CHEMISTRY MADE SIMPLE

Revised Edition

Revised by
John T. Moore, Ed.D.

A Made Simple Book
Broadway Books
New York

Produced by The Philip Lief Group, Inc.

CHEMISTRY MADE SIMPLE.

For information, address: Broadway Books, a division of Random House, Inc., 1745 Broadway, New York, NY 10019.

Previous editions of this book were originally published in a different form in 1955 and 1984 by Doubleday, a division of Random House, Inc., and are here reprinted by arrangement with Doubleday.

Produced by The Philip Lief Group, Inc.
Managing Editors: Judy Linden, Jill Korot.
Design: Annie Jeon.

First Broadway Books trade paperback edition published 2004

Library of Congress Cataloging-in-Publication Data

Chemistry made simple.—Rev. ed. / revised by John T. Moore.
p. cm. – (Made simple)
Rev. ed. Of: Chemistry made simple / by Fred C. Hess. Rev. ed. / rev. by Arthur L. Thomas. c1984.
ISBN 0-7679-1702-2

1. Chemistry—Popular works. I. Moore, John T. II. Hess, Fred C. Chemistry made simple. III. Made Simple (Broadway Books)

QD37.H4 2004
540—dc22 2004045856

10 9 8 7 6 5 4 3 2

CONTENTS

LIST OF TABLES

EXPERIMENTS

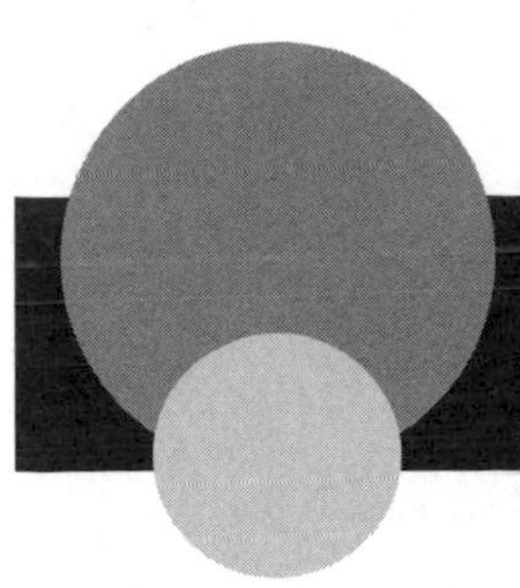

INTRODUCTION

Look around for a moment. What do you see? A painted wall? Is it made of plaster? Or laminated plywood? Or bonded panelboard? Is there a photograph hanging on it? Or a lithographed print? Or an oil painting? Do you see a glass window with synthetic wood shutters or with rayon drapes? Is that table by the wall varnished, stained, or coated with shellac? Are your chairs in leather or vinyl? What metal is that lamp made of?

Look at your clothes. What kinds of dyes produced those colors? Are your clothes fade-resistant, stain-resistant, or permanent press? Will bleach turn them white or yellow them? What do the cleaners use to clean clothes? What detergent do you use in your washing machine? Are the soles of your shoes leather or synthetic material? Do you have a plastic raincoat?

Look at your kitchen, with its stainless steel appliances and ceramic tile floor. Does your cooktop burn gas? Or does electricity pass through metal or ceramic elements to produce heat for cooking? Is your kitchen table surface varnished wood or glass? Is it set with ironstone dishes? Are your eating utensils made of an alloy called sterling silver? Or of another alloy called stainless steel? Or are the utensils silver-plated?

As you look around, it rapidly becomes apparent that practically everything you can see is either a product from some chemical industry or has in some way been treated by chemicals. Even the ink and paper of this page are chemical products. Many of your drugs and foods are either prepared in chemical laboratories or analyzed by chemists for purity and safety.

Even the human body is an astounding chemical factory. It is so complex that chemists have barely begun to understand the secrets of the many strange and wonderful processes that take place within it. The body is made up of chemicals that grow! Perhaps just as remarkable, some of those chemicals grow just so far and then seem to stop. Of course, all living things contain chemicals that grow, but the fundamental differences between things that are alive and things that are inanimate still await explanation by scientists.

Given this chemical complexity, it is little wonder, then, that people are curious about this branch of science that affects their daily lives. As the products of chemistry become even more numerous, people need a basis for selecting the products that best suit their needs. For example, choosing a suitable motor oil for their cars, or safe pesticides for their gardens, makes people aware that an understanding of chemistry will help them deal more effectively with those questions.

Chemistry Made Simple was written to give you a basic understanding of chemistry, whether you are a student taking your first chemistry course or simply someone wanting to learn a bit more about the chemical world around you. The book is divided into 26 chapters, each covering a specific aspect of chemistry. The chapters incorporate simple experiments that you can do at home to give you a hands-on experience with chemistry. These experiments require no special laboratory environment—a kitchen will do just fine—and no toxic chemicals are involved; *however, you must be careful when doing these experiments—and always wear your safety goggles.*

Chapter 1 is a brief introduction to chemistry and its place in science. You will learn about the scientific method and the *Système International* (SI), or international system, of measurements used by scientists. You will also be given some tips on working chemistry problems.

In Chapter 2 you will learn about the physical and chemical properties of matter, and you will encounter one way to classify matter—as an element, a compound, or a mixture. Finally, you will be shown the general types of chemical reactions that you are likely to encounter in your study of chemistry.

Chapter 3 introduces you to the world of matter and models of matter. You will learn about the atom and the particles that are contained within it, and you will be introduced to the periodic table, that marvelous classification system that chemists (and chemistry students) use daily.

Chapter 4 discusses chemical bonding. You will learn the two general types of bonding—ionic and covalent—and how to write the formulas for such compounds. You will also learn how to name them.

Chapter 5 covers some of the laws of chemistry. You will learn how to write chemical equations and mass relationships in chemical reactions, and you will be shown how to find the answers to two important questions: How much will it take, and how much can I make?

Chapter 6 focuses on the gaseous state of matter. We will look at the temperature–pressure–volume–amount relationships in real gases and study the models used in the study of the world of gases. You will learn to use the combined gas equation and ideal-gas equation in gas-law calculations.

Chapter 7 covers the other two states of matter, liquids and solids. You will learn what happens when matter goes from one state to another as well as the structures found in the liquid and solid states.

Chapter 8 focuses on the mixtures that we call solutions. You will learn ways of expressing the concentrations of solutions as well as some of the unique properties of solutions.

Chapter 9 is another chapter on solutions, but especially solutions of salts and similar substances. You will examine the properties of acids and bases in this chapter, and you will be introduced to the pH concept.

Chapters 10 and 11 discuss chemical reactions that involve electron exchange. In Chapter 10, redox reactions are introduced, and you will learn how to balance those equations. Chapter 11 focuses on a specific application of redox reactions, the kinds that occur in electrochemical cells. You will learn the difference between a cell and a battery and will learn how to calculate the voltage produced by a given cell.

Chapter 12 focuses on the gases contained in the atmosphere, such as nitrogen and oxygen, and touches on air pollution, the topic of Chapter 25.

Chapters 13 through 21 examine the properties of various families of elements, giving the occurrence, preparation, physical and chemical properties, uses, and important compounds of the members of these families. Along the way you will learn a little about such topics as metallurgy, the making of steel, and photography.

Chapter 22 focuses on the chemistry of just one element, carbon. This chapter on organic chemistry discusses the different classes of organic compounds and introduces polymers.

Chapter 23 focuses on nuclear chemistry. You will learn about the different types of nuclear reactions and the radiation they produce. The discussion of fusion reactions explains why they hold great promise for meeting humankind's future energy needs.

Chapter 24 then presents the chemistry of energy. You will learn about the sources of energy, both renewable and nonrenewable, and why scientists are studying alternative sources of energy.

Chapter 25 discusses the chemistry of the environment and related topics like air pollution, heavy-metal contamination, water pollution, thermal pollution, and radioactive waste disposal.

Chapter 26 focuses on the chemistry of living organisms, specifically the building blocks of the human body—proteins, carbohydrates, fats, and nucleic acids—and ends with a brief look at biotechnology.

In most chapters you will find at least one set of problems. If you are taking a chemistry course, be sure to do the problems; chemistry is not a spectator sport! To learn to do chemistry problems, you need to practice.

When I was in school, I loved chemistry. Seeing one substance change into another was magical to me. I wanted to know why and how the transformation occurred. After more than 30 years of teaching, I am still amazed by chemistry. I hope that in using this book you not only learn something about chemistry but also gain a sense of wonder and excitement about what we chemists call the *central science*.

SCIENCE AND CHEMISTRY

KEY TERMS

matter, phenomenon, hypothesis, theory, law, mass, weight, *Système International d'Unités* (SI)

Science is knowledge gained from the study of the behavior of nature. Chemistry is that branch of science that deals with the composition of all forms of matter and with the change of one form of matter into another. (*Matter* is anything that takes up space.) Chemistry is concerned not only with what is, and what happens, in nature, but also with how changes take place. It is this knowledge of how change occurs that leads to our ability to control such changes by

1. imitating them to make greater quantities of better products;
2. using them as clues to find new processes and to create new products; and
3. inhibiting them to preserve useful products.

Science is knowledge gained from the study of the behavior of nature.

Chemistry is that branch of science that deals with the composition of all forms of matter and with the change of one form of matter into another.

THE SCIENTIFIC METHOD

In all branches of science, including chemistry, knowledge is obtained fundamentally by careful observation of the behavior of nature. A particular natural event is called a phenomenon. Scientists, in seeking an understanding of how a phenomenon takes place, first use their training and background to offer a possible explanation of what is happening. That "best guess" explanation is called a hypothesis. Scientists then undertake a series of experiments in which they permit the phenomenon to occur repeatedly under carefully controlled conditions. Based on these experiments, scientists gather evidence (observations or facts) that either supports the hypothesis or causes it to be rejected or modified. This cycle—hypothesis, experimentation, and hypothesis modification—continues repeatedly. In fact, in good science, it never stops. As scientists develop new and better ways of extending their senses, allowing them to make increasingly better observations, they constantly check their old hypotheses and modify them as needed.

When the weight of evidence seems to indicate that a hypothesis is sound, scientists then announce their findings as a *theory*, a hypothesis that has withstood the test of time. If the theory is subsequently proven to be a nonvarying phenomenon in nature, scientists are said to have discovered a natural *law*. Theories and laws are then applied to other phenomena to increase understanding of them and to adapt them to refine or to create products for human use.

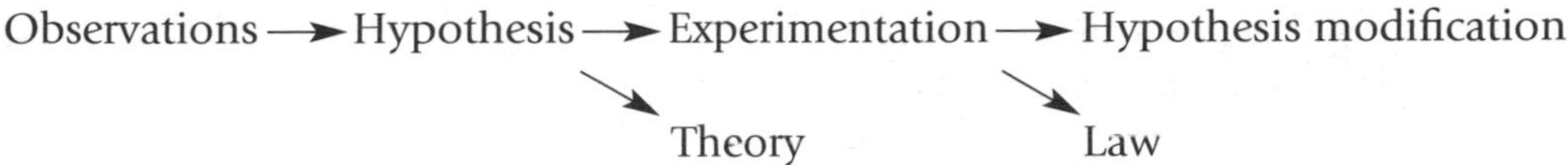

In the early days of chemistry, it was common for one person to carry out this entire process alone. The contributions of early chemists to our knowledge and understanding of nature were monumental, yet because of the limitations of the conditions under which chemists worked, erroneous theories and even false laws were generally accepted for a time. As the number of skilled scientists increased, errors became less common, and knowledge of nature rapidly unfolded. It is now much more common for teams of skilled chemists to work together on a single project, with each member contributing to the solution of the problem from his or her own special background. The work of these teams is called *research*. Basic research is concerned both with the checking and rechecking of apparently sound theories and with the persistent search for deeper understanding of the behavior of nature. Applied research, in contrast, is concerned with the application of scientific knowledge to the development of new and better products for consumers.

Scientists believe that no single theory or law represents absolute truth. Rather, they feel that each new discovery brings them nearer to truth, so they are always ready to modify their concept of the behavior of nature if/when new evidence emerges. Thus, scientific progress can be described as:

Careful observation + Persistent search for meaning + Intelligent thought → Progress

This equation represents the scientific method to approaching problems that has been credited with the tremendous growth of science in the past three centuries.

Although the scientific method is responsible for many great discoveries, serendipity also has played a part in the advancement of humankind's knowledge. Serendipity is an accidental, fortuitous discovery. For example, Teflon®, microwave cooking, Post-it notes, and many other products and processes were discovered by accident; however, it takes a well-trained scientist, with an open and inquiring mind, to take full advantage of luck.

Just as observation is the starting point that leads to new scientific discoveries, so can it serve as a springboard for studying a field of science.

TRIVIA: The first synthetic dye was created by an 18-year-old British chemistry student, William Perkin, in 1856. Perkin, who was home from school for Easter, was trying to make quinine in his home lab but instead made a purple dye. With the financial backing of his parents, Perkin built a factory to make this dye and became successful and wealthy. The Germans used many of Perkin's ideas and techniques to build the world's first extensive chemical industry.

MEASUREMENT

One of the basic components of observation is measurement. Modern science simply did not exist until people learned to measure precisely such quantities as distance, volume, weight, temperature, pressure, and time. The invention of suitable devices for measuring these quantities not only enabled scientists to gather quantitative data but also permitted the use of mathematics for evaluating their observations.

In the case of chemistry, the invention of the balance was a critical development because it could be used to demonstrate the most important fact discovered to that time—namely, that all changes in nature from one form to another take place on a definite weight or mass basis. Until this relationship was recognized, there simply was no science of chemistry. (We explain the difference between mass and weight in the next section.)

SYSTÈME INTERNATIONAL D'UNITÉS (SI)

Just as the behavior of nature is independent of national boundaries, so science, the study of natural behavior, is international in scope. The international system of units, or *Système International d'Unités* (abbreviated SI), adopted in 1960, is a standardized system of weights and measures used all over the world for scientific work. Prior to the adoption of the SI, the earliest measurements were based on dimensions of the body, such as the foot, inch, and cubit. The metric system was adopted by most scientists in the eighteenth century. The metric system is convenient to use because all units are based on powers of 10. The SI established certain basic metric units, and units derived from them, as preferred units to be used in scientific measurements. SI units mean the same thing to all scientists. The English system of units (feet, pounds, slugs, gallons, miles, etc.) is still used in the United States, even though the metric system has been formally adopted.

The fundamental SI unit of length is the *meter*, which is about 3.28 feet, or about 10 percent longer than a yard. The meter was first defined as one ten-millionth of the distance from the equator to the North Pole. A bar of platinum–iridium alloy was inscribed with two marks 1 meter apart. This bar was carefully preserved at Sèvres, France, near Paris. This definition changed with the onset of improved technology when it was discovered that the distance from the equator to the North Pole was not precise enough for universal applications. The meter is currently defined as the distance traveled by a ray of electromagnetic (EM) energy in a vacuum in 1/299,792,458 of a second. As with all scientific and mathematical units, measurements are standardized for universal use and conversion, and then they are modified and made more precise as new technologies are developed.

The SI uses the following seven base units (and their symbols) for measurement:

length: meter (m)

mass: kilogram (kg)

time: second (s)

electric current: ampere (A)

thermodynamic temperature: kelvin (K)

amount of substance: mole (mol)

luminous intensity: candela (cd)

The measures we will use most often in our exploration are length, mass, and amount of substance.

NOTE: The kilogram is the SI base unit for mass. This unit is an exception among base units, because it carries a prefix ("kilo").

The liter (L), although not an SI base unit, is for liquid measure. A cubic centimeter (abbreviated

cm^3 or cc) is, for all practical purposes, the same volume as a milliliter (mL), and 1000 milliliters equals one liter. A liter is slightly larger than a quart; it is equal to 1.0567 quarts. One milliliter of water at 4.08°C (39.2°F) has a mass of 1 gram. A nickel coin has a mass of about 5 grams.

Mass is proportional to weight. You will often hear the term "weight" when the actual quantity being measured is mass. *Mass* refers to the amount of matter in a body, whereas *weight* refers to the gravitational attraction of Earth on the body. The original meaning of weight, still in general use today, is equivalent to mass. A definition common in scientific studies uses weight for a particular kind of force. To understand the differences between mass and weight, note the following comparisons:

1. Mass is a measurement of the amount of matter an object contains; weight is the measurement of the pull of gravity on that object.

2. Mass is measured by comparing a known amount of matter with an unknown amount of matter. Weight is measured with a scale.

3. The mass of an object does not change when an object's location changes. Weight, however, does change with location, when the location change brings about a change in gravitational attraction.

Prefix	Symbol	Factor
pico-	p	1×10^{-12}
nano-	n	1×10^{-9}
micro-	μ	1×10^{-6} (0.000001)
milli-	m	1×10^{-3} (0.001)
centi-	c	1×10^{-2} (0.01)
deci-	D	1×10^{-1} (0.1)
deka-	da	1×10^{1} (10)
hecto-	H	1×10^{2} (100)
kilo-	K	1×10^{3} (1000)
mega-	M	1×10^{6} (1,000,000)
giga-	G	1×10^{9} (1,000,000,000)

Table 1.1 SI Prefixes

Table 1.1 lists the prefixes that are used with units in the SI system. For example, the prefix milli (m) means 0.001. Thus, we can have a millimeter (0.001 meter) or a milliliter (0.001 liter) or a millisecond (0.001 second). We may also have a kilometer (1000 meters) or a kilogram (1000 grams). The most commonly used prefixes appear in boldface in Table 1.1.

The powers-of-10 SI prefixes allow for converting between units by moving the decimal point. As with dollars and cents, one can change from one unit to another—

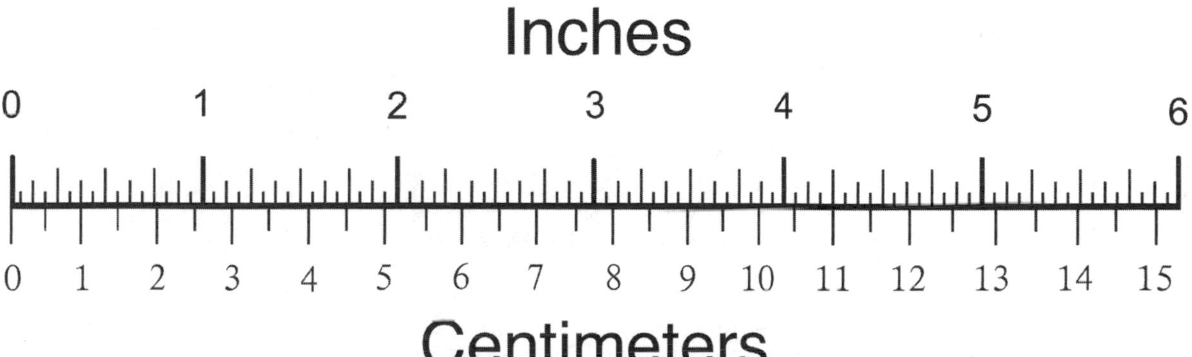

Figure 1.1

SI to English	English to SI
LINEAR MEASURE	
1 millimeter (mm) = 0.03937 inch (in.)	1 in. 0 = 25.4 mm
1 centimeter (cm) = 0.3937 inch (in.)	**1 in. = 2.54 cm**
1 meter (m) = 39.37 inches (in.)	1 ft = 30.48 cm
1 meter (m) = 3.281 feet (ft)	1 ft =0.3048 m
1 meter (m) = 1.0936 yards (yd)	1 yd = 0.9144 m
1 kilometer (km) = 3281 feet (ft)	1 mi = 1609 m
1 kilometer (km) = 0.6214 mile (mi)	**1 mi = 1.609 km**
SQUARE MEASURE	
1 square centimeter (cm^2) = 0.155 square inches (sq in. or in.^{2})	1 in.^{2} = 6.4516 cm^2
1 square meter (m^2) = 10.764 square feet (sq ft or ft^2)	1 ft^2 = 0.0929 m^2
1 square meter (m^2) = 1.196 square yards (sq yd or yd^2)	1 yd^2 = 0.8361 m^2
CUBIC MEASURE	
1 cubic centimeter (cm^3 or cc) = 0.061 cubic inches (in.^{3})	1 in.^{3} = 16.387 cm^3
1 cubic meter (m^3) = 35.320 cubic feet (cu ft or ft^3)	1 in.^{3} = 0.0283 m^3
1 cubic meter (m^3) = 1.308 cubic yards (cu yd or yd^3)	1 in.^{3} = 0.7645 m^3
CAPACITY	
1 milliliter (mL) = 0.0338 fluid ounce (fl oz)	1 fl oz = 29.573 mL
1 liter (L) = 2.1134 pint (pt) liquid	1 pt (liquid) = 0.4732 L
1 liter (L) = 1.0567 quart (qt) liquid	**1 qt (liquid) = 0.9463 L**
1 liter (L) = 0.9081 quart (qt) dry	1 qt (dry) = 1.1012 L
1 liter (L) = 0.2642 gallon	1 gallon = 3.7853 L
WEIGHT	
1 gram (g) = 15.43 grains (gr)	1 gr = 0.0648 g
1 gram (g) = 0.0353 ounce (oz)	1 oz = 28.35 g
1 kilogram (kg) = 2.2046 pounds (lb)	**1 lb = 453.6 g**
	1 lb = 0.4536 kg

Table 1.2 Conversions Between SI and English Units of Measure

for example, meters to centimeters—simply by moving the decimal point. One dollar (1.00) is equal to 100 cents. In converting from dollars to cents, we move the decimal point two places to the right. Similarly, in converting from meters to centimeters, we move the decimal point two places to the right (2.45 m = 245 cm).

Table 1.2 gives the factors for converting the most common measurements between the SI and the English system. Four SI–English conversion factors are set in boldface; all other values in this table are calculated from these standards. As you study this table try to relate the various quantities to things you are familiar with. For example: Which of your fingernails is 1 centimeter wide? Where is the palm of your hand 10 centimeters wide? It may help you to know that a 1/4-teaspoon measuring spoon holds 0.8 milliliter, and that three aspirin tablets (325 mg each) have a mass of just less than 1 gram.

THE MATHEMATICS OF CHEMISTRY

Chemistry is a quantitative science. Much of the meaning of chemistry is lost if mathematical relationships are omitted in its study. The mathematics of introductory chemistry involves only simple arithmetic: addition, subtraction, multiplication, division, simple proportion, exponentiation (raising to powers), and taking roots. We shall see that even these last two operations can be simplified. Don't be frightened by numbers.

You will be shown how to go about solving typical math problems in chemistry. Practice exercises will give you an opportunity to apply the solution techniques. Answers to each problem will be supplied. Chemistry is not a spectator sport. The way to succeed is through practice. You will find that the problems are not too difficult, and they will lead you to a much better understanding of chemistry.

Before we get to the problems, we present two concepts that are important in doing calculations.

The Factor-Label Method

All quantities or measurements have a numeral associated with a unit—for example, 5.3 grams, 3.6 meters. The factor–label method is a system for solving numerical problems by using known relationships among units (factors) to find an unknown quantity. A problem is set up using the necessary mathematical quantities and operations to obtain the desired units. Then, the setup is solved using the numbers associated with the units to obtain the answer.

For example, suppose we want to convert exactly 1 mile to inches. We do not have a direct conversion factor, so we use related factors: we know that there are 5280 ft in a mile, and 12 in. in a foot:

1 mile:

$$\frac{1 \text{ mile}}{1}$$

then we convert to feet:

$$\frac{1 \text{ mile}}{1} \times \frac{5280 \text{ ft}}{1 \text{ mile}}$$

and then finally we convert to inches:

$$\frac{1 \text{ mile}}{1} \times \frac{5280 \text{ feet}}{1 \text{ mile}} \times \frac{12 \text{ inches}}{1 \text{ foot}}$$

This expression is the setup. It tells us what relationship and mathematical operations we are to use. Now, we can carry out the math and arrive at the final answer:

$$\frac{1\ \cancel{\text{mile}}}{1} \times \frac{5280\ \cancel{\text{feet}}}{1\ \cancel{\text{mile}}} \times \frac{12\ \text{inches}}{1\ \cancel{\text{foot}}} = 63{,}360\ \text{inches}$$

Notice that the units must cancel properly, leaving the desired unit for the final answer. When we have the same a unit of measure (e.g., mile) in the numerator of one factor and in the denominator of another, the units cancel each other. If the problem has been set up correctly, after all identical units are canceled, the desired unit will be left. Several different units and conversion factors can be used in the same equation, as necessary, as in the example.

Significant Digits

When any quantity is measured the result can be known only to within a certain accuracy, depending on the quality of the instrument being used and the skill of the person making the measurement. For example, if you measure a length with a ruler marked in millimeters and report the result as 18.3 cm, the three digits are known with certainty. The value of the second decimal place is uncertain, because you would have to estimate the distance between the third and fourth tick marks on the ruler. The number of *significant figures* in a quantity is the number of digits in it that are known with certainty. Thus, the value 18.3 has three significant figures.

In counting the number of significant digits in the quantities you are using in a mathematical operation, use only quantities that are considered measured values, not exact ones. There are exactly 12 inches in 1 foot, 5280 feet in one mile, and 2.54 cm in 1 inch (by definition). Then, when counting significant digits in a measured value, use the following four rules to determine whether a digit is significant:

1. All nonzero digits are significant.

   Example: 23.45 has 4 significant digits, and 3.67×10^{-3} has 3 significant digits.

2. All zeros between nonzero digits are significant.

   Example: 2305.6 has 5 significant digits.

3. All zeros to the left of the first nonzero digit are *not* significant.

   Example: 0.00234 has 3 significant digits, and 0.0002034 has 4 (the zero between 2 and 3 is significant).

4. Zeros that appear after the last nonzero digit are significant only if there is a decimal point in the number.

   Example: The number 23000 has only 2 significant digits; 230.00 has 5 significant digits.

The following are a couple of simple rules that scientists use in determining the number of significant digits to report:

Addition and subtraction. In addition and subtraction problems, the final answer is reported to the number of decimal places in the quantity with the *fewest* number of decimal places. For example, in the problem 23.1 + 18.45 + 27.563, the answer, 69.113, should be reported to the tenths place: 69.113 is rounded to 69.1. In rounding off, if the first number to be dropped is 5 or greater, the last retained number is rounded up by 1. If the first number to be dropped is less than 5, the last number retained remains the same.

NOTE: Round off only the *final* answer, not intermediate values.

> 5 or greater: round it up
> Less than 5: let it stay

Multiplication and division. In multiplication and division problems, the final answer is reported to the same number of significant digits as the number with the *fewest* number of significant digits. For example, in the multiplication problem 39.31 (4 significant digits) × 27.8531 (6 significant digits) = 1094.905361, the answer should be rounded to 4 significant digits: 1095.

EXAMPLE 1. How many inches are there in 15 centimeters?

SOLUTION: In Table 1.2 we find that 1 in.= 2.54 cm. Therefore:

$$\frac{15\ \not{cm}}{1} \times \frac{1\ in}{2.54\ \not{cm}} = 5.90551181\ in. = 5.9\ in.$$

(Remember that 2.54 cm = 1 in. is exact, by definition, so we use the 15 cm to determine the number of significant figures in the final answer.)

EXAMPLE 2. Express the exact measurement 1273.000 grams in terms of milligrams, centigrams, decigrams, grams, dekagrams, hectograms, and kilograms.

SOLUTION:

1,273,000	milligrams (mg)
127,300.0	centigrams (cg)
12,730.00	decigrams (dg)
1,273.000	grams (g)
127.3000	dekagrams (dag)
12.73000	hectograms (hg)
1.273000	kilograms (kg)

Notice that in moving in either direction, up or down the prefix scale shown in Table 1.1, the decimal point moves *one place* for each change in prefix. This is also true if meters or liters are used in place of grams. Notice also that the number of significant figures does not change; this is because we assume an exact measurement.

EXAMPLE 3. Express 9255 square meters in terms of square millimeters, square centimeters, square decimeters, square meters, square dekameters, square hectometers, and square kilometers.

SOLUTION:

9,255,000,000	square millimeters (mm^2)
92,550,000	square centimeters (cm^2)
925,500	square decimeters (dm^2)
9,255	square meters (m^2)
92.55	square decameters (dam^2)
0.9255	square hectometers (hm^2)
0.009255	square kilometers (km^2)

NOTE: In calculations with square measure, the decimal point is moved *two places* for each incremental change in prefix. For example, in the preceding example, 9255 m^2 = 92.55 dam^2. The decimal point moved two places to the left in the conversion from the smaller to the larger unit.

EXAMPLE 4. Express 625 cubic meters in terms of each of the prefixes in Table 1.1.

SOLUTION:

625,000,000,000	cubic millimeters (mm^3)
625,000,000	cubic centimeters (cm^3)
625,000	cubic decimeters (dm^3)

625	cubic meters (m^3)
0.625	cubic dekameters (dam^3)
0.000625	cubic hectometers (hm^3)
0.000000625	cubic kilometers (km^3)

NOTE: In calculations with cubic measure, the decimal point is moved *three places* for each incremental change in prefix. For example, in the preceding example, 625 m^3 = 0.625 dam^3. The decimal point moved three places to the left in the conversion from the smaller to larger unit.

EXAMPLE 5. Convert 85.0 mi/h to km/s. The following is one of many possible setups to this problem:

$$\frac{85.0\ \cancel{\text{mi}}}{1\ \cancel{\text{h}}} \times \frac{5280\ \cancel{\text{ft}}}{1\ \cancel{\text{mi}}} \times \frac{12\ \cancel{\text{in}}}{1\ \cancel{\text{ft}}}$$

$$\times \frac{2.54\ \cancel{\text{cm}}}{1\ \cancel{\text{in.}}} \frac{1\ \text{m}}{100\ \cancel{\text{cm}}} \times \frac{1\ \text{km}}{1000\ \cancel{\text{m}}}$$

$$\times \frac{1\ \cancel{\text{h}}}{60\ \cancel{\text{min}}} \times \frac{1\ \cancel{\text{min}}}{60\ \text{s}}$$

$$= 0.0379984\ \text{km/s}$$

The final answer is rounded to 3 significant figures, based on the value 85.0 mi/h, because all other quantities are exact. The final answer is then 0.0380 km/s.

This problem set has many possible setups, depending on which conversion factors you choose to use (or remember).

PROBLEM SET NO. 1

1. How many milliliters are in 3 fluid ounces?
2. Convert 1 pound 3 ounces to kilograms.
3. Convert 528 square inches to square meters.
4. Convert 22,400 cubic centimeters to cubic feet.
5. Convert 250 grams to ounces.
6. Convert 25 square yards to square centimeters.
7. How many liters are in 1 measuring cup (8 fluid ounces)?
8. How many milliliters are in a 1-tablespoon measure (1 tablespoon = 3 teaspoons = 1/16 cup = 1/2 fluid ounce)?
9. A gasoline tank has a capacity of 18 gallons. Express this capacity in liters.
10. Assuming the dimension of a book to be 8.5 in. × 11 in. × 0.625 in., find the volume of the book in cubic decimeters.

CHAPTER 2 MATTER

KEY TERMS

physical properties, chemical properties, density, specific gravity, chemical reaction, element, compound, mixture

Strike a match. Any kind of match will do. Watch it carefully. What do you see? What do you hear? What do you smell? What do you feel? Blow out the match. Did it go out completely? Repeat the process, this time holding the match in a horizontal position. Notice the shape of the flame. Do you see the liquid creeping just ahead of the flame? Light a wooden toothpick with the match, and then blow them both out. Blow harder on the toothpick. Blow on it again. What happens? Do you have any evidence that match manufacturers are safety-conscious? What differences in the properties of the match before and after burning can you find? Can the charred remnants of the match still be called a match?

A tremendous amount of chemistry was illustrated by the phenomena that you have just observed. You will notice that in making observations we use not only our eyes but also our other senses. The more senses we can use in observation, the more thorough our findings will be. We will use our observations to get acquainted with some of the fundamental terms and ideas of chemistry. The observations will help us visualize chemical ideas and help explain other phenomena of nature.

PROPERTIES OF MATTER

We distinguish one form of matter from another by characteristics called *properties*. When you were asked to handle the match and the toothpick, you knew just what was meant because you were familiar in a general way with the properties of those objects. You are aware, of course, that a wooden match has more properties in common with a toothpick than it has with a paper match. The wood is a common substance of the two objects. A *substance* is a definite variety of matter, and all specimens of a substance have the same properties. Aluminum, iron, rust, salt, and sugar are all examples of substances.

NOTE: Each of these is homogeneous, or uniform, in its makeup. In contrast, granite and concrete cannot be called substances, because they are not homogeneous; each is made up of several different substances.

Substances have two major classes of properties: physical and chemical. *Physical properties* describe the characteristics of a substance as it exists, that is, such observable and measurable characteristics as size, shape, color, density, and mass. *Chemical properties* describe the ability of a substance to change into a new and completely different substance.

PHYSICAL PROPERTIES

Substances have two kinds of physical properties: intensive and extensive.

Intensive Physical Properties

Intensive physical properties include those features that do not depend on the quantity or amount of the substance. The following are some of the important intensive physical properties:

Density is the mass per unit volume of a substance. Density is usually expressed in grams per cubic centimeter (g/cm^3) or, in the units you may be accustomed to using, in pounds per cubic foot (lb/cu ft). Because 1 cm^3 of water has a mass of 1 g, the density of water (at 4 °C) is g/cm^3. A cubic foot of fresh water has a mass of 62.4 lbs. The density of water in the English system is 62.4 lb/cu ft. Multiplying a metric density by 62.4 gives the density of the substance in English units. Table 2.1 lists the densities of some common substances.

Substance	Density (g/cm^3)	(lb/cu ft)
aluminum	2.7	168.5
brass	8.6	536.6
copper	8.9	555.4
cork	0.22	13.7
diamond	3.5	218.4
gold	19.3	1204.3
ice	0.917	57.2
iron	7.9	493.0
lead	11.3	705.1
magnesium	1.74	108.6
mercury	13.6	848.6
rust	4.5	280.8
salt	2.18	136.0
sugar	1.59	99.2
steel	7.83	488.6
sulfur	2.0	124.8
water, fresh	1.0	62.4
water, sea	1.025	64.0
zinc	7.1	443.0

Table 2.1 Density of Common Substances

These values can be used to work chemistry problems:

EXAMPLE 1. How much does 400. cm^3 of mercury weigh?

SOLUTION: From Table 2.1: the density of mercury is 13.6 g/cm^3.

Therefore:

$$\frac{400.\ \text{cm}^3}{1} \times \frac{13.6\ \text{g}}{1\ \text{cm}^3} = 5440\ \text{g}$$

(3 significant digits)

Specific gravity is the ratio of the mass of a given volume of a substance to the mass of the same volume of water at the same temperature. Because 1 cm^3 of water has a mass of 1 g, specific gravity is numerically equal to the metric density of a substance. Both density and specific gravity have to do with the "lightness" or "heaviness" of a substance. Aluminum is "lighter" than lead. Water is "lighter" than mercury. The term *density* is used more often with solids, whereas specific gravity is used more often with liquids or solutions (acid in a car battery, or antifreeze in its radiator). An advantage of specific gravity is that it does not depend on the temperature—the specific gravity of ethyl alcohol is the same at 20 °C as at 100 °C. This is not true of its density; however, the difference is so slight that most chemists use the term density.

Specific gravity can be used to solve problems:

EXAMPLE 2. A particular solution has a specific gravity of 1.20. How much volume (cm^3) does 10. g of this solution occupy?

SOLUTION: A specific gravity of 1.20 = a density of 1.20 g/cm^3.

Therefore:

$$\frac{10.\text{ g}}{1} \times \frac{1\text{ cm}^3}{1.20\text{ g}} = 8.3\text{ cm}^3$$

(2 significant digits)

Hardness is the ability of a substance to resist scratching. A substance will scratch any other substance softer than itself. The Mohs hardness scale is used to compare the hardness of substances. The scale lists various minerals of different hardness (Table 2.2), but because so few of these minerals are commonly known, Table 2.2 also gives the approximate hardness of some familiar substances and objects. Low hardness numbers indicate soft substances; the higher the number, the harder the substance.

Odor. Many substances have characteristic odors. Some have pleasant odors, like methyl salicylate (oil of wintergreen); some have pungent odors, like ammonia or sulfur dioxide (a gas that forms when the head of a match burns); some have disagreeable odors, like hydrogen sulfide (a gas that forms in rotten eggs).

Color. You are familiar with the color of such substances as gold and copper. White substances are usually described as colorless. Normally, it takes a combination of several specific physical properties to identify a given substance. The color of gold is not unique, as many unfortunate prospectors found out. Their strikes of "fool's gold" looked like gold but turned out to be pyrite (also known as iron sulfide), a far less valuable substance.

Familiar Substance (representative sample)	Hardness	Familiar Substance (representative sample)	Hardness
talc	1	graphite	0.7
gypsum	2	asphalt	1.3
calcite	3	fingernail	1.5
fluorite	4	rock salt	2.0
apatite	5	aluminum	2.6
feldspar	6	copper	2.8
quartz	7	brass	3.5
topaz	8	knife blade	5.4
corundum	9	file	6.2
diamond	10	glass	6.5

Table 2.2 Mohs Hardness Scale

Extensive Physical Properties

Extensive physical properties are such features as weight, dimensions, and volume. Unlike the intensive properties, these properties depend on the amount of matter present. A large chunk of gold and a small chunk of gold have the same color, density, hardness, and so forth (intensive properties), but they have different masses and volumes (extensive properties). Objects, particularly manufactured objects, may possess similar extensive properties, but these properties are in no way fundamentally related to the substances that make up the objects. Thus, matches and toothpicks are objects. Each is made according to a pattern of extensive properties. But toothpicks may be made of wood or of plastic, two completely different substances with totally different intensive physical properties.

CHEMICAL PROPERTIES

The chemical properties of a substance describe its ability to form new substances under given conditions. A change from one substance to another is called a chemical change, or a *reaction*. Hence, the chemical properties of a substance may be thought of as a listing of all the chemical reactions of a substance and the conditions under which the reactions occur.

The striking of a match illustrates several chemical properties of the substances in a match. Examine Figure 2.1 carefully. Notice the various substances present in each type of match. When you strike a safety match, the heat of friction of the head of the match rubbing on the glass is sufficient to cause the phosphorus on the scratching area to burn. This burning then generates enough heat to cause the substances in the head of the match to ignite. The burning of these, in turn, produces the heat necessary for the matchstick to catch fire. All the substances in the match burn (chemical property), but individual substances do so at a different temperature (condition). None of the substances burns at room temperature! Because the phosphorus is contained only on the scratching area of the box or cover of the matches, matches can be

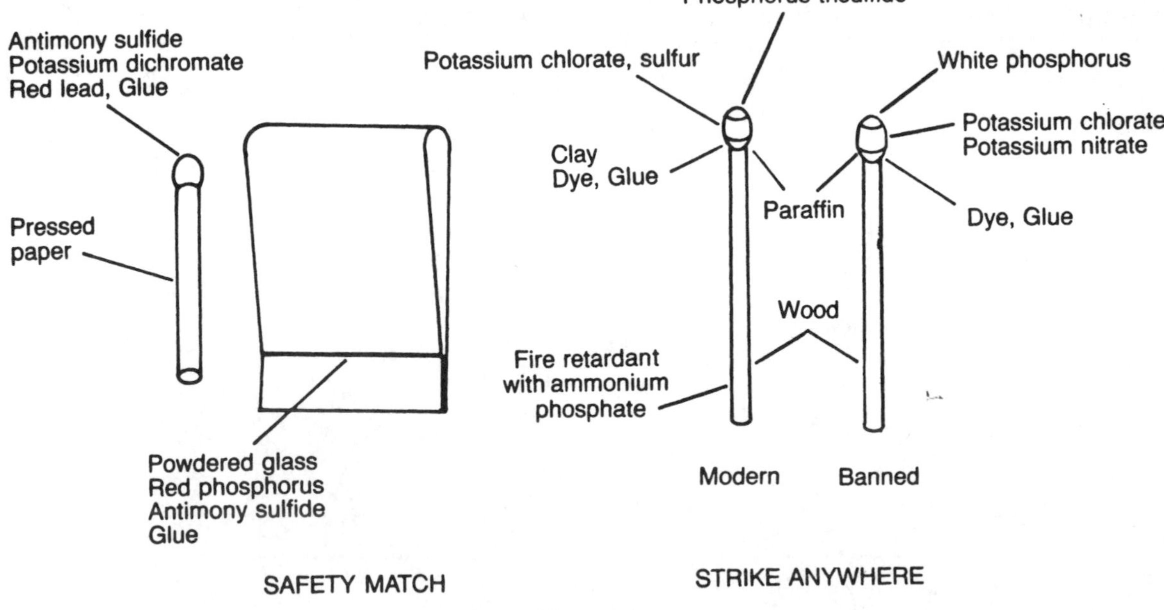

Figure 2.1

"struck" only on this area. (Occasionally, safety matches can be struck on glass or linoleum, where rubbing produces sufficient heat to cause the head to start burning.)

The phosphorus trisulfide in the tip of the "strike-anywhere" match is very sensitive to heat. Rubbing this tip on almost any moderately hard surface will produce sufficient frictional heat to cause this substance to burn. The other substances in the tip, and finally the matchstick, are then ignited as the temperature rises.

> White phosphorus was formerly used in the tip of this type of match. This substance likewise bursts into flame at temperatures slightly above room temperature. However, the men who worked with white phosphorus and inhaled its fumes frequently contracted a disease known as phosphorous necrosis, more commonly known as "phossy jaw." This is one of the most severe occupational diseases known. It began with a toothache and painful swelling of the gums and jaw; then abscesses would form, accompanied by a rancid discharge; finally, the victim's jawbone would literally rot and give off a greenish white glow in the dark. The only recourse for those afflicted was to have their jawbones surgically removed, resulting in agony and disfiguration. Laws were eventually passed prohibiting the use of white phosphorus; matches were then made of red phosphorous, which is harmless.

The charred remnants of the matchstick and toothpick consist principally of carbon, one of the new substances formed when wood or paper burns. The "afterglow" you observed in the toothpick is a chemical property of carbon. You have seen the same phenomenon in a charcoal fire. The matchstick, however, exhibited no afterglow because it had been treated with a solution of a fire retardant substance that soaked into the wood. A substance called borax was formerly used for this purpose, but ammonium phosphate is generally considered to be more effective and is now widely used, not only in matchsticks but also in draperies, tapestries, and other types of decorations, as well as in children's sleepwear.

KINDS OF MATTER

As you look at the different objects about you, you are perhaps impressed by the almost endless variety of matter. Classification of the kinds of matter into fundamental groups was an impossible task until chemists began to probe into the composition of matter. Knowledge of composition quickly led to the discovery that all matter is made up of either pure substances or mixtures of pure substances. Pure substances, in turn, are of two types: elements or compounds. Figure 2.2 diagrammatically shows the kinds of matter based on composition.

Elements

Elements are the basic constituents of all matter. An element is the simplest form of matter. It cannot be formed from simpler substances, nor can it be decomposed into simpler

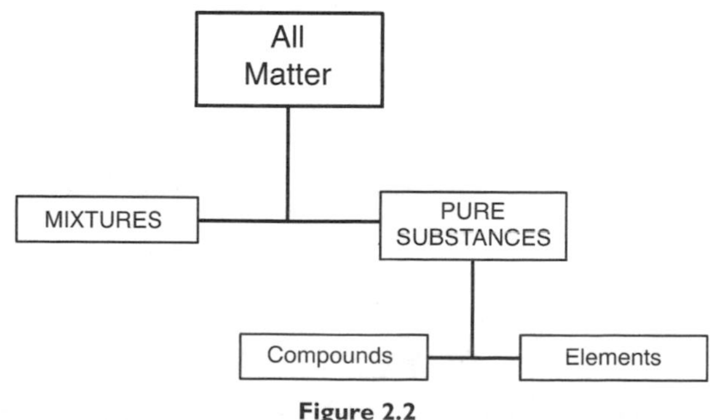

Figure 2.2

varieties of matter. Some elements exist free in nature; others are found only in combination. Free or combined, elements are the building blocks that make up every different variety of matter in the universe. Table 2.3 lists some commonly known elements together with the symbol chemists use for each.

How many of these names have you seen? How many of them have you heard of? A complete list of the 116 known elements appears in Chapter 3.

In general, the symbols are made up of the principal letter or letters in the name of the element. The symbols of elements that were known in antiquity are taken from their Latin names: copper (*cuprum*) Cu; gold (*aurum*) Au; iron (*ferrum*) Fe; lead (*plumbum*) Pb; mercury (*hydrargyrum*) Hg; potassium (*kalium*) K; silver (*argentum*) Ag; sodium (*natrium*) Na; tin (*stannum*) Sn.

If all matter were to be broken down into its constituent elements, the percentage of each element in nature would be as shown in Figure 2.3. The elements in your body can easily be remembered from the advertising sign shown in Figure 2.4. The symbols of the most common elements in the human body are contained in the sign. Use Table 2.3 to look up the names

Element	Symbol	Element	Symbol
aluminum	Al	neon	Ne
argon	Ar	nickel	Ni
arsenic	As	nitrogen	N
bromine	Br	oxygen	O
calcium	Ca	phosphorus	P
carbon	C	platinum	Pt
chlorine	Cl	plutonium	Pu
copper	Cu	potassium	K
fluorine	F	radium	Ra
gold	Au	silicon	Si
helium	He	silver	Ag
hydrogen	H	sodium	Na
iodine	I	sulfur	S
iron	Fe	tin	Sn
lead	Pb	uranium	U
magnesium	Mg	zinc	Zn
mercury	Hg		

Table 2.3 Common Elements with Their Chemical Symbols

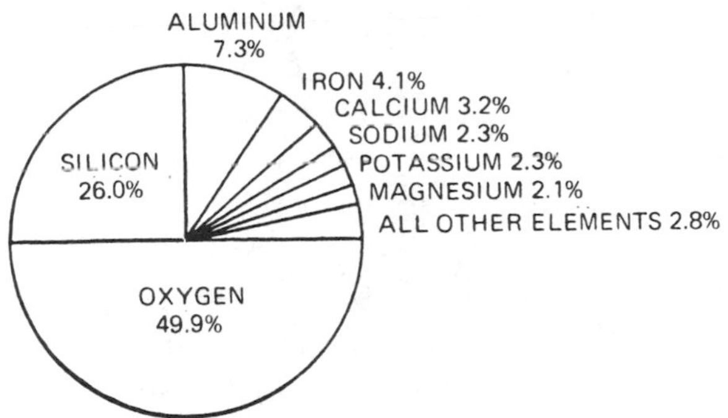

Figure 2.3 Distribution of Elements

of the 12 elements represented in the figure. The last two symbols in the sign, Na and Cl (combined as NaCl), stand for sodium chloride—ordinary table salt.

Compounds

A *compound* is a pure substance made up of elements that are chemically combined in a specific ratio. A compound is perfectly homogeneous (the same throughout) and has a definite composition regardless of its origin, location, size, or shape. A compound can be decomposed into its elements only by a chemical change. The elements in a compound cannot be separated by any physical means.

Figure 2.4

Compounds are much more abundant than elements. Millions of compounds are known. Water, sand, ammonia, sugar, salt, alcohol, and hydrogen peroxide are all examples of familiar compounds. Bear in mind that when elements combine to form a compound, the elements lose all their properties, and a new set of properties unique to the compound is created. For example, if you were to eat sodium or inhale chlorine gas, you would quickly die, for both of these elements are poisonous. When these two elements combine with each other, however, they form ordinary table salt, a substance eaten as part of a regular diet.

Mixtures

Most natural forms of matter are mixtures of pure substance. A *mixture* is a combination of pure substances (elements, with or without one or more compounds) held together by physical rather than chemical means. Soil and most rock, plants and animals, coal and oil, air and cooking gas, rivers and oceans—these are all mixtures. Mixtures can be of two types: homogeneous or heterogeneous. Homogeneous mixtures have the same composition throughout. Dissolving sugar in water forms a homogeneous mixture called a *solution*. Heterogeneous mixtures vary in composition at different locations within the sample. A mineral is a heterogeneous mixture. You can actually see the differences in composition from spot to spot within the mineral sample.

Mixtures differ from compounds in the following ways:

1. The components of a mixture retain their own properties. If you examine a fragment

of concrete you will observe that the grains of sand or gravel held together by the cement retain their identity and can be picked free. Their substance has not been changed in the formation of the concrete.

2. Unlike compounds, which have a definite, fixed composition, mixtures have widely varying compositions. Thus, solutions are mixtures. An infinite number of different salt–water solutions can be made simply by varying the amount of salt dissolved in the water.

3. Mixtures can be separated into their components by physical means. This separation can be accomplished by taking advantage of the differences in the physical properties of the ingredients. For example, no matter how completely you mix or grind salt and pepper together, the salt can be separated from the pepper by dissolving the salt in water. The insoluble pepper will remain unaffected. The separation is completed by straining or filtering the liquid through a piece of cloth or paper that will hold back the pepper, and then evaporating the liquid to dryness to recrystallize the salt.

Perhaps you would like to try this separation for yourself. Read the following procedure fully, and gather your materials before you start. Then, proceed with the experiment.

EXPERIMENT 1: Be sure to wear your safety goggles! Mix 1/4 teaspoon of salt and about 1/8 teaspoon of ground black pepper in a small drinking glass. Stir until a good mixture is obtained. Fill the glass about half full with water and stir until the salt is dissolved. Place a handkerchief or small piece of cloth loosely over the top of a small saucepan. Filter the liquid into the pan by pouring it carefully through the cloth. Notice that all the pepper remains on the cloth and that the filtrate (the salt solution) in the pan is perfectly clear. Taste the clear filtrate to see if the salt is really there. Over very low heat, boil away the water in the pan, being sure to remove the pan from the heat just as the last bit of liquid disappears. The white sediment left in the pan is recrystallized salt. Taste it to make sure. (Normally in the chemistry lab we never taste—we test—but in this case testing is permitted because you are working with simple foods—water, salt, and pepper.)

The separation of mixtures into ingredients is an important operation. Almost every industry that uses natural products as raw materials uses one or more of the basic methods of separating mixtures. All the methods take advantage of differences in the physical properties of the components. Some of the important methods of separating mixtures are as follows:

1. *Filtration*. As you saw in Experiment 1, passing a solution through a filter can separate substances. Larger or insoluble components stay on the filter while smaller or soluble components pass through its holes. Water is often filtered for use in factories and power plants. Water filtration systems reduce the cost of chemical treatment programs and save wear and tear on valves, pumps, heat exchangers, and cooling-tower nozzles.

2. *Sorting*. This process involves separating the desired component from the other components in a fragmented mixture. Some sorting may be done by hand; other sorting may be performed by machine. The mining of coal is an example of this process. In mining, the coal is blasted loose from within the earth and is then separated by sorting from the rock that accompanies the coal.

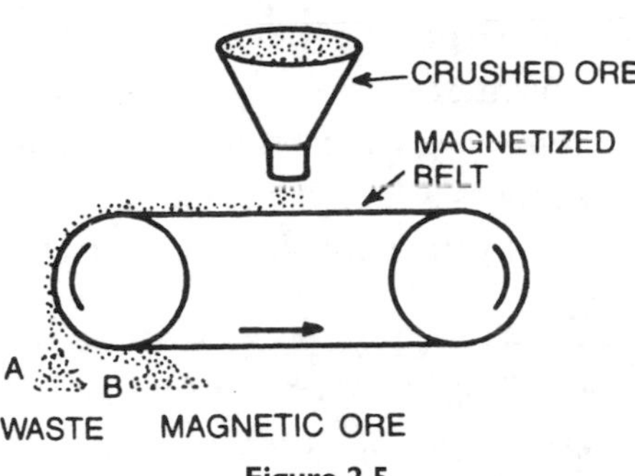

Figure 2.5

Figure 2.6

3. *Magnetic separation.* Some iron ore is magnetic. This ore is scooped up in giant shovels from the earth; it is crushed and poured onto a magnetized belt, as shown in Figure 2.5. The nonmagnetic waste material drops off the belt at A, but the magnetic ore clings to the belt until it reaches B, and is thus separated.

4. *Distillation.* This process takes advantage of the differences in boiling points of the components of a solution. *Boiling point* is the temperature at which a substance changes from a liquid to a gas or vapor. In distillation, the component with the lowest boiling point boils away first, leaving behind the components with higher boiling points. The component with the lower boiling point is said to be more *volatile* than the other components. The component that boils off as a gas is then trapped and (changed back to a liquid) by cooling and is collected in a new container.

   EXPERIMENT 2: Dissolve a teaspoon of sugar in a cup of water and place the solution in a teakettle. Taste the solution to be sure it is sweet. Heat the solution to boiling. Wearing your safety goggles and oven mitts, hold a large plate vertically with the far edge just in front of the spout of the kettle so that the steam strikes it (see Figure 2.6). *Be very careful—steam can severely burn you!* Let the condensed moisture (condensate) run down the plate into a clean cup. Taste the condensate. Is it sweet? Where is the sugar? Which is more volatile, water or sugar? Remove the kettle from the burner or turn it off before the solution boils completely to dryness.

   Simple distillation effectively separates water and sugar because the boiling points of these two substances are relatively far apart. When the boiling points of components to be separated are close together, a process known as *fractional distillation* is used. In this process, a large tower or column is erected above the boiling pot and fitted with cooling coils or a cooling jacket (Figure 2.7). This arrangement provides efficient condensation of the less volatile component and permits the more volatile ones to escape to a new container. The separation of crude petroleum into components used for such products as gasoline, lubricating oil, and fuel oil is accomplished by fractional distillation.

5. *Extraction.* The process of extraction involves dissolving a component out of a mixture using a suitable solvent. Water was the solvent used to extract salt from the salt and pepper mixture in Experiment 1. Water is also used to extract the flavor of coffee from ground coffee beans in a coffee maker.

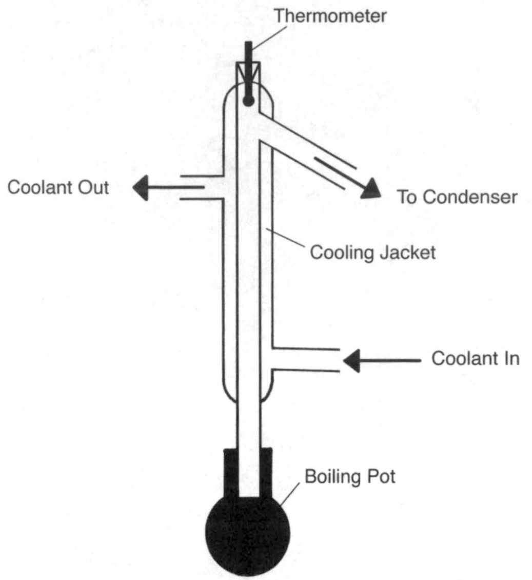

Figure 2.7 Fractional Distillation Column

Alcohol is used to extract vanillin (vanilla flavor) from vanilla beans. Sulfuric acid is used to extract copper from ore in the mining industry.

6. *Gravitation*. This process takes advantage of differences in density or specific gravity of the components in a mixture. In the panning of gold, the gold grains settle to the bottom of the pan because of their high density, and the lighter rocks are washed over the edge of the pan with water. In wheat harvesting, the light chaff is blown away from the denser wheat grains.

CHEMICAL CHANGE

TRIVIA: The fastest chemical reaction known occurs when light enters your eye and rhodopsin, a protein, starts a chemical chain reaction that allows you to see.

A *physical change* is an alteration of the properties of a substance that does not affect the substance itself. Hammering a piece of metal modifies its shape and increases its hardness, but the substance of the metal remains unchanged. Freezing water to ice, or boiling it to steam, causes a drastic change in physical properties, but the substance remains water.

In a chemical reaction, one or more substances change into another substance. A reaction forms one or more new substances with a new set of physical properties. Thus, physical change accompanies chemical change.

In contrast, a *chemical change*, or *reaction*, involves such a thorough change in a substance that an entirely new substance is formed in the process. The new substance that is created has its own set of properties, so physical change accompanies chemical change. Do you remember how completely the match was transformed as it burned? That was a chemical change. All burning involves chemical change, as does the rusting of iron, the toasting of bread, the taking of a photograph, and the digestion of food. Chemical change is a common occurrence.

There are four principal types of chemical change: combination, decomposition, single displacement, and double displacement. All chemical changes involve one or a combination of these basic varieties. Let's examine each type more thoroughly.

Combination

Combination is the direct joining of two or more simple substances—either elements or simple compounds or both—to form a more complex compound. Sometimes, burning or combustion reactions are placed in a separate reaction category, but combustion is simply a combination reaction that involves the oxygen in the air. For example, copper will join with oxygen in the air, when heated, to form the compound copper oxide.

EXPERIMENT 3: *Be sure to wear your safety goggles.* Remove about 2 in. of insulation from a 6-in. length of copper wire. Clean the exposed metal with sandpaper to a bright copper color. Heat the copper to redness in the upper part of a gas flame for about 1 minute. Let the wire cool. Notice the black coating on the copper. This is copper oxide. Scrape it off with a knife. This exposes copper metal again, as indicated by the color. Repeat this experiment until you are satisfied that the copper is really reacting or changing.

The reaction demonstrated in Experiment 3 can be stated in words, where the arrow means "yields":

Copper + **Oxygen** → **Copper oxide**

an element + an element → a compound of the two elements

The arrow indicates that the substances on the left are transformed into the substances on the right during the chemical change. Furthermore, many chemical changes are reversible, which means that the substances on the right can sometimes re-form the substances on the left.

Decomposition

Decomposition is the breaking down of a compound into simpler compounds or into its elements. For example, hydrogen peroxide decomposes in strong light or on contact with blood. Hydrogen peroxide is a compound of hydrogen and oxygen. It decomposes into water (a simpler compound of hydrogen and oxygen) and oxygen (an element).

EXPERIMENT 4: Sprinkle a few grains of dried yeast (the kind used for baking) on a plate and pour a small amount of hydrogen peroxide solution onto it. Watch the solution closely. The bubbles that form are bubbles of oxygen gas. The rest of the peroxide forms water.

Hydrogen peroxide → **Oxygen** + **Water**

a compound → an element + a compound

Single Displacement

Single displacement involves the substitution of one element for another in a compound. For example, if a piece of iron is dropped into a solution of sulfuric acid (the solution in a car battery), hydrogen gas bubbles out of the solution. Sulfuric acid is a compound of hydrogen, sulfur, and oxygen. The iron displaces the hydrogen, freeing it as an element, and forms a new compound, iron sulfate (iron, sulfur, and oxygen), in solution. This reaction can be stated as follows:

Iron + **Sulfuric Acid** → **Hydrogen** + **Iron sulfate**

an element + a compound → an element + a compound

Double Displacement

In double displacement reactions, two compounds react to form two new compounds by exchanging parts. To observe a reaction of this type we need a special solution. Let's make that solution first.

EXPERIMENT 5: We are going to extract a chemical indicator from purple cabbage. Take a head of purple cabbage and chop up a portion of it. Put the chopped cabbage into a glass or ceramic pot with enough water to cover the cabbage. Heat the cabbage to boiling and continue to boil it for 5 minutes. Cool the solution and then strain the purple liquid into a bottle. Store it in the refrigerator.

Keep this purple-cabbage "indicator solution" in a stoppered bottle. We will use this solution several times. This solution has the property of turning red to violet in acid solutions but is blue to blue-green in alkaline (basic) solutions. An alkali is a compound that is the opposite of an acid. An alkali neutralizes an acid to water and salt solution. Such a reaction is a double displacement type. Let's observe one of these reactions.

EXPERIMENT 6: *Be sure to wear safety goggles and gloves during this experiment.* Dissolve a few crystals of caustic soda (sodium hydroxide, commonly called lye) in one-quarter cup of water. Add 2 or 3 drops of the purple-cabbage solution prepared in Experiment 5. The blue color shows that sodium hydroxide is an alkali. Add vinegar (acetic acid) drop by drop, while stirring, to the sodium hydroxide solution. When the solution turns red, the reaction is completed. The acid has neutralized the alkali.

All the substances involved in the reaction in Experiment 6 are compounds. The reaction may be stated thus:

Sodium hydroxide	+	**Acetic acid**	→	**Sodium acetate**	+	**Water**
an alkali	+	an acid	→	a salt	+	a compound

Energy is the ability to do work. Heat, light, sound, and electricity are some of the many forms of energy. In every chemical change, energy is either given off (an exothermic reaction) or absorbed (an endothermic reaction). For example, water decomposes into its elements, hydrogen and oxygen, by absorbing electrical energy. It is an endothermic reaction. On the other hand, hydrogen and oxygen combine explosively to form water. That is an exothermic reaction.

It is important not to confuse the energy change in a reaction with the conditions under which a reaction occurs. Wood burns to produce heat, but not at room temperature. The wood must first be heated to a point considerably above room temperature before it will begin to burn. The high initial temperature is a condition under which the reaction of burning takes place. The production of heat by burning wood is a result of the reaction itself.

Many reactions take place only in the presence of a *catalyst*. A catalyst is a substance that alters the speed or rate of a chemical reaction without becoming permanently changed itself. A catalyst that slows a reaction is called a negative catalyst or *inhibitor*. Water is a catalyst for many reactions. Perfectly dry iron will not rust in dry air. Dry crystals of acetic acid will not react as in Experiment 6 with dry crystals of sodium hydroxide.

EXPERIMENT 7: *Be sure to wear safety goggles and hold the sugar cube with tongs. Be careful—sometimes the moisture in sugar cubes causes them to break apart and splatter.* With a match, try to burn a cube of sugar. Notice that the sugar melts but does not burn. Dip the other end of the sugar cube into some cigarette or cigar ashes. (Bear in mind that these ashes have already been burned!) Apply a flame to the ash-covered end of the cube. It now burns because of the presence of a catalyst.

SUMMARY

Matter is made up of elements, compounds, and mixtures. It has both chemical and physical properties. Physical change involves modification of properties without changing the substance. Physical properties describe matter as it exists. Chemical properties describe the ability of a form of matter to change to new substances. Chemical change involves forming new substances. Energy changes accompany chemical changes. Catalysts change the speed of a chemical reaction.

PROBLEM SET NO. 2

1. Which of the following physical properties of water are intensive and which are extensive?
   (a) Water freezes at 32 °F.
   (b) A sample of spring water has a temperature of 46 °F
   (c) Water dissolves in alcohol.

2. A sample of aluminum has a mass of 5.4 g. What is its volume?

3. How much cork (in cm^3) has the same mass as 1 cm^3 of gold?

4. The acid in a car battery has a specific gravity of 1.28. How much does 1 ft^3 of this acid solution weigh?

5. (a) Which substances in Table 3 will float on water?
   (b) Which substances in Table 3 will sink in mercury?

6. Name the elements in the human body that are listed in Figure 5.

7. Which of the following are mixtures and which are pure substances?
   (a) ocean water
   (b) a snowflake
   (c) ink
   (d) paint
   (e) sugar

8. Which of the following are physical changes and which are chemical changes?
   (a) snow melting
   (b) milk souring
   (c) baking soda neutralizing an "acid" stomach
   (d) gas bubbling out of ginger ale
   (e) the disappearance of sugar when it dissolves in water

9. Which type of chemical change is illustrated by each of the following?
   (a) the rusting of iron
   (b) the explosion of dynamite
   (c) acid "eating" a piece of zinc
   (d) boric acid or vinegar solutions poured on a lye spill on the floor

STRUCTURE OF MATTER

KEY TERMS

atom, electron, proton, neutron, nucleus, atomic number, atomic mass, quantum mechanical model, energy level, orbital, Aufbau principle, electron configuration, isotope, periodic table, group or family, period, metal, nonmetal, metalloid

We have seen that chemical change involves a complete transformation of one substance into another. Early chemists reasoned that such a thorough change must in some way be related to the way matter is constructed. They sought to find out the nature of the building blocks that make up the different varieties of matter. They hoped that once they could create some sort of model of the fundamental particles of matter, they could then explain not only the various ways that matter is constructed but also the behavior of substances during the process of chemical change.

As early as 450 B.C.E. some Greek philosophers reasoned that all matter was built up of tiny particles called atoms. This idea was slow to develop, but in 1802, the English chemist John Dalton suggested that all matter could be broken down into elements, the smallest particles of which he referred to as atoms. By 1895 the theory that atoms exist was extended to account for particles of matter even smaller than atoms. By 1913 evidence of the presence of several subatomic particles had been gathered. The work of probing into the structure of matter continues today. We have not yet learned the full story, and many features of the behavior of matter are still unexplained.

From a chemical point of view an atomic model has been developed that is quite satisfactory. We will use it to explain all common phenomena. We will also look at some of its weaker points to show that science is not static but rather is constantly changing as people progress toward a better understanding of nature.

ATOMS

If we were to take a strip of aluminum (an element) or a piece of copper (an element) and subdivide it into increasingly smaller pieces, we would eventually come to a tiny particle that if further subdivided would no longer show the properties of the element. We call the smallest particle of an element that has all the properties of the element an *atom*. Atoms are really quite small, but microscopes have been developed to see them. It would take about 100 million atoms to make a line 1 in. long. You can thus picture that a 1-in. cube would contain a huge number of atoms. The important thing to remember is that atoms are both tiny and numerous.

ATOMIC STRUCTURE

JOKE: A neutron goes into a bar and asks the bartender the price of a glass of wine. The bartender replies, "No charge for you."

In 1913 Niels Bohr, a Danish scientist, suggested an atomic model that serves chemists well to the present day. He pictured the atom as consisting of three basic kinds of particles: electrons, protons, and neutrons. The *electron* is a particle possessing a negative (−) electrical charge. The *proton* is a particle consisting of a positive (+) electrical charge equal in magnitude (but opposite in type) to the charge on the electron. The *neutron* is a particle with no electrical charge. The proton and neutron have essentially the same mass. A mass of 1 *atomic mass unit* (1 amu) has been assigned to each. The electron is much smaller; it weighs only about 1/1848 as much as the proton or the neutron. From a chemical point of view, we can consider the mass of the electron to be negligible. Table 3.1 summarizes the properties of these three particles that make up an atom.

In the Bohr model of the atom, protons and neutrons are considered to be packed together in the center of the atom to form what is known as the *nucleus*. The average nucleus occupies about one ten-thousandth of the total volume of an atom. Electrons travel around this nucleus in orbits that are relatively far away from the nucleus. The Bohr model describes the electrons as occupying orbits, much like the orbits of the planets circling the sun; however, current understanding is that the electrons occupy electron clouds of different sizes and shapes, at different distances from the nucleus.

Subatomic Particle	Electrical Charge	Mass (amu)
Electron	−1	0
Proton	+1	1
Neutron	0	1

Table 3.1 Constituents of an Atom

At this point, three important characteristics of atoms can be stated. These three atomic characteristics are summarized in Table 3.2.

1. Despite the presence of electrically charged particles in atoms, all elements are observed to be electrically neutral. Therefore, the number of protons (positive) in the nucleus of an atom must be equal to the number of electrons (negative) surrounding the nucleus.

2. Because elements differ from one another, their atoms must differ structurally. Each element has an atomic number. The atomic number is more than just a catalog number; it is a special characteristic of each element. The *atomic number* is equal to the number of protons in the nucleus of the atom. Thus, each atom of hydrogen (atomic number 1) has a single proton in its nucleus. Each atom of uranium (atomic number 92) has 92 protons in its nucleus. Because atoms are electrically neutral, the atomic number also equals the number of electrons in the atom.

3. Equal numbers of atoms of different elements weighed under the same conditions have a different mass. Therefore, the atoms of different elements have different atomic masses. The *atomic mass* of an atom is equal to the sum of the number of protons and the number of neutrons in the nucleus of the atom. (Remember that the mass of the electron is negligible in comparison with that of a proton or neutron.) Thus, essentially all the mass of an atom is contained in its nucleus. Atomic masses are relative,

Characteristic	Structural Explanation
neutral atoms	number of protons = number of electrons
atomic number	number of protons = number of electrons = atomic number
atomic mass	number of protons + number of neutrons = atomic mass

Table 3.2 Summary of Atomic Characteristics

which is to say they do not give the number of grams or pounds that an atom weighs; they merely tell how much heavier or lighter an atom of one element is compared with another. For example, the atomic mass of oxygen is 16, and the atomic mass of helium is 4. This means that each atom of oxygen has a mass 16/4, or 4 times as much as each atom of helium.

Let's pause here to see how our atomic model is shaping up. Can you visualize a nugget of kernels like a popcorn ball stuck in the center of a very large ball of cotton candy? Imagine that the popcorn ball also has peanuts in it, giving it two different kinds of kernels (particles). We can think of the popcorn kernels as protons and the peanuts as neutrons, all tightly held together in the nucleus. The cotton candy ball represents the space in which the electrons are most likely to be found. However, if the nucleus of an actual atom were the size of our popcorn–peanut ball, then the ball of cotton candy would be miles in diameter! A model of a hydrogen atom would consist of a single piece of popcorn with a large cotton candy ball surrounding it. A uranium model would contain quite a lot of popcorn (92 pieces) and many peanuts (146), surrounded by a large cotton candy sphere. The sum of the particles in the uranium nucleus is 238, which is the atomic mass of uranium. Figure 3.1 gives us another picture of the modern-day atomic model.

DISTRIBUTION OF ELECTRONS

The model of the atom that we have been considering so far is too simplistic. To better explain experimental evidence and to provide a clearer description of the atom, the quantum mechanical model of the atom was developed. The *quantum mechanical model* describes the electrons as being found in volumes of space at definite average distances from the nucleus (*energy levels*) and being contained in "electron clouds" or *orbitals*, which are volumes of space of different shapes in which the probability of finding the electron is high. Each energy level contains a definite number of orbitals of specific types (the different types being called *subshells* or sublevels), and each orbital or subshell contains a specific number of electrons. Here are the first four energy levels and the orbitals and subshells they contain:

Energy level 1 — s orbital (spherical with nucleus at center)

Energy level 2 — 2 subshells: s and p
- s orbital (larger sphere than 1s)
- p subshell (3 orbitals—dumbbell shaped)

Energy level 3 — 3 subshells: s, p, and d
- s orbital (larger sphere than 2s)
- p subshell (3 orbitals—dumbbells larger than 2p)

d subshell (5 orbitals—various shapes)

Energy level 4 — 4 subshells—s, p, d, and f
- s orbital (larger than 3s)
- p subshell (3 orbitals—dumbbells larger than 3p)
- d subshell (5 orbitals—larger than 3d)
- f subshell (7 orbitals—various shapes)

Figure 3.1 shows the shapes of the s, p, and d orbitals.

When electrons fill the various energy levels, they follow a certain set of rules called the *Aufbau principle*:

1. The electrons fill the lowest energy level first.

2. Each orbital can hold a maximum of two electrons.

The s, p, and d Orbitals

(a.) s Orbital

p_z p_x p_y

(b.) p Orbitals

$d_{x^2-y^2}$ d_{xz} d_{z^2}

d_{xy} d_{yx}

(c.) d Orbitals

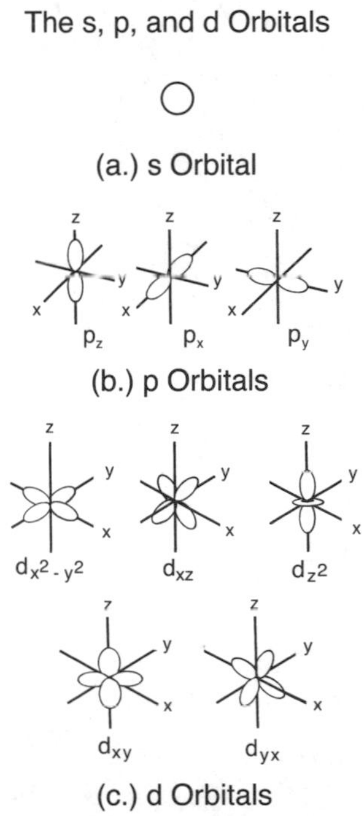

Figure 3.1 The s, p, and d Orbitals

3. If there are two or more vacant orbitals of the same energy level (as in the three p orbitals of the p subshell), each orbital accepts one electron before the orbitals in the subshell begin filling with the second electron. This is called *Hund's rule.*

Figure 3.2 (a) shows an energy diagram of these subshells and orbitals. Here the orbitals, represented by dashes, hold a maximum of two electrons each. Electrons are shown as up or down arrows, paired in an orbital. Figure 3.2 (b) is the energy diagram for nitrogen, which has 7 electrons. Notice that the 1s orbital fills first, then the 2s orbital, and finally one electron goes into each of the 2p subshells. The next electron that would be added (to represent oxygen with 8 electrons) would pair up with an electron in any of the half-filled 2p orbitals.

To help you remember the pattern of orbital filling, follow the pattern of arrows in Figure 3.3.

As useful as energy level diagrams are, they are awkward to use. As a shortcut, chemists often write the electron configuration of an atom. The *electron configuration* is a notation that shows the type of orbital at each energy level that contains electrons and the number of electrons in each orbital. The orbitals at a particular energy level are denoted by the number of the energy level (1, 2, 3) and the letter of the type of orbital (s, p, d). Thus, 3p would represent a p orbital at energy level 3.

Look at Figure 3.2 (b), the energy level diagram for nitrogen. It has 2 electrons in an s orbital at energy level 1, 2 electrons in an s orbital at energy level 2, and 3 electrons in a p orbital at energy level 3. Nitrogen's electronic configuration, therefore, is:

Nitrogen $1s^2 2s^2 2p^3$

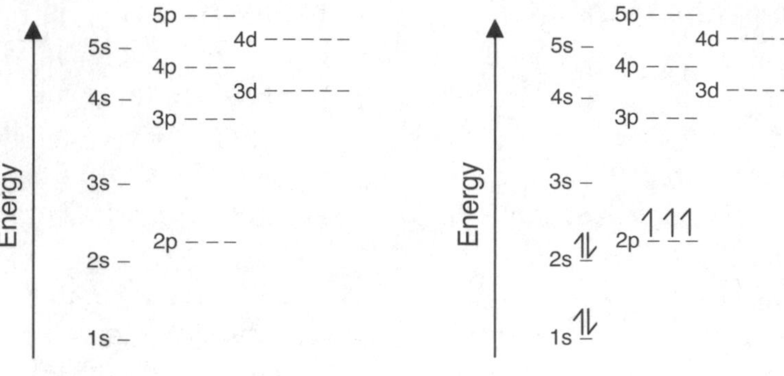

(a.) Relative Position of Energy Levels

(b.) Energy Level Diagram for Nitrogen

Figure 3.2 Energy Level Diagrams

NOTE: We did not distinguish among the three p subshells. It can be done, but most chemists like to keep things as simple as possible and use the preceding notation.

An even more condensed form configuration uses brackets to represent the electron configuration of the preceding noble gas element (group 8A (18) in the periodic table; see Figure 3.4). (Remember that the noble gas elements include helium, neon, argon, krypton, xenon, and radon.) In the case of nitrogen, helium (He) is the preceding noble gas. You can therefore use [He] to represent helium's electron configuration of $1s^2$. The electron configuration of nitrogen then becomes:

Nitrogen [He] $2s^2 2p^3$

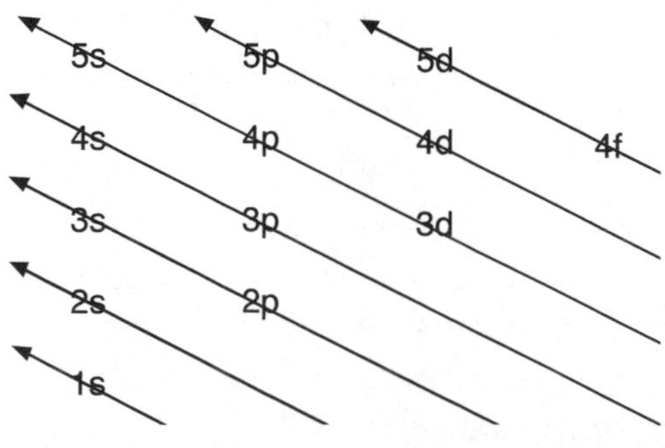

Figure 3.3 Orbital Filling Pattern

Chlorine (Cl, 17 electrons) has the electron configuration:

Cl $1s^2 2s^2 2p^6 3s^2 3p^5$ or [Ne] $3s^2 3p^5$

Table 3.3 gives the condensed electron configuration of all the elements that are currently known.

Table 3.3 Distribution of Electrons

Atomic Number	Element Name	Element Symbol	Electron Configuration
1	hydrogen	H	$1s^1$
2	helium	He	$1s^2$
3	lithium	Li	$[He]2s^1$
4	beryllium	Be	$[He]2s^2$
5	boron	B	$[He]2s^2 2p^1$
6	carbon	C	$[He]2s^2 2p^2$
7	nitrogen	N	$[He]2s^2 2p^3$
8	oxygen	O	$[He]2s^2 2p^4$
9	fluorine	F	$[He]2s^2 2p^5$
10	neon	Ne	$[He]2s^2 2p^6$
11	sodium	Na	$[Ne]3s^1$
12	magnesium	Mg	$[Ne]3s^2$
13	aluminum	Al	$[Ne]3s^2 3p^1$
14	silicon	Si	$[Ne]3s^2 3p^2$

(continued)

Table 3.3 (Continued)

Atomic Number	Element Name	Element Symbol	Electron Configuration
15	phosphorus	P	$[Ne]3s^{2}3p^{3}$
16	sulfur	S	$[Ne]3s^{2}3p^{4}$
17	chlorine	Cl	$[Ne]3s^{2}3p^{5}$
18	argon	Ar	$[Ne]3s^{2}3p^{6}$
19	potassium	K	$[Ar]4s^{1}$
20	calcium	Ca	$[Ar]4s^{2}$
21	scandium	Sc	$[Ar]3d^{1}4s^{2}$
22	titanium	Ti	$[Ar]3d^{2}4s^{2}$
23	vanadium	V	$[Ar]3d^{3}4s^{2}$
24	chromium	Cr	$[Ar]3d^{5}4s^{1}$
25	manganese	Mn	$[Ar]3d^{5}4s^{2}$
26	iron	Fe	$[Ar]3d^{6}4s^{2}$
27	cobalt	Co	$[Ar]3d^{7}4s^{2}$
28	nickel	Ni	$[Ar]3d^{8}4s^{2}$
29	copper	Cu	$[Ar]3d^{10}4s^{1}$
30	zinc	Zn	$[Ar]3d^{10}4s^{2}$
31	gallium	Ga	$[Ar]3d^{10}4s^{2}4p^{1}$
32	germanium	Ge	$[Ar]3d^{10}4s^{2}4p^{2}$
33	arsenic	As	$[Ar]3d^{10}4s^{2}4p^{3}$
34	selenium	Se	$[Ar]3d^{10}4s^{2}4p^{4}$
35	bromine	Br	$[Ar]3d^{10}4s^{2}4p^{5}$
36	krypton	Kr	$[Ar]3d^{10}4s^{2}4p^{6}$
37	rubidium	Rb	$[Kr]5s^{1}$
38	strontium	Sr	$[Kr]5s^{2}$
39	yttrium	Y	$[Kr]4d^{1}5s^{2}$
40	zirconium	Zr	$[Kr]4d^{2}5s^{2}$
41	niobium	Nb	$[Kr]4d^{4}5s^{1}$
42	molybdenum	Mo	$[Kr]4d^{5}5s^{1}$
43	technetium	Tc	$[Kr]4d^{5}5s^{2}$
44	ruthenium	Ru	$[Kr]4d^{7}5s^{1}$
45	rhodium	Rh	$[Kr]4d^{8}5s^{1}$
46	palladium	Pd	$[Kr]4d^{10}$

Atomic Number	Element Name	Element Symbol	Electron Configuration
47	silver	Ag	$[Kr]4d^{10}5s^{1}$
48	cadmium	Cd	$[Kr]4d^{10}5s^{2}$
49	indium	In	$[Kr]4d^{10}5s^{2}5p^{1}$
50	tin	Sn	$[Kr]4d^{10}5s^{2}5p^{2}$
51	antimony	Sb	$[Kr]4d^{10}5s^{2}5p^{3}$
52	tellurium	Te	$[Kr]4d^{10}5s^{2}5p^{4}$
53	iodine	I	$[Kr]4d^{10}5s^{2}5p^{5}$
54	xenon	Xe	$[Kr]4d^{10}5s^{2}5p^{6}$
55	cesium	Cs	$[Xe]6s^{1}$
56	barium	Ba	$[Xe]6s^{2}$
57	lanthanum	La	$[Xe]5d^{1}6s^{2}$
58	cerium	Ce	$[Xe]4f^{1}5d^{1}6s^{2}$
59	praseodymium	Pr	$[Xe]4f^{3}6s^{2}$
60	neodymium	Nd	$[Xe]4f^{4}6s^{2}$
61	promethium	Pm	$[Xe]4f^{5}6s^{2}$
62	samarium	Sm	$[Xe]4f^{6}6s^{2}$
63	europium	Eu	$[Xe]4f^{7}6s^{2}$
64	gadolimium	Gd	$[Xe]4f^{8}6s^{2}$
65	terbium	Tb	$[Xe]4f^{9}6s^{2}$
66	dysprosium	Dy	$[Xe]4f^{10}6s^{2}$
67	holmium	Ho	$[Xe]4f^{11}6s^{2}$
68	erbium	Er	$[Xe]4f^{12}6s^{2}$
69	thulium	Tm	$[Xe]4f^{13}6s^{2}$
70	ytterbium	Yb	$[Xe]4f^{14}6s^{2}$
71	lutetium	Lu	$[Xe]4f^{14}5d^{1}6s^{2}$
72	hafnium	Hf	$[Xe]4f^{14}5d^{2}6s^{2}$
73	tantalum	Ta	$[Xe]4f^{14}5d^{3}6s^{2}$
74	tungsten	W	$[Xe]4f^{14}5d^{4}6s^{2}$
75	rhenium	Re	$[Xe]4f^{14}5d^{5}6s^{2}$
76	osmium	Os	$[Xe]4f^{14}5d^{6}6s^{2}$
77	iridium	Ir	$[Xe]4f^{14}5d^{7}6s^{2}$
78	platinum	Pt	$[Xe]4f^{14}5d^{9}6s^{1}$

(continued)

Table 3.3 (Continued)

Atomic Number	Element Name	Element Symbol	Electron Configuration
79	gold	Au	$[Xe]4f^{14}5d^{10}6s^1$
80	mercury	Hg	$[Xe]4f^{14}5d^{10}6s^2$
81	thallium	Tl	$[Xe]4f^{14}5d^{10}6s^26p^1$
82	lead	Pb	$[Xe]4f^{14}5d^{10}6s^26p^2$
83	bismuth	Bi	$[Xe]4f^{14}5d^{10}6s^26p^3$
84	polonium	Po	$[Xe]4f^{14}5d^{10}6s^26p^4$
85	astatine	At	$[Xe]4f^{14}5d^{10}6s^26p^5$
86	radon	Rn	$[Xe]4f^{14}5d^{10}6s^26p^6$
87	francium	Fr	$[Rn]7s^1$
88	radium	Ra	$[Rn]7s^2$
89	actinium	Ac	$[Rn]6d^17s^2$
90	thorium	Th	$[Rn]6d^27s^2$
91	protactinium	Pa	$[Rn]5f^26d^17s^2$
92	uranium	U	$[Rn]5f^36d^17s^2$
93	neptunium	Np	$[Rn]5f^46d^17s^2$
94	plutonium	Pu	$[Rn]5f^67s^2$
95	americium	Am	$[Rn]5f^77s^2$
96	curium	Cm	$[Rn]5f^76d^17s^2$
97	berkelium	Bk	$[Rn]5f^97s^2$
98	californium	Cf	$[Rn]5f^{10}7s^2$
99	einsteinium	Es	$[Rn]5f^{11}7s^2$
100	fermium	Fm	$[Rn]5f^{12}7s^2$
101	mendelevium	Md	$[Rn]5f^{13}7s^2$
102	nobelium	No	$[Rn]5f^{14}7s^2$
103	lawrencium	Lr	$[Rn]5f^{15}6d^17s^2$
104	rutherfordium	Rf	$[Rn]5f^{14}6d^27s^2$
105	dubnium	Db	$[Rn]5f^{14}6d^37s^2$
106	seaborgium	Sg	$[Rn]5f^{14}6d^47s^2$
107	bohrium	Bh	$[Rn]5f^{14}6d^57s^2$
108	hassium	Hs	$[Rn]5f^{14}6d^67s^2$
109	meitnerium	Mt	$[Rn]5f^{14}6d^77s^2$
110	darmstadtium	Ds	$[Rn]5f^{14}6d^97s^1$
111	unununium	Uuu	$[Rn]5f^{14}6d^{10}7s^1$
112	ununbium	Uub	$[Rn]5f^{14}6d^{10}7s^2$
113*	ununtrium	Uut	$[Rn]5f^{14}6d^{10}7s^27p^1$
114†	ununquadium	Uuq	$[Rn]5f^{14}6d^27s^27p^2$
115*	ununpentium	Uup	$[Rn]5f^{14}6d^{10}7s^27p^3$
116‡	ununhexium	Uuh	$[Rn]5f^{14}6d^{10}7s^27p^4$

* Experimentally formed in February 2004; not yet ratified.
† Reported informally in January 1999 but not yet ratified.
‡ Isolated in December 2002.

One other important fact about the arrangement of electrons in atoms is that the electrons in the outermost energy level (particularly the s and p electrons) are those involved in chemical reactions and bonding. These are called the *valence electrons.*

ISOTOPES

Evidence shows that not all the atoms of a given element are identical; they can vary in atomic mass. Atoms of an element with different atomic masses are called *isotopes* of the element. About 2000 isotopes are known of an estimated 8000 believed to exist.

For example, there are three different kinds of hydrogen atoms. The first has an atomic mass of 1, the second has an atomic mass of 2, and the third has an atomic mass of 3. The only difference is the number of neutrons in the nucleus of each isotope. All three isotopes have but one proton, because all are atoms of hydrogen and have atomic number 1. Similarly, all three isotopes have a single electron in the 1s orbital, because each must remain electrically neutral. Isotopes, then, are atoms of the same element that possess *different numbers of neutrons in their nuclei*. These isotopes can be distinguished from one another using the following notation:

$$^{A}_{Z}X$$

where X is the atomic symbol of an element, Z is the atomic number (number of protons), and A is the mass number (the number of protons + neutrons). In the neutral atom, the atomic number also equals the number of electrons. Therefore, if the atomic number is subtracted from the mass number (A minus Z), the number of neutrons can be determined.

Number of neutrons = Mass number − Atomic number

The three isotopes of hydrogen are then represented as follows:

$^{1}_{1}H$	$^{2}_{1}H$	$^{3}_{1}H$
1 proton	1 proton	1 proton
1 electron	1 electron	1 electron
0 neutrons	1 neutron	2 neutrons

Another way of representing isotopes is to give the element name followed by a hyphen and then the mass number. The three isotopes of hydrogen are then written as hydrogen-1, hydrogen-2, and hydrogen-3.

ATOMIC MASS

Almost all the elements have isotopes. The relative abundance of each isotope of a given element in nature varies considerably. For example, the element chlorine has two principal isotopes, one of atomic mass 35 amu and one of atomic mass 37 amu. If you picked up a container of chlorine, about 75% of the chlorine atoms in the container would have atomic mass 35 amu and the other 25% would have atomic mass 37 amu. The average mass of all the atoms in the container would be about 35.5. The atomic mass of chlorine can be found in the periodic table, Figure 3.4. You will see that it is 35.45 amu. This number is a weighted average atomic mass—that is, it is the average of the atomic masses of the isotopes of the element, taking into account the relative abundance of each isotope in nature.

$$(\text{percent})(\text{mass}) + (\text{percent})(\text{mass}) = (\text{weighted})\ \text{average}$$

$$(0.75)(35) + (0.25)(37) = 35.5$$

Actual atomic masses are given in grams, but the relative values used in the periodic table are given based on a comparison of the mass of a given number of atoms of the element to the mass of the same number of atoms of the isotope: carbon-12, whose mass is 12 amu. Thus, 1 amu is the mass of a single atom of carbon-12.

SYMBOLS

Table 2.3 (Chapter 2) listed the symbols of some common elements. Symbols are very important in chemistry, because they represent three things:

1. the name of an element
2. one atom of an element
3. a quantity of the element equal in mass to its atomic mass

For example, when we write the symbol "O," we mean not only the name, oxygen, but we also represent a single oxygen atom with this symbol. What is perhaps most important of all, because oxygen has an atomic mass of 16 amu, the symbol "O" stands for 16 units of mass of this element.

THE ATOMIC MODEL REVISITED

Our modern atomic model is now sufficiently developed to explain chemical phenomena. The model contains a nucleus, composed of positive protons and neutral neutrons, that constitutes the mass of the atom. Surrounding the nucleus are clouds of electrons carrying sufficient negative charge to offset the positive charge on the nucleus.

If you are familiar with the properties of electricity, you know that opposite charges attract one another, and like charges repel one another. Why, then, aren't the negative electrons attracted into the positive nucleus, making the atom collapse? and why doesn't the nucleus fly apart as a result of the repulsions among the protons?

The answer to the first question is in the idea that the electrons are moving about the nucleus. If you attach a piece of string to a ball and whirl the ball around you, the string will get tight. The whirling motion would make the ball fly away from your hand, but the string keeps it from doing so. In the atom, the electrical attraction between the positive nucleus and the negative electrons balances the tendency of the whirling electrons to escape from the nucleus.

The nucleus does not fly apart as a result of the repulsions among the protons because an attractive force known as the strong nuclear force holds the nucleus together. The exact nature of the attractive force is not well understood, but it is known to be the strongest force in the universe, stronger than both the electromagnetic force and gravity, but it acts only at a distance comparable to the diameter of a proton or neutron. Another nuclear force, the weak nuclear force, is involved in radioactive decay processes.

THE PERIODIC TABLE

In 1869 two scientists, Julius Lothar Meyer, a German, and Dmitri Mendeleev, a Russian, independently came up with a classification system for the elements based on all the chemical information known to that time. This system arranged all the known elements according to their atomic number in what is known as the *periodic table.* This table (Figure 3.4) shows the atomic number, symbol, and atomic mass of each element. Where atomic masses have not yet been accurately measured, the approximate value is given in parentheses.

The vertical columns in the table are called *groups* or *families*. These groups are numbered on the periodic table in two ways. The newer way is to number the groups (columns) from left to right sequentially, 1–18; however, many chemists prefer the older method, in which a number and a letter are used (e.g., IIA, VIIIB).

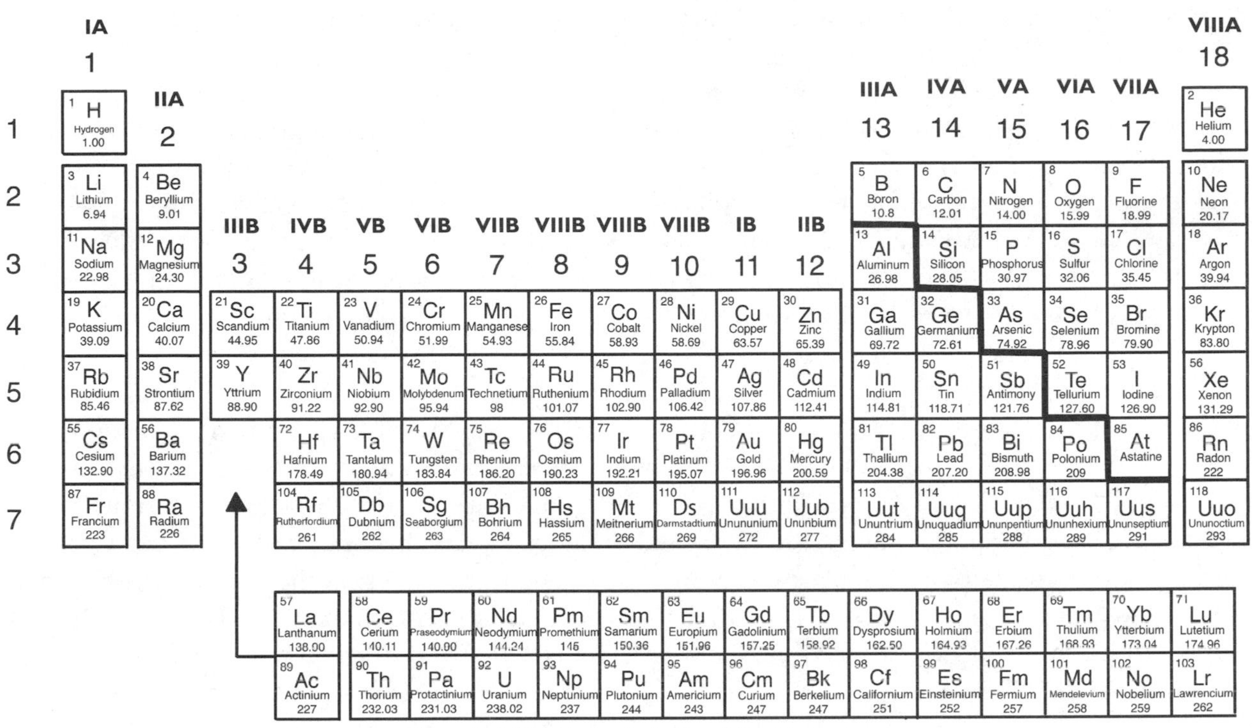

Figure 3.4 The Periodic Table of the Elements

All of the elements in a group have the same electronic structure in their outermost subshell and thus have similar chemical properties. For example, all the elements in group IA have 1 electron in the outermost subshell (check this with Table 3.3). Elements in group IIA have 2 outermost electrons, elements in group IIIA have 3 outermost electrons, and so on. For the A elements, the number preceding the A gives the number of electrons in the outermost energy level. The noble gas elements (VIIIA) at the far right of the table have 8 outermost electrons. The B elements, which are called the *transition elements*, can be thought of as arranged in subgroups, and all these have 2 outermost electrons with the exception of the copper-silver-gold subgroup, which has only 1 outermost electron.

The horizontal rows of elements are called *periods*. There are a total of seven periods on the periodic table. As we move from left to right in a period, electrons are being added to the same energy level. Elements within a period have different chemical properties. It is important to note that the last element of each period is a noble gas element. The lanthanide elements of the inner transition series are part of period 6, and the actinide elements of the inner transition series are part of period 7.

All four structural types of elements are shown in the table. The simple elements are found in groups 1A through 7A, and because their logical conclusion, the completely filled shell, is found in the noble gases, they can be said to be found in groups 1A though 8A. The transition elements are at the center; the lanthanide and actinide elements are extracted from the table and listed at the bottom. The noble gas elements form a group at the extreme right.

Notice the stair-stepped line on the right-hand side of the periodic table in Figure 3.4. Elements to the left of that line are commonly called *metals*. They have physical and chemical properties that we normally associate with metals in everyday life: they have a luster, they are malleable (able to be hammered into a thin sheet) and ductile (able to be drawn out into a thin wire), are good conductors of electricity and heat, and in chemical reactions they tend to lose electrons. Elements to the right of the stair-stepped line are called *nonmetals* and have properties opposite those of the metals: they don't have a luster, they are not malleable or ductile, they are nonconductors of electricity and heat, and in chemical reactions they tend to accept electrons. A group of elements that border the stair-stepped line are called the *metalloids*. They include B, Si, Ge, As, Sb, Te, and At. These elements have properties of both metals and nonmetals and are economically very important in the semiconductor industry because of their unique electrical conductivity properties.

The chemical behavior of elements is based on the electronic structure of their atoms, particularly the structure of the outer energy level. Because the members of a group of elements have the same structure in their outer subshell, we can expect the members of that group to show similar chemical behavior. For this reason, we can expect to find many uses for the periodic table as we explore the chemical behavior of elements.

SUMMARY

An atom is the smallest particle of an element capable of showing the properties of the element. There are two principal regions within an atom: a central nucleus and the electron clouds that surround it. In the nucleus are positively charged protons and neutral neutrons. Each of these particles has a mass of 1 amu. Each electron has a negative charge and negligible mass.

The atomic number of an atom is the number of protons in the nucleus of the atom. The atomic mass of an atom is the sum of the number of protons and neutrons in the nucleus. Isotopes are atoms of the same element possessing different atomic masses owing to differing numbers of neutrons.

The atomic mass of an element is the average atomic mass of all the isotopes of the element taking into account the relative natural abundance of each isotope of the element. Chemical symbols represent the name of an element, one atom of the element, and one atomic mass's worth of the element. The periodic table is a structural classification of the elements based on the distribution of electrons in the energy levels.

PROBLEM SET NO. 3

1. Write the electron configuration of the following elements:
   (a) carbon
   (b) silicon
   (c) bromine

2. Write the isotope representation ($^{A}_{Z}X$) for the following elements:
   (a) atomic number 5; atomic mass 11
   (b) atomic number 12; atomic mass 24
   (c) atomic number 18; atomic mass 40
   (d) atomic number 34; atomic mass 79
   (e) atomic number 50; atomic mass 93

COMPOUNDS

KEY TERMS

noble gas, bonding, ion, ionic bonding, activity, covalent bonding, molecule, oxidation number, formula, structural formula, Lewis structural formula, electronegativity, formula mass, mole

Elements in the free or uncombined state make up only a small fraction of matter. Most matter occurs as compounds or mixtures of compounds. Let's put our atomic model of the atom to the test to see whether it is useful in explaining how elements can combine to form all the various compounds.

THE NOBLE GAS ELEMENTS

TRIVIA: The element helium was first discovered on the sun in 1868 by examining spectral lines given off by the sun. But it wasn't until 1895 that helium was found on Earth.

Take another look at Table 3.3 in Chapter 3. Find the *noble gas* elements helium, neon, argon, krypton, xenon, and radon. Notice that except for helium, all these elements have eight valence (s and p) electrons in the outermost energy level. Then, notice that in each case the next element in the table has one electron in a new energy level. The energy levels below that level are filled, and that single electron must occupy the lowest unfilled energy level.

Now, look at these same elements on the periodic table (Figure 3.4 in Chapter 3). You will find all of them in the column at the far right under the heading group VIIIA or 18. Each noble gas occurs at the end of a period. Among the noble gas elements, krypton, xenon, and radon react with reluctance to form compounds, but so far, it has not been possible to form compounds of helium, neon, and argon.

Property	Helium	Neon	Argon	Krypton	Xenon	Radon
Symbol	He	Ne	Ar	Kr	Xe	Rn
Atomic number	2	10	18	36	54	86
Atomic mass	4.003	20.179	39.948	83.80	131.29	222
Density, g/cm^3	0.00018	0.0009	0.00179	0.00374	0.0058	0.0099
Solubility, mL/100 mL water	1.49	1.5	5.6	6.0	28.4	0.000002
Boiling point, °C	−268.9	−246.0	−185.7	−152.3	−107.1	−61.8
Melting point, °C	−272.2	−248.7	−189.2	−156.6	−111.9	−71.0

Table 4.1 Properties of Noble Gas Elements

You might ask why we bring up this group of elements that form very few compounds in a chapter devoted to the formation of compounds. Well, these elements possess a structure so stable that they resist compound formation. This is because they have a filled valence energy level. We observe in nature that this condition produces stability due to a lower energy state than that possessed by most all the other elements. Other elements are less stable because they do not have a filled valence energy level. When elements combine to form compounds they undergo a rearrangement of their electron structures to gain an electron configuration similar to that of a noble gas element (that is, a filled valence energy level). Such a rearrangement causes these elements to become structurally more stable.

BONDING

The combining of elements to form compounds through a shift of electronic structure is known as *bonding*. The major ways in which elements achieve noble gas configuration are by the loss or gain of electrons or the sharing of electrons. These two ways correspond to two types of bonding: ionic and covalent.

Ionic Bonding

Consider for a moment the structure of an atom of sodium. It has an electron configuration of $1s^2 2s^2 2p^6 3s^1$. Sodium has one electron in its valence energy level (energy level 3) and eight electrons in energy level 2. If that lone electron in energy level 3 is removed from the sodium atom, the remaining electron configuration is identical with the structure of neon, the noble gas element that immediately precedes sodium in the periodic table. The removal of the electron changes the nature of the atom by causing it to have one excess positive charge. It is no longer a neutral sodium atom because, although its nucleus is still that of sodium, it possesses one fewer electron than the neutral sodium atom. Nor is the atom neon, because its nucleus has too many protons. An atom that acquires an electrical charge owing to the loss or gain of electrons is called an *ion*. If that ion has a positive charge, due to the loss of one or more electrons, it is called a *cation*. If it has a negative charge, due to the gain of one or more electrons, it is called an *anion*. The ion we are discussing is a sodium cation, because it lost one electron and now has a positive charge. Ions possess properties that are totally different from those of the atoms from which they come.

The idea of forming an ion from an atom is reasonable enough, but where does the electron go that is lost? The elements near sodium in the periodic table, like potassium or calcium or magnesium, will not accept an additional electron, for it will not bring them nearer a stable electron configuration; however, at the other end of the periodic table are the nonmetal elements—for example, chlorine. Chlorine has seven electrons in its outermost subshell. The addition of one more electron will give it the same stable configuration as argon, a noble gas element. The gain of the electron by the chlorine atom means that now there is one more electron than protons, one more negative charge than positive charges. This chemical species is neither a chlorine atom nor an argon atom. It is a *chloride ion*, and because it has an overall negative charge, it is an anion. Once the electron has been transferred from the sodium atom to the chlorine atom, these are two oppositely charged ions (sodium = positive, chlorine = negative) that are capable of attracting each other electri-

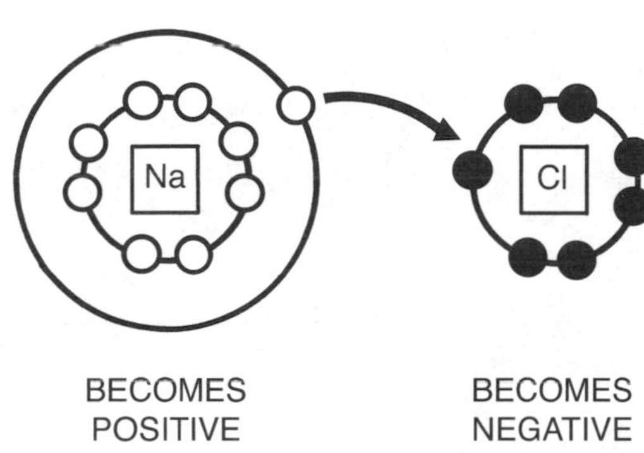

Figure 4.1 Ionic Bonding

cally. (Recall that unlike charges attract and like charges repel.) The ions do so and form the familiar compound sodium chloride: ordinary table salt. Figure 4.1 shows the formation of this compound. The process of forming a compound through the transfer of electrons is called *ionic bonding*.

A careful consideration of the periodic table will lead to the discovery of the elements that are involved in ionic bonding. Elements in groups IA, IIA, and IIIA give up one, two, or three electrons, respectively, to form positive ions. Elements in groups VA, VIA, and VIIA accept three, two, or one electron, respectively, to form negative ions. In general, metals give up electrons to form cations, and nonmetals accept electrons to form anions. Table 4.2 gives the symbols for typical ions formed by elements in each of these groups. Also, it has been observed that the loss and gain of one or two electrons is quite common, with the loss or gain of three electrons being far less common. Loss or gain of more than three electrons is not common at all in the A family elements.

Because electromagnetic forces must be overcome, it takes energy to remove an electron from a sodium atom or to force an electron into a chlorine atom. Similarly, it seems reasonable that it takes more energy to strip the two electrons from a magnesium atom than to remove the one electron from a sodium atom. The ease with which an element loses or gains electrons is a measure of its *activity*. Based on energy considerations, we can state the following: elements in groups IA and VIIA are more active than those in groups IIIA and VA. Of the elements in group IVA, only elements near the bottom form ions. Thus, it can be seen that activity decreases as we consider elements toward the center of the periodic table.

- Elements in groups 1A, 2A, and 3A give up 1, 2, or 3 electrons, respectively, to form positive ions.
- Elements in groups 5A, 6A, and 7A accept 3, 2 or 1 electron, respectively, to form negative ions.
- In general,
    - metals give up electrons to form cations;
    - nonmetals accept electrons to form anions.

Within a given group, there is also a range of chemical activity. In group IA, the negative electron to be removed from the lithium atom is much closer to the positive nucleus than the electron to be removed from the cesium atom. The closer the electron is to the nucleus, the greater the attractive force and the more energy it takes to remove the electron. Thus, it takes less energy to form a cesium ion than to form a lithium ion. Therefore, cesium is much more active than lithium. If we look at group VIIA, similar reasoning tells us that it requires

Monatomic ions	1A Family	2A Family	3A Family	5A Family	6A Family	7A Family
	Li^+	Mg^{2+}	Al^{3+}	N^{3-}	O^{2-}	F^-
	Na^+	Ca^{2+}			S^{2-}	Cl^-
	K^+	Sr^{2+}				Br^-
	Rb^+	Ba^{2+}				I^-
	Cs^+					

Table 4.2 Symbols of Selected Typical Ions

more energy to force an electron into a bromine atom, where the positive nucleus is buried within a cloud of negative electrons, than to force an electron into a fluorine atom, where the positive nucleus is relatively close to the outermost subshell and can help attract the extra electron. Based on energy considerations, we can state that the most active elements in terms of loss and gain of electrons are to be found in the lower left and upper right areas of the periodic table. Table 4.2 shows the common monatomic (one-atom) ions you will encounter.

The transition and inner transition elements (the B elements on the periodic table) also form ions. However, because both their outermost and second outermost subshells are unfilled, they give up not only their outermost electrons to form ions, but they may also give up some electrons from their second outermost subshell as well. Thus, it is common to find these elements forming two or more different positive ions.

The compounds formed by ionic bonding consist of oppositely charged ions packed and held together by electrical attraction. Such compounds are called *ionic agglomerates* or *salts*.

Covalent Bonding

Examining the guidelines on loss or gain of electrons, we would expect an element like carbon, which is in group 4A, to be fairly inert (unreactive) and to form few compounds. Yet, this element forms more compounds than all the other elements together. Obviously, then, there must be some other bonding mechanism that allows these compounds to form.

Carbon has four electrons in its outermost energy level, and hydrogen has one electron in its energy level. Suppose that four hydrogen atoms were to approach a carbon atom so closely that the electron cloud of each hydrogen atom penetrated the electron cloud of the carbon atom. The electrons in these overlapping clouds would be influenced by the nuclei of both types of atoms. Both atoms would essentially be sharing these electrons. What would be the net effect? Figure 4.2 shows us. The electron of each hydrogen atom is indicated by *x*, and the carbon electrons are indicated by dots (the inner electrons of carbon are not shown). We can see that two electrons are now associated with each hydrogen atom, giving them the stable helium configuration, and eight electrons are associated with the carbon atom, giving it the stable neon configuration. Both types of

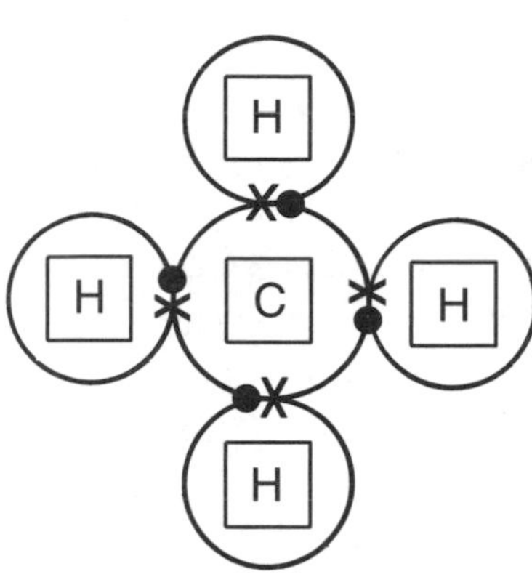

Figure 4.2 Covalent Bonding

atoms have attained stable structures through this sharing process. The compound that results is methane, the principal ingredient of natural gas used in cooking. The process of forming a compound through the sharing of pairs of electrons is called *covalent bonding*.

A pair of electrons shared between two atoms is a *bond*. In methane, carbon is joined to four hydrogen atoms by single bonds. Many compounds exist in which two or even three pairs of electrons are shared by two atoms. Figure 4.3 shows the bonding in carbon dioxide, a gas that bubbles out of carbonated water, and in acetylene, a gas commonly used in welding.

Two pairs of electrons are shared between each oxygen atom and the central carbon atom in carbon dioxide, giving each of the three atoms eight electrons. Three pairs of electrons are shared between the two carbon atoms in acetylene, and a single pair is shared between the carbon and hydrogen atoms. Carbon dioxide is said to have two double bonds, and acetylene is said to have a triple bond between its two carbon atoms.

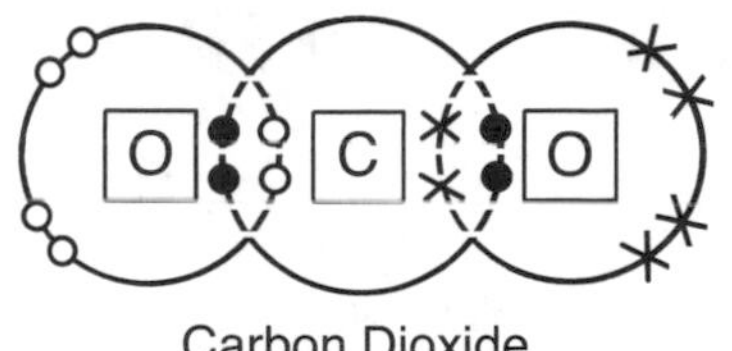

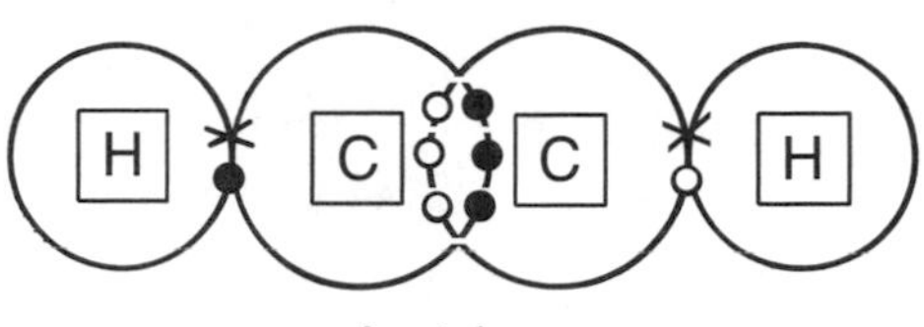

Figure 4.3 Other Covalent Bonds

The net effect of covalent bonding is to form a compound containing a definite number of atoms. These individual particles, which possess all the properties of the compound, are called *molecules*. Molecules are present only in covalent compounds. Ionic compounds are not molecules but are made up of ions packed together. Table 4.3 summarizes the differences between ionic bonding and covalent bonding.

Bonding	Ionic	Covalent
process	complete transfer of electrons to form ions	sharing of pairs of electrons
via	electrostatic attraction	overlap of electron clouds
product	ionic agglomerates (salts)	molecules

Table 4.3 Comparison of Ionic and Covalent Bonding

OXIDATION NUMBER

The *oxidation number* of an element is the charge that would be present on an atom if the element or compound in which the atom is found were ionic. It is a "bookkeeping" number that can be used sometimes to predict a compound's *formula*, the composition of a substance indicated by symbols of each element present and subscript numbers showing the number of each type of atom involved. *Elements in the free (uncombined) state have an oxidation number of zero.* Most elements exhibit a variety of oxidation numbers depending on the particular compound of which they are a part. The following general rules can help you determine the oxidation number of an element in a compound:

1. Elements of group 1A of the periodic table normally have an oxidation number of +1.

2. Elements of group 2A normally have an oxidation number of +2.

3. Elements of group 7A normally have an oxidation number of −1 in binary compounds (compounds that contain only two elements).

4. In ionic compounds in general:
   a. The oxidation number of an ion is numerically equal to the charge on the ion. (An ion that has a 2+ charge, like Mg^{2+}, would have an oxidation number of +2.)
   b. Positive ions have positive oxidation numbers.
   c. Negative ions have negative oxidation numbers.

5. In covalent compounds in general:
   a. The oxidation number of an atom in a covalent compound is numerically equal to the number of its electrons shared with atoms different from itself. For example, the carbon atom in carbon dioxide shares all four of its outermost electrons with oxygen atoms, so its oxidation number is 4 (refer to Figure 4.3); but in acetylene, three carbon electrons are shared with another carbon atom, and one electron is shared with a hydrogen atom. The carbon-to-carbon bonds don't count because the atoms are identical, so the oxidation number of carbon in acetylene is 1.
   b. Oxygen always has an oxidation number of −2 in its compounds (except for peroxides, in which its oxidation number is −1).
   c. Elements like carbon, silicon, nitrogen, phosphorus, sulfur, and chlorine, when they are centrally located in covalent molecules, normally have positive oxidation numbers.

6. The net sum of all of the oxidation numbers exhibited in a given compound must be zero.

These rules generally apply to most chemical compounds. Where exceptions occur in this book, they are pointed out.

FORMULAS

The *formula* of a compound is a ratio of the number of atoms of each element present in the compound. In ionic compounds, the formula gives the simplest ratio of constituents in whole numbers. In covalent compounds, the formula simply gives the lowest whole-number ratio of elements in the compound (empirical formula) or the exact number of atoms of each element present in a molecule of the compound (molecular or true formula).

The symbols of the elements present are used in writing formulas. If more than one atom of an element is required in the formula, a subscript numeral is written after the symbol of the element to indicate the number of its atoms in the formula. For example, the formula for water is H_2O. This means that in every molecule of water two atoms of hydrogen and one atom of oxygen are present. (*Note*: When only one atom of an element is present in the formula, the subscript 1 is understood and not written.) The formula for sodium chloride, NaCl, tells us that this compound contains equal numbers of sodium and chlorine atoms. We know from previous discussions that in this compound the atoms are actually present as ions. A formula gives no indication as to whether a compound is ionic or covalent. This characteristic must be ascertained from the properties of the compound.

Hints:

1. If the compound is composed of simply a metal and a nonmetal, it is probably ionic.

2. Combinations of nonmetals indicate covalent bonding.

Writing Formulas of Ionic Compounds

If we know the electrical charge of each element in a compound, we can easily write its formula. Let's look at a few examples.

EXAMPLE 1. A compound consists solely of magnesium and chlorine. What is its formula?

SOLUTION: Magnesium (Mg), a group 2A element, forms a cation with a charge of +2. Chlorine (Cl), a group 7A element, forms an anion with a charge of −1.

Remember that an ionic compound, like any compound, is electrically neutral (it must have the same number of positive charges as negative charges). To form a compound in which the sum of the charges is zero, it will take two chloride anions to counteract each magnesium cation.

$$1\ Mg^{2+} + 2\ Cl^-$$

$$2 \text{ positive charges} + 2 \text{ negative charges}$$

$$= 0$$

The formula of this compound is

$$MgCl_2$$

The name of this compound is magnesium chloride. The suffix *-ide* is used with the root of the name of the negative element in binary compounds. The terms *oxide, sulfide, nitride, phosphide, carbide, fluoride, bromide*, and *iodide* appear in the names of compounds in which these negatively charged elements are combined with one other positively charged element to form a binary compound.

Look carefully at the formula of magnesium chloride, $MgCl_2$. Following Mg, the subscript 1 is understood (and not shown). Following Cl is the subscript 2. Do you see that the electrical charge of each element has been "crisscrossed" and then written as a subscript after the symbol of the other element? Let's try this idea with another example.

EXAMPLE 2. What is the formula for aluminum oxide?

SOLUTION: Aluminum (Al), a group 3A element, forms a cation with an electrical charge of +3. Oxygen (O), a group 6A element, forms an anion with a charge of −2.

Crisscrossing the charges and using them as subscripts, we obtain the formula

$$Al_2O_3$$

Note that the + and − signs of the charges are ignored when we write formulas.

Does this formula for aluminum oxide satisfy the rule of zero net charge for compounds? Let's check it.

For Al: $2 \times (+3) = +6.$

For O: $3 \times (-2) = -6.$

Net valence (sum) = 0.

EXAMPLE 3. What is the formula for calcium sulfide?

SOLUTION: Calcium (Ca), a group 2A element, forms a cation with a charge of +2. Sulfur (S), a group 6A element, forms an anion with a charge of −2. Crisscrossing the charges and writing them as subscripts, we obtain the formula

$$Ca_2S_2$$

However, this is *not* the simplest formula for this compound. This formula tells us that the ratio of calcium to sulfur atoms is 2:2. This ratio, of course, is the same as a ratio of 1:1. Therefore, to write this formula in its simplest form, we reduce the subscripts to 1, and the formula becomes

$$CaS$$

Now let's look at the relationship between formulas and charges the other way around. Suppose we have the formula of a compound and we want to find the charges of the elements present. Let's look at some examples.

EXAMPLE 4. What are the charges of the elements in sulfur dioxide, SO_2?

SOLUTION: Because oxygen tends to gain two electrons to reach noble gas configuration, it normally has a charge of −2. Because there are two oxygen atoms in the formula, the total negative charge is −4. Therefore, to satisfy the requirement of a zero net charge, the charge on sulfur must be +4.

EXAMPLE 5. What are the oxidation numbers associated with the elements in sulfuric acid, H_2SO_4? (We'll use oxidation numbers here because sulfuric acid is a covalently bonded compound, so it is more appropriate to talk about oxidation numbers instead of charges.)

SOLUTION: According to the rules for oxidation numbers presented earlier in this chapter, oxygen almost always has an oxidation number of −2. Hydrogen, a Group 1A element, has an oxidation number of +1. The four oxygen atoms give us a total oxidation number of $4 \times (-2) = -8$. The two hydrogen atoms give us an oxidation number sum of $2 \times (+1) = +2$. Therefore, in order to make the net sum of the oxidation numbers in the compound zero, sulfur in H_2SO_4 has an oxidation number of +6.

Polyatomic Ions

In many chemical compounds, clusters of elements behave as if they were a single element. Such a group of elements, which carries either a positive or negative charge, is known as a *polyatomic* (composed of many atoms) ion (or radical). Consider the following series of compounds:

Series A	
sodium chloride	NaCl
sodium hydroxide	NaOH
sodium nitrate	$NaNO_3$
Series B	
sodium sulfide	Na_2S
sodium sulfate	Na_2SO_4
sodium carbonate	Na_2CO_3

In series A, the hydroxide (OH^-) and the nitrate (NO_3^-) groups behave toward sodium in exactly the same way as a single chlorine atom does. Similarly, in series B, the sulfate (SO_4^{2-}) and carbonate (CO_3^{2-}) groups behave toward sodium in exactly the same way as a single sulfur atom does. All forms of these groups are polyatomic ions.

The atoms within a polyatomic ion (or radical) are held together by covalent bonds, but in each case they contain either an excess or deficiency of electrons, causing the radical to possess an electrical charge. Thus, polyatomic ions combine as a unit with other ions to form ionic compounds.

Polyatomic ions possess a net oxidation number equal in magnitude and sign to the net charge on the polyatomic ion, just like any other ion. Table 4.4 gives the names, formulas, and charges of the common polyatomic ions.

Because ammonium is a polyatomic cation, it forms ionic compounds with all the polyatomic anions. Notice how the formulas of these compounds are written:

ammonium acetate: $NH_4C_2H_3O_2$

ammonium carbonate: $(NH_4)_2CO_3$

ammonium phosphate: $(NH_4)_3PO_4$

NOTE: It takes 2 ammonium ions to counteract the charge of the carbonate ion, and 3 ammonium ions to counteract the charge of the phosphate ion. (Remember ammonium phosphate from the matchstick?)

Charge of +1	Charge of −1	Charge of −2	Charge of −3
ammonium, NH_4^+	acetate, $C_2H_3O_2^-$	carbonate, CO_3^{2-}	phosphate, PO_4^{3-}
	bicarbonate, HCO_3^-	chromate, CrO_4^{2-}	
	hypochlorite, ClO^-	dichromate, $Cr_2O_7^{2-}$	
	chlorate, ClO_3^-	sulfate, SO_4^{2-}	
	hydroxide, OH^-	sulfite, SO_3^{2-}	
	cyanide, CN^-		
	nitrate, NO_3^-		
	nitrite, NO_2^-		
	permanganate, MnO_4^-		

Table 4.4 Polyatomic Ions

Look carefully at the names and formulas of the radicals. The suffixes *-ite* and *-ate* occur repeatedly. These suffixes are used only with polyatomic ions that contain oxygen atoms. Notice that "-ite" radicals contain less oxygen than "-ate" radicals. For example:

sulfite, SO_3^{2-} sulfate, SO_4^{2-}

nitrite, NO_2 nitrite, NO_3

NOTE: No definite number of oxygen atoms is specified for either type. The formulas of each radical must be learned individually through repeated use.

Writing Structural Formulas

When we work with covalent compounds it is often not enough to know the molecular formula for a compound, because more than one compound can have a particular molecular formula. For example, two different compounds have the molecular formula C_2H_6O. To distinguish between these compounds, we need to write a *structural formula*. A structural formula shows the actual number of each type of element in a compound, and it also indicates the bonding pattern. Quite often, we show these in a condensed form in which we do not show every individual bond. For example, the structural formulas of the two compounds that have the molecular formula of C_2H_6O are CH_3CH_2OH and CH_3OCH_3.

Sometimes, however, we need more detail than condensed structural formulas provide. In these situations we use what is called a *Lewis structural formula*. The Lewis structural formula indicates the bonding pattern with a line to represent a shared pair of bonding electrons in the molecule and dots to represent electrons not involved in bonding. Thus, we are drawing a picture of the molecule. For example, let's write the Lewis structural formula for water, H_2O.

1. First, using whatever information is available, arrange the atoms in the compound relative to each other. Many times this information will be given, or you may have to try several arrangements to see which one works. (Often, if there is a single atom of one element and multiples of another, as in water, the single atom is the central atom and the others are bonded to it.)

HOH

You don't have to be concerned about the relative positions of the hydrogen atoms—just be sure it is understood that both are bonded to the oxygen.

2. Next, find the total number of valence electrons available for all atoms. In this case: $2 \times 1 = 2$ for the hydrogen atoms and 6 for the oxygen atoms, giving a total of 8. If the molecule that you are "drawing" is a polyatomic ion, add or subtract electrons to account for the charge.

3. Now, place a single line (corresponding to a shared pair of electrons) between the elements that are covalently bonded.

H–O–H

This accounts for 4 of the 8 valence electrons available (2 electrons per line $\times$ 2 lines = 4 electrons).

4. Finally, place the remaining electrons in pairs around the appropriate atom so that every atom except hydrogen has 8 electrons (either shared or not shared) around it. Hydrogen should have 2 electrons.

$$H{-}\ddot{\underset{\cdot\cdot}{O}}{-}H$$

5. Although electrons have correctly been placed around each atom, the water molecule is not linear but rather is angular because the lone electron pairs on the oxygen repel each other, forcing the hydrogen atoms closer to each other. Other factors involved in the shape of the water molecule are beyond the scope of this book.

Sometimes more than one pair of electrons is shared, and a double or triple bond is shown using multiple lines (2 for a double bond and 3 for a triple bond). If a polyatomic ion is being represented, the Lewis structural formula is drawn and then enclosed in brackets, with the ionic charge shown as a superscript outside the bracket in the upper right-hand corner. For example, the cyanide ion, CN^-, is represented as

$$[:C \equiv N:]^-$$

BOND POLARITY

In covalent compounds, electron pairs are shared; however, often this sharing is not equal. Unless the two atoms involved in the bond are the same element, one of the atoms attracts the bonding pair more than the other does, and it pulls the electron pair closer to itself. When this happens, the atom that has the greater attraction acquires a partial negative charge because it is feeling the effect of the extra electron, and the other atom acquires a partial positive charge because one of its valence electrons is pulled slightly away. A separation of charge results, with one part of the molecule having a partial negative charge and the other part a partial positive charge. This type of covalent bonding, in which there is an unequal sharing of the electron pair involved in bonding, is called a *polar covalent bond*. The direction in which the bonding pair of electrons will shift depends on the electronegativity of the atoms involved in the bond.

Electronegativity is a measure of the attraction that an atom has for a bonding pair of electrons. Electronegativity is correlated with the position of an element in the periodic table. Electronegativity values increase in a period from left to right and decrease in a family from top to bottom (Figure 4.4).

Therefore, fluorine is the most electronegative element. (It is not helium, because helium is a noble gas and is relatively inert.) The element with the higher electronegativity value attracts the bonding electrons slightly toward it and acquires a partial negative charge, whereas the other atom takes on a partial positive charge. Overall, the molecule remains neutral, but there are areas of partial charge within the structure.

Molecules in which one end has a partial positive charge and the other end has a partial negative charge are called *polar molecules*. Because these molecules have partially charged ends, they can attract other polar molecules; the positive end of one polar molecule can attract the negative end of another polar molecule. An extreme case of this attraction occurs when a hydrogen atom is bonded to an extremely electronegative element (F, O, N).

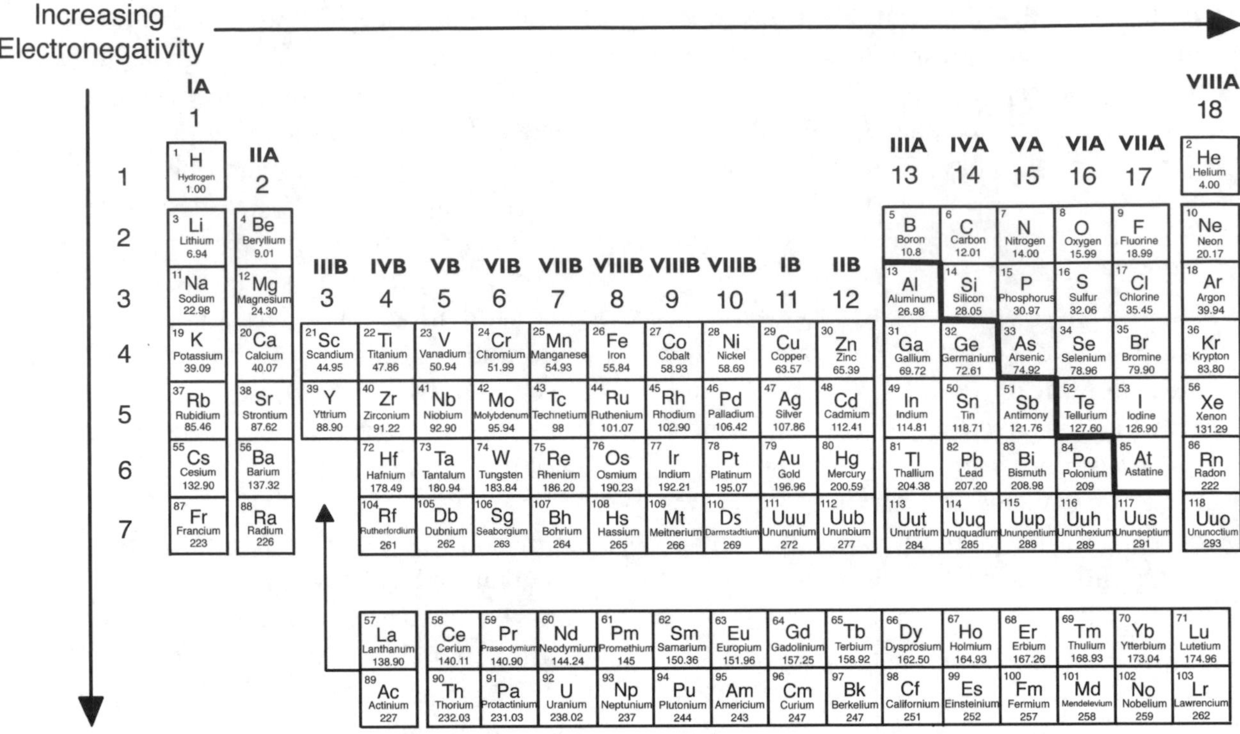

Figure 4.4

This bond is highly polar, and the hydrogen on one molecule attracts the F, O, or N on another molecule. This strong attraction is called a *hydrogen bond* (H-bond). Because hydrogen bonds tend to hold molecules somewhat closer and more strongly than normal they affect the physical properties of the substance, such as melting point (the temperature at which a substance changes from a solid to a liquid), boiling point (the temperature at which a substance changes from a liquid to a gas), and so forth. For example, hydrogen bonding explains the unusually high melting point, boiling point, and surface tension (the attractive force at the surface of a liquid that makes it behave as if there were a thin film over the liquid surface) of water.

FORMULA OR MOLECULAR MASSES

Just as symbols represent more than the name of an element, formulas stand for more than the name of a compound. A formula stands for:

1. the name of a compound;
2. the molecule of the compound (if it is covalently bonded); and
3. a quantity of the compound equal in mass to its formula mass.

> The formula mass of a compound is the sum of all the atomic masses of the elements present in the formula of the compound.

The concept of the formula mass of a compound is one of the most important ideas in chemistry. Its definition is very simple: the *formula mass* of a compound is the sum of all the atomic masses of the elements present in the formula of the compound.

The formula mass of sodium chloride, NaCl, is found as follows:

Atomic mass of Na	22.990 amu
Atomic mass of Cl	35.453 amu
Formula mass of NaCl	58.443 amu

The formula mass of water, H_2O, is found similarly.

Atomic mass of H (×2)	2.016 amu
Atomic mass of O	15.999 amu
Formula mass of H_2O	18.015 amu

Because the formula of a covalent compound represents the constituents of a molecule of the compound, the formula mass is usually referred to as the "molecular mass" of the compound. Actually, because a formula gives no indication as to the type of bonding, ionic or covalent, present in a compound, the term "molecular mass" is commonly used even with ionic compounds, although no molecule is present in these compounds. Thus, the terms "formula mass" and "molecular mass" are commonly used interchangeably.

An amount of a compound expressed in grams equal to its formula mass is called a *mole* (abbreviated *mol*). For example, 18.015 of water is 1 mol of water. (The quantity 18.015 is equal to the formula mass of water [18.015 amu] found in the preceding example.) The number of moles of a compound is found by using the following expression:

$$\frac{\text{Mass of compund (g)}}{\text{Formula mass (g/mol)}} = \text{Number of moles (mol)}$$

SUMMARY

Noble gas elements resist formation of compounds because of the great stability of the electronic configurations in their atoms in having a filled valence energy level. In forming compounds, elements gain, lose, or share electrons in order to have their distribution of electrons rearranged to match that of the noble gas elements. Compound formation that results from the process of transfer of electrons from one atom to another is called ionic bonding. The ions produced by this transfer pack together electrically to form ionic agglomerates or salts. Metals normally lose electrons to form positively charged ions called cations. Nonmetals normally gain electrons to form negatively charged ions called anions. These ions may simply be of a single atom (monoatomic) or of groups of atoms (polyatomic).

Compound formation through the process of sharing electrons between atoms is called covalent bonding. This process produces covalently bonded particles called molecules.

A formula is a ratio of the number of atoms of each element present in a compound. It contains the symbols of each element present, with subscripts after each symbol to indicate the number of atoms of the element present. The Lewis structural formula visually represents the bonding pattern and electron distribution in the compound.

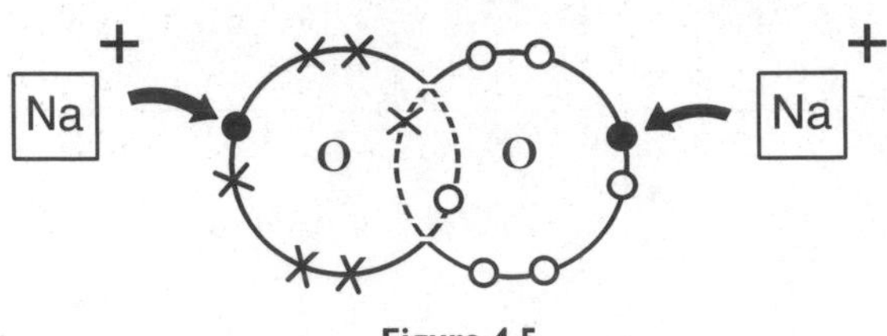

Figure 4.5

One of the central concepts in chemistry is the mole concept. A mole is a quantity of a compound equal in mass to the formula mass, or molecular mass, of the compound. The formula mass is the sum of the atomic masses of the elements present.

PROBLEM SET NO. 4

1. Figure 4.5 is a diagram of the electronic configuration in a compound (only the outermost electrons of each atom are shown). Determine the charge of each atom in the compound. What is the name of the compound?
2. A compound consists solely of aluminum and sulfur. Work out its formula. What is its name?
3. A compound consists solely of magnesium and nitrogen. Work out its formula. What is its name?
4. What are the oxidation numbers of each atom in sodium carbonate, Na_2CO_3?
5. What are the oxidation numbers of each atom in calcium nitrate, $Ca(NO_3)_2$?
6. What is the formula of ammonium sulfate?
7. What is the formula of magnesium phosphate?
8. Compute the formula mass of sugar, $C_{12}H_{22}O_{11}$.
9. Compute the formula mass of aluminum sulfate, $Al_2(SO_4)_3$.
10. A 200-g measure of sodium hydroxide, NaOH, is to be used in preparing a solution. How many moles of sodium hydroxide are to be used?
11. Draw the Lewis structural formula of carbon dioxide, CO_2.

LAWS OF CHEMISTRY

KEY TERMS

law of conservation of matter, law of definite proportions, Avogadro's law, equation, empirical formula

The model of an atom has given us a reasonable explanation of how atoms are combined in compounds. Now let's look at some of the basic laws of chemistry. These laws were discovered only after years of painstaking observation of the behavior of nature.

NOTE: They were all known before our atomic model was created. Each one contributed to the development of our model. Our main concern with these laws now is that they help us understand chemical change.

CONSERVATION OF MATTER

The *law of conservation of matter* states that matter is neither created nor destroyed during chemical change. This means that the sum of the mass(es) (total mass) of the substance(s) entering a chemical change is precisely equal to the sum of the mass(es) (total mass) of the substance(s) formed as a result of chemical change. This law has been verified by repeated study of chemical changes using delicate balances to measure the mass of *reactants* (the substances that one starts with in a chemical reaction) and *products* (the substances formed as a result of a chemical reaction. Actually, because mass and energy are related, it is more accurate to say that *mass and energy* are neither created nor destroyed during chemical change.

Chemical change involves a redistribution of electrons, either by transfer or by sharing, but no new electrons are formed in chemical change, nor are any destroyed. The nuclei of atoms, which possess all the mass, remain unchanged and enter into new combinations solely as a result of the redistribution of electrons. Thus, our atomic model is consistent with the law of conservation of matter.

DEFINITE PROPORTIONS

The *law of definite proportions* states that a given compound always contains the same elements combined in the same proportions by mass. The determination of the composition of a substance is a process known as *analysis* of the compound. Repeated analyses of samples of a compound always show that it contains the same elements in the same mass proportions.

For example, water always contains 8 parts by mass of oxygen to 1 part by mass of hydrogen. Let's see if these results are consistent with our concepts of atomic structure and compound formation. Oxygen, with atomic mass 16.0 g/mol, has six electrons in its outermost shell. Hydrogen, with atomic mass 1.0 g/mol, has one electron in its shell. Oxygen needs two electrons to fill its valence (outermost) shell. Our concept of compound formation tells us that two hydrogen atoms are required to provide enough electrons to fill the valence shell of oxygen. Furthermore, this gives us a mass proportion of 16 parts by mass of oxygen (one atom) to 2 parts by mass of hydrogen (two atoms). This proportion, 16 to 2, is consistent with the 8 to 1 proportion always found in the analysis of water.

The law of definite proportions has further significance. The process of causing elements or more simple compounds to combine to form compounds is known as *synthesis* of compounds. The law of definite proportions dictates that a compound formed by synthesis must contain the same mass (proportions of its elements) as any other samples of this compound. Thus, water produced in a laboratory by combining oxygen and hydrogen must contain 8 parts by mass of oxygen to 1 part by mass of hydrogen, the same as any other water sample. Now, suppose that we took 8 parts by mass of oxygen and 2 parts by mass of hydrogen and attempted to combine them. What would happen? We can see that there is too much hydrogen. The 8 parts of oxygen would combine with 1 part of hydrogen, and the rest of the hydrogen would remain unchanged. In that case, the oxygen is said to be the *limiting reactant,* because the amount of water formed is based on the amount of oxygen present. Likewise, in this case, there is said to be an *excess* (an amount of a reactant beyond that theoretically required for a reaction) of hydrogen present, because there is more hydrogen present than oxygen can combine with.

Similarly, if we were to begin with 10 parts by mass of oxygen and 1 part by mass of hydrogen, 8 parts of oxygen would combine with the 1 part of hydrogen to form water, and the rest of the oxygen would be left unchanged. Here, the hydrogen is the limiting reactant, and an excess of oxygen is present. This concept of limiting and excess reactants is very important, because in most chemical changes that involve two or more reactants, one of the reactants will be the limiting reactant, and the others will be in excess.

AVOGADRO'S LAW

The nineteenth-century Italian scientist Amedeo Avogadro was a pioneer in the study of quantitative relationships in science, especially with gases. *Avogadro's law* states that equal volumes of gases measured at the same temperature and pressure contain equal numbers of molecules. All gases exist as molecules. By finding the ratio of mass of equal volumes of various gases, we can find the ratio of their molecular masses. For example, let's look again at the compound water and its elements hydrogen and oxygen. We can easily convert water to a gas (steam) and weigh a given volume of it. Likewise, the same volume of hydrogen and oxygen, both gases, can be brought to the same temperature as the steam and weighed. The mass ratios found by this procedure always turn out to be as follows:

hydrogen	1 part by mass
oxygen	16 parts by mass
water	9 parts by mass

The sample of oxygen weighs 16 times as much as the sample of hydrogen, and the sample of steam weighs 9 times as much as the hydrogen. Because Avogadro's law states that each of these samples contains the same number of molecules, the individual molecules of each of these substances must possess these same proportions by mass. Now, we know the mass of one of these molecules. The formula of water is H_2O, and its molecular mass, obtained by adding the atomic masses in the formula (1 + 1 + 16), is 18 g/mol. Because we used a mass of 9 parts for water, we need to recalculate the mass ratio we found using the basis of 18 g/mol for water, by multiplying each component by 2:

hydrogen:	2 parts by mass
oxygen:	32 parts by mass
water:	18 parts by mass

Because this ratio is a ratio of molecular masses, and because the actual molecular mass of water is 18 g/mol, the actual molecular mass of hydrogen must be 2 g/mol, and the actual molecular mass of oxygen must be 32 g/mol.

Therefore, the molecule of hydrogen gas must contain two atoms of hydrogen, because the atomic mass of hydrogen is 1/mol. Similarly, the molecule of oxygen must contain two atoms of oxygen, because the atomic mass of oxygen is 16 g/mol, half the molecular mass. The formula of hydrogen gas is therefore written H_2 to indicate the two atoms in the molecule. The formula for oxygen gas is O_2. Both of these molecules are covalently bonded.

JOKE: Seen on a bumper sticker: Everbody has Avogadro's number.

Avogadro's law is thus very useful in finding the molecular mass and formula of a gaseous substance, provided that its mass can be compared with the mass of a substance whose formula is known. Experimental and mathematical studies of Avogadro's law have indicated its accuracy beyond reasonable doubt.

> Because equal volumes of gases contain equal numbers of molecules, finding the mass of equal volumes of different gases lets us calculate their molecular masses.

EQUATIONS

An *equation* is a concise statement of a chemical reaction using the symbols and formulas of the reactants and products. For example, when sulfur or any other substance burns in air, it combines with oxygen in air to produce an oxide. Let's look at this reaction in the form of a chemical equation.

S(s)	+	O_2(g)	→	SO_2(g)
sulfur		oxygen		sulfur dioxide

The (s) indicates that the sulfur is in the solid state; the (g) indicates that the oxygen and sulfur dioxide are in the gaseous state. If any of the substances had been a liquid, that state would have been indicated by (l). Sometimes the specific reaction a substance undergoes depends on its state, so if it makes a difference, the state should be indicated using these notations.

Examine the equation closely. Is it consistent with the law of conservation of matter? In other words, are there equal numbers of each type of atom on each side of the equation? Yes. This equation therefore is said to be *balanced*. An equation is almost meaningless unless it is balanced.

This equation tells us more than merely that sulfur combines with oxygen to produce sulfur dioxide. It has a quantitative significance just as symbols themselves do. It tells us that one atomic mass's worth of sulfur reacts with one molecular mass's worth of oxygen to produce one molecular mass's worth of sulfur dioxide. If we use grams as units, we get

$S(s)$	+	$O_2(g)$	→	$SO_2(g)$
32.1 g		32.0 g		64.1 g

In other words, this equation tells us that 1 mol of sulfur combines with 1 mol of oxygen to produce 1 mol of sulfur dioxide.

Let's look at another reaction. You will recall from Experiment 3 that when copper is heated in air, black copper oxide is formed. This reaction is indicated as follows:

$Cu(s)$	+	$O_2(g)$	→	$CuO(s)$
copper		oxygen		copper oxide

What about our law of conservation of matter now? Do you see that we have apparently destroyed some oxygen? This equation is not balanced. It is called a *skeleton* equation; it indicates only the names of the substances involved.

This equation would be balanced if we could put a subscript 2 following the O of CuO to make it CuO_2. But this would violate the law of definite proportions, because black copper oxide always has the formula CuO. In balancing equations, the subscripts in the formulas of compounds may not be changed.

A skeleton equation is balanced by placing numbers, called *coefficients*, in front of the formulas of the substances in the reaction. Look again at our skeleton equation. An even number of oxygen atoms appears on the left side of the equation. By placing the coefficient 2 in front of CuO, we would have two oxygen atoms on each side of the equation, because the coefficient multiplies all the symbols in the formula immediately following it. This would change our equation to read

$$Cu(s) + O_2(g) \rightarrow 2\ CuO(s)$$

Now, we have too much copper on the right. We can remedy this by placing another coefficient 2 in front of the Cu, giving us the following.

$$2\ Cu(s) + O_2(g) \rightarrow 2\ CuO(s)$$

Now, the equation is balanced. We have two copper atoms and two oxygen atoms on each side of the equation. The balanced equation now reads 2 mol of copper combines with 1 mol of oxygen to produce 2 mol of copper oxide. The following expression shows how the mass of each of the substances in the balanced equation may be indicated:

$2\ Cu(s)$	+	$O_2(g)$	→	$2\ CuO(s)$
2 mol × 63.6 g/mol		1 mol × 32.0 g/mol		2 mol (63.6 + 16.0 g/mol)
127.2 g	+	32.0 g	=	159.2 g

Thus, 127.2 units of mass of copper combine with 32.0 units of mass of oxygen to form 159.2 units of mass of copper oxide. These units of mass may be grams, pounds, tons, or another unit, just so long as all three masses are expressed in the same units. This mass relationship also tells us that copper and oxygen combine in a mass ratio of 127.2 parts by mass of copper to 32.0 parts by mass of oxygen. Similarly, 159.2 parts by mass of copper oxide are formed for every 32.0 parts by mass of oxygen or every 127.2 parts by mass of copper.

Let's look at one more example. Butane gas (C_4H_{10}) is commonly used as a bottled gas in rural areas. It burns with oxygen to form carbon dioxide and water. The skeleton equation is

$$C_4H_{10}(g) + O_2(g) \rightarrow CO_2(g) + H_2O(g)$$

Now let's balance this skeleton equation using the "even numbers" technique described in the previous example.

1. Starting with oxygen, we see an even number of oxygen atoms on the left, and an odd number on the right. The CO_2 has an even number of oxygen atoms, so we have to work with the 1 oxygen atom in H_2O. Let's try a coefficient of 2:

$$C_4H_{10}(g) + O_2(g) \rightarrow CO_2(g) + 2\ H_2O(g)$$

That gives us an even number of oxygen atoms, but we need 10 hydrogen atoms to balance the hydrogen atoms on the left side of the equation, and the coefficient of 2 gives us only 4 (2×2). Therefore, we need a larger coefficient.

2. A coefficient of 5 gives us the right amount of hydrogen (5×2), but 5 is an odd number, so we must go to the next even multiple of 5, which is 10. This will do, but it gives us 20 hydrogen atoms on the right. Moving to the left side of the equation, we can place another coefficient of 2 in front of the C_4H_{10}, giving us 20 hydrogen atoms on the left. We now have

$$2\ C_4H_{10}(g) + O_2(g) \rightarrow CO_2(g) + 10\ H_2O(g)$$

The hydrogen atoms are balanced, and we have an even number of oxygen atoms on each side.

3. Now, we look at the carbon. We have eight carbon atoms on the left (2×4), so we need a coefficient of 8 in front of the CO_2 to balance the carbon. This gives us

$$2\ C_4H_{10}(g) + O_2(g) \rightarrow 8\ CO_2(g) + 10\ H_2O(g)$$

We still have an even number of oxygen atoms on each side.

4. Now, we are finally ready to balance the oxygen. There are a total of 26 oxygen atoms on the right side of the equation. A coefficient of 13 in front of the O_2 will give us 26 oxygen atoms on the left side. Now, our equation is balanced and looks like this:

$$2\ C_4H_{10}(g) + 13\ O_2(g) \rightarrow 8\ CO_2(g) + 10\ H_2O(g)$$

This equation reads: 2 mol of butane combines with 13 mol of oxygen to produce 8 mol of carbon dioxide and 10 mol of water. The mass proportions involved are as follows:

Reactants: butane: 2(48.0 + 10.0) g = 116.0 g
oxygen: 13(32.0) = 416.0 g
Total mass of reactants **532.0 g**

Products: carbon dioxide: 8 (12.0 + 32.0) g = 352.0 g
water: 10(2.0+16.0) g = 180.0 g
Total mass of products **532.0 g**

The characteristics of a balanced equation may be summarized as follows:

1. It obeys the law of conservation of matter.

2. It obeys the law of definite proportions.

3. Its coefficients give the molar proportions of reactants and products involved in the reaction.

Symbols, formulas, and equations all have definite quantitative meanings. We are now ready to look at some numerical applications based on these ideas.

PERCENTAGE COMPOSITION

If we know the formula of a compound, we can easily find the percentage by mass of each element present. A statement of the percentage of each element present in a compound is called its *percentage composition*. In chemistry, this composition is always on a mass basis unless specifically stated otherwise. (Sometimes the composition of mixtures of gases is given on a volumetric basis.)

The computation of percentage composition from the formula of a compound is based on the meaning of symbols and formulas. Each symbol stands for one atomic mass's worth of the element it represents, and each formula stands for one molecular mass's worth of the compound it represents. Let's see how percentage composition calculations are carried out.

EXAMPLE 1. What is the percentage composition of water, H_2O?

SOLUTION:

	Number of Atoms	Atomic Mass (g)	Total Mass (g)
hydrogen	2	1.0	2.0
oxygen	1	16.0	16.0
	Molecular mass of H_2O:		18.0

$$\text{Percentage of hydrogen} = \frac{2.0\text{ g}}{18.0\text{ g}} \times 100$$

$$= 11.1\%\text{ hydrogen}$$

$$\text{Percentage of oxygen} = \frac{16.0\text{ g}}{18.0\text{ g}} \times 100$$

$$= 88.9\%\text{ oxygen}$$

Note that the percentage of each element is found from the expression

$$\frac{\text{Total mass of element present}}{\text{Molecular mass of compound}} = \%\text{ of element}$$

EXAMPLE 2. What is the percentage composition of sulfuric acid, H_2SO_4?

SOLUTION:

	Number of Atoms	Atomic Mass (g)	Total Mass (g)
hydrogen:	2	1.0	2.0
sulfur:	1	32.1	32.1
oxygen:	4	16.0	64.0
	Molecular mass of H_2SO_4:		98.1

$$\text{Percentage of hydrogen} = \frac{2.0\text{g}}{98.1\text{g}} \times 100$$

$$= 2.0\%\text{ hydrogen}$$

$$\text{Percentage of sulfur} = \frac{32.1\text{g}}{98.1\text{g}} \times 100$$

$$= 32.7\%\text{ sulfur}$$

$$\text{Percentage of oxygen} = \frac{64.0\text{g}}{98.1\text{g}} \times 100$$

$$= 65.2\%\text{ oxygen}$$

EXAMPLE 3. Find the percentage of oxygen in calcium nitrate, $Ca(NO_3)_2$

	Number of Atoms	Atomic Mass (g)	Total Mass (g)
calcium:	1	40.1	40.1
nitrogen:	2	14.0	28.0
oxygen:	6	16.0	96.0
	Molecular mass of $C(NO_3)_2$:		164.1

$$\text{Percentage of oxygen} = \frac{96.0\,\text{g}}{164.1\,\text{g}} \times 100$$

$$= 58.5\%\ \text{oxygen}$$

Note particularly how the number of atoms of each element was obtained.

EXAMPLE 4. An iron ore field contains ferric oxide, Fe_2O_3, also known as hematite, mixed with rock that bears no iron. Naturally, both hematite and rock are scooped up in the gigantic shovels used in mining the ore. Samples taken at various spots in the ore field show that the field contains 80% hematite and 20% rock. Find the mass of pure iron in one ton of this ore and the percentage of iron in the ore field.

SOLUTION:

1. Mass of Fe_2O_3 per ton of ore: 2000 lb × 0.80 = 1600 lb of Fe_2O_3 per ton of ore.

2. Percentage of iron in Fe_2O_3:

	Number of Atoms	Atomic Mass (g)	Total Mass (g)
iron	2	55.8	111.6
oxygen	3	16.0	48.0
Molecular mass of Fe_2O_3:			159.6

$$\text{Percentage of iron} = \frac{111.6\,\text{g}}{159.6\,\text{g}} \times 100$$

$$= 69.9\%\ \text{iron}$$

3. Mass of iron per ton of ore:

$$1600\ \text{lb} \times 0.699 = 118.4\ \text{lb of iron per ton of ore}$$

4. Percentage of iron in the field:

$$\frac{1118.4\ \text{lb}}{2000\ \text{lb}} \times 100 = 55.9\%\ \text{iron in the ore field}$$

Example 4 shows how percentage composition problems may be a part of many different varieties of practical problems. Such fields as analytical chemistry, metallurgy, mining, mineralogy, and geology all make use of calculations of this type.

COMPUTATION OF FORMULAS

If we know the percentage composition of a compound, we can compute the *empirical* (simplest) *formula* of the compound. As we have seen, a formula is a ratio of the number of atoms of each element present in the compound. The empirical formula indicates which atoms are present in the lowest whole-number ratio.

For example, the true formula of hydrogen peroxide is H_2O_2. Its empirical formula is HO. In general, the empirical formula is the true formula for all ionic compounds. In covalent compounds, for which the formula represents the composition of the molecule of the compound, the true formula is either the same as the empirical formula or is some whole-number multiple of it. Let's find the empirical formula of a compound.

EXAMPLE 5. A compound is analyzed and found to contain 75% carbon and 25% hydrogen. Find its empirical formula.

SOLUTION: Because each different type of atom contributes to the total mass of the compound in parcels of mass equal to its own atomic mass, we can divide the mass percentage of a given element by its atomic mass to get the relative number of atoms of the element that contribute to the total mass percent. For the compound under consideration this would be:

$$\text{carbon:} \quad \frac{75 \text{ (mass percent)}}{12 \text{ (atomic mass)}} = 6.25$$

$$\text{hydrogen:} \quad \frac{25 \text{ (mass percent)}}{1 \text{ (atomic mass)}} = 25.0$$

Thus we have 6.25 carbon atoms for every 25 hydrogen atoms in this compound. To reduce these numbers to the empirical whole numbers, we divide each by the smaller. The entire calculation is then as follows:

$$\text{carbon:} \quad \frac{75}{12} = 6.25; \ \frac{6.25}{6.25} = 1$$

$$\text{hydrogen:} \quad \frac{25}{1} = 25.0; \ \frac{25.0}{6.25} = 4$$

Therefore, the empirical formula of this compound is CH_4.

EXAMPLE 6. A compound contains 21.6% sodium, 33.3% chlorine, and 45.1% oxygen. Find its empirical (simplest) formula.

SOLUTION:

$$\text{sodium:} \quad \frac{21.6}{23.0} = 0.94; \ \frac{0.94}{0.94} = 1$$

$$\text{chlorine:} \quad \frac{33.3}{35.5} = 0.94; \ \frac{0.94}{0.94} = 1$$

$$\text{oxygen:} \quad \frac{45.1}{16.0} = 2.82; \ \frac{2.82}{0.94} = 3$$

Therefore, the empirical formula of this compound is $NaClO_3$.

EXAMPLE 7. In some crystalline solids, molecules of water form part of the crystal structure. Such solids are known as *hydrates*. Ordinary household washing soda, made up of sodium carbonate and water, is a typical hydrate. The percentage of water present can be found by measuring the loss in mass of a hydrate sample dried in a hot oven. A 20.00-g sample of washing soda is dried in an oven. After drying it is found to weigh 7.57 g. Compute

(a) the percentage of water in washing soda;

(b) the formula of washing soda (expressed in moles).

SOLUTION:

(a) Percentage of $H_2O = \frac{\text{loss in mass}}{\text{original mass}} \times 100 -$

$$\frac{20.00\text{ g} - 7.57\text{ g}}{20.00\text{ g}} \times 100 = 62.15\%$$

(b) The percentage of Na_2CO_3 is 100% − 62.15% = 37.85%

$$Na_2Co_3\text{:} \quad \frac{37.85}{106} = 0.357; \ \frac{0.357}{0.357} = 1$$

$$H_2O\text{:} \quad \frac{62.15}{18.0} = 3.45; \ \frac{3.45}{0.357} = 10$$

(to the nearest whole number)

Therefore, the formula of washing soda must indicate 1 part of Na_2CO_3 and 10 parts of H_2O. Its formula is written as follows: $Na_2CO_3 \cdot 10\ H_2O$. This is the standard method of writing the formula of a hydrate. It indicates that the crystal contains 10 mol of water for every mole of sodium carbonate. Note that because a molar ratio of constituents was sought, the molecular mass of each constituent was used in finding the molar ratio.

MASS RELATIONSHIPS IN EQUATIONS

Chemical equations tell us the number of moles of each substance involved in a given reaction. For example, the equation for the rusting of iron

$$4\ Fe(s) + 3\ O_2(g) \rightarrow 2\ Fe_2O_3(s)$$

tells us that iron combines with oxygen in a ratio of 4 mol of iron to 3 mol of oxygen, and that 2 mol of iron oxide is produced for every 4 mol of iron entering the reaction.

These molar ratios, in turn, indicate the ratio of mass of each substance involved. The equation tells us that iron and oxygen combine in a ratio of (4×55.8) parts by mass of iron to (3×32.0) parts by mass of oxygen, and that (2×159.8) parts by mass of iron oxide are thereby produced in this reaction. We could "read" the equation either in terms of moles (represented by the coefficients in the balanced chemical equation) or in terms of mass (grams).

If we multiply the coefficient of a substance in a balanced equation (the number of moles) by the formula mass of the substance (grams/mole), we obtain a quantity known as the *equation mass* of the substance. The actual masses of substances involved in a chemical reaction are in the same ratio as their equation masses.

Because of these relationships, if we know the balanced equation for a reaction and the actual mass of any one substance involved in the reaction, we can find the actual mass of any other substance participating in the reaction. We use the following proportion:

$$\frac{\text{Actual mass of one substance}}{\text{Its equation mass}} = \frac{\text{Unknown actual mass}}{\text{Its equation mass}}$$

Let's look at an example that involves finding actual mass.

EXAMPLE 8. 27.95 g of iron is oxidized completely.

(a) What mass of oxygen combined with the iron?

(b) What mass of iron oxide was produced?

SOLUTION: First, we write the balanced equation for the reaction, and then we write the equation mass of each substance below its formula as follows:

$4\ Fe(s)$	+	$3\ O_2(g)$	$\rightarrow$	$2\ Fe_2O_3(s)$
4 mol ×		3 mol ×		2 mol ×
55.8 g/mol		32.0 g/mol		159.6 g/mol

(a) Use substitution in the preceding expression to find the actual mass of oxygen (let x represent the unknown mass.):

$$\frac{27.95\ g}{223.3\ g} = \frac{x}{96.0\ g}$$

$$x = \frac{27.95\ g \times 96.0\ g}{223.2\ g}$$

$$x = 12.0\ \text{g of oxygen}$$

(b) Use substitution in the expression to find the actual mass of iron oxide:

$$\frac{27.95\ g}{223.3\ g} = \frac{x}{319.2\ g}$$

$$x = \frac{27.95 \text{ g} \times 319.2 \text{ g}}{223.2 \text{ g}}$$

$$x = 39.97 \text{ g of iron oxide}$$

The steps, then, in solving this type of problem are as follows:

1. Write the balanced equation for the reaction.

2. Find the equation mass of the substances concerned.

3. Set up a proportion involving the actual mass and the equation mass, and solve for the unknown.

EXAMPLE 9. Sodium hydroxide, NaOH, can be prepared by treating sodium carbonate, Na_2CO_3, with calcium hydroxide, $Ca(OH)_2$, according to the following skeleton equation:

$Na_2CO_3(aq) + Ca(OH)_2(aq)$
$\rightarrow NaOH(aq) + CaCO_3(s)$

What mass of NaOH can be produced from 74.2 g of Na_2CO_3?

SOLUTION: Balanced equation:

$Na_2CO_3(aq) + Ca(OH)_2(aq)$
1 mol × 106.0 g/mol
$\rightarrow 2\ NaOH(aq) + CaCO_3(s)$
2 mol × 40.0 g/mol

Therefore:

$$\frac{74.2 \text{ g}}{106.0 \text{ g}} = \frac{x}{80.0 \text{ g}}$$

$$x = \frac{74.2 \text{ g} \times 80.0 \text{ g}}{106.0 \text{ g}}$$

$$x = 56.0 \text{ g of NaOH}$$

NOTE: It is assumed that there is sufficient calcium hydroxide present to react with all the sodium carbonate. If any excess calcium hydroxide is present, it will remain unchanged, because the sodium carbonate is the limiting reactant in this case.

SUMMARY

The basic laws of chemistry, such as the law of conservation of matter and the law of definite proportions, give us a way to interpret chemical change in term of masses. The coefficients in chemical equation may represent individual atoms and molecules consumed or formed, or they may represent the number of moles of reactants and products involved in the reaction. If you know the number of moles of each substance and their formula masses, you can determine the mass of each substance that reacts and is formed (the equation masses). Using the equation masses then lets you calculate the actual mass of substances consumed or formed. You can answer the questions, How much will it take? and How much will I make?

PROBLEM SET NO. 5

1. The formula of hydrogen gas is known to be H_2. Equal volumes of hydrogen and nitrogen gas are weighed at the same temperature and pressure. The nitrogen gas weighs 14 times as much as the hydrogen. What is the formula of nitrogen gas?

2. Find the percentage composition of sugar, $C_{12}H_{22}O_{11}$.

3. Find the percentage of copper in hydrated copper sulfate, $CuSO_4 \cdot 5\ H_2O$.

4. A sample of impure NaCl is found to contain 58% Cl. What is the percent purity of the sample?

5. A compound contains 52.9% Al and 47.1% O. Find its formula.

6. 50.88 g of copper combines with 12.84 g of sulfur. Find the formula of the compound that is formed.

7. Balance the following equations:
   (a) $NaCl + H_2SO_4 \rightarrow Na_2SO_4 + HCl$
   (b) $NH_3 + O_2 \rightarrow N_2 + H_2O$
   (c) $ZnS + O_2 \rightarrow ZnO + SO_2$
   (d) $C_3H_8 + O_2 \rightarrow CO_2 + H_2O$
   (e) $Ca_3(PO_4)_2 + SiO_2 + C$
   $\rightarrow CaSiO_3 + CO + P$

8. Lime, CaO, is prepared commercially by heating limestone, $CaCO_3$. The equation is

$$CaCO_3 \rightarrow CaO + CO_2$$

What mass of lime could be obtained by heating 500 lb of limestone?

9. Potassium nitrate, KNO_3, decomposes when heated, according to the following skeleton equation:

$$KNO_3 \rightarrow KNO_2 + O_2$$

   (a) Balance the equation.
   (b) How many moles of O_2 will be formed from 12 mol of KNO_3?
   (c) What is the name of KNO_2? (See Table 4.2.)
   (d) What mass (g) of KNO_2 will be formed by heating 12 mol of KNO_3?

10. Nitrogen and hydrogen combine in the presence of a catalyst to form ammonia, NH_3, according to the equation

$$N_2 + 3H_2 \rightarrow 2\ NH_3$$

280 g of N_2 and 100 g of H_2 are placed a reaction chamber.

   (a) Which of the two substances is the limiting reactant?
   (b) How many moles of the limiting reactant are present?
   (c) How many moles of excess reactant are present in excess?
   (d) How many moles of NH_3 can be produced?
   (e) What mass of NH_3 can be produced?

CHAPTER 6

GASES

KEY TERMS

Boyle's law, Charles's law, Dalton's law of partial pressures, ideal gas law, kinetic-molecular theory, Graham's law of diffusion, equivalent mass, specific heat

Matter exists in three physical states: gaseous, liquid, and solid. A gas has no internal boundary. It expands to fill any container completely regardless of the size or shape of the container. A liquid has one internal boundary, its surface, and fills its container below its surface regardless of the shape of the container. A solid is rigid—that is, it bounds itself internally in all dimensions. It needs no external container.

The properties of these three states of matter are related to our concept of the structure of matter, and a study of them will help us become more familiar with the chemical behavior of matter. Let's first turn our attention to the gaseous state.

PRESSURE

The *pressure* of a gas is the force it exerts on a unit surface area. Pressure is measured with an instrument known as a *barometer*. The first barometers were made of a glass tube about 1 m long, sealed at one end, and filled completely with liquid mercury. The open end was then immersed in a dish of mercury by temporarily sealing the open end, inverting the tube, dipping the temporarily sealed end under the surface of mercury in the dish, and removing the temporary seal. The mercury then flowed into the tube until the pressure of the column of mercury in the tube exactly equaled the pressure of air on the surface of the mercury in the dish. Today's barometers are made of different materials, but they work in the same way.

Because the height of the column of mercury depends on the pressure of the air, the measurement of its height (see Figure 6.1) gives an indication of the air pressure. At sea level, the average air pressure supports a column of mercury 760 mm, or 29.92 in., high. The SI unit of pressure is the pascal (Pa), which is a force of one Newton exerted on an area of one square meter (N/m^3). One atmosphere equals

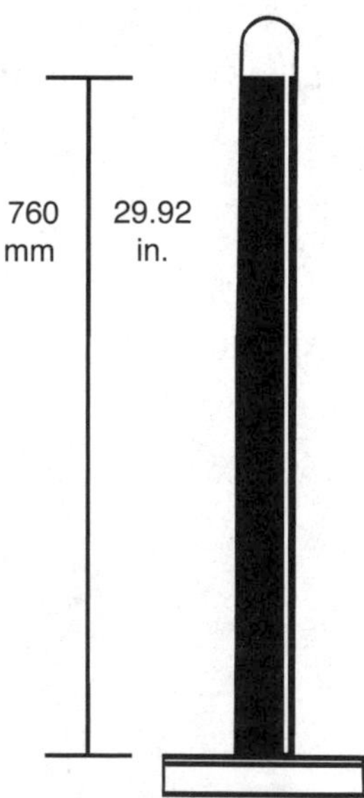

Figure 6.1 Mercury Barometers

101,325 Pa, or 101 kPa. This pressure is called *standard pressure.*

One millimeter of mercury pressure is also called a *torr,* in honor of Evangelista Torricelli, who invented the first barometer in 1643. If, at a given temperature, a gas is compressed, the volume of the gas will decrease. The English scientist Robert Boyle studied this phenomenon carefully and found that at a given temperature the volume occupied by a gas is inversely proportional to its pressure. This relationship is known as *Boyle's law,* first stated in 1663. In equation form Boyle's law is

$$P = k\frac{1}{V}$$

where

P = pressure of a gas sample
V = volume of a gas sample
k = a constant needed to make the two sides equal

Therefore,

$$P \times V = k$$

Thus, at a given temperature, the product of the pressure and volume of a gas must be constant. If the pressure is increased, the volume must decrease to maintain the constant product. For a given gas sample to be studied under different pressures, the following expression must hold:

$$P_1 V_1 = P_2 V_2$$

where

P_1 = original pressure of a gas sample
V_1 = original volume of the sample
P_2 = new pressure of the sample
V_2 = new volume of the sample

If we know the volume of a gas at one pressure, we can find the volume at any other pressure by that equation. We can also use it to find a pressure if we know both volumes and the other pressure.

TEMPERATURE

Temperature is measured with an instrument known as a *thermometer.* A thermometer consists of a glass tube of small, uniform bore with a bulb blown in one end. The tube is filled completely with mercury (or some similar material) at a temperature slightly above the temperature at which it is to be used, and then sealed off. As the mercury cools, it contracts in the tube, leaving a vacuum (void or empty space) in the upper part of the tube. The thermometer must then be calibrated (marked accurately) with known reference temperatures. The freezing point and boiling point of water are standard reference points. The thermometer is first immersed in melting ice, and the position of the end of the mercury column is marked with a line on the glass tube. The thermometer is then suspended in steam rising from water that is boiling at an atmospheric pressure of 760 mm, and the position of the end of the mercury column is again marked with a line. The distance between these two lines is then divided into equal segments called *degrees.* The size of each degree depends on the scale of temperature to be used.

There are two common temperature scales: Celsius (formerly called Centigrade) and Fahrenheit. On the Celsius scale, denoted by °C, the freezing point of water is called 0°C, and the boiling point of water is called 100°C. Thus, this scale has 100 degrees between the two standard reference points. The Celsius scale is most often used in scientific work.

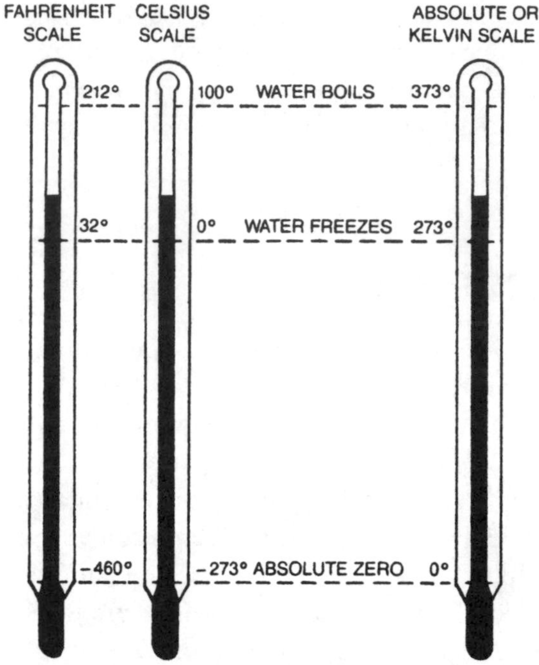

Figure 6.2 Comparison of Temperature Scales

On the Fahrenheit scale, denoted by °F, the freezing point of water is 32°F, and the boiling point of water is 212°F. Thus, in this scale there are 180 degrees between the two standard reference points. The Fahrenheit scale is often used in the home and in industry. The two temperature scales are related by the following expressions:

$$°F = \left[\frac{9}{5}°C\right] + 32 \quad \text{and} \quad °C = \frac{5}{9}[°F - 32]$$

These expressions are used to convert a temperature from one scale to the other.

The French scientist Jacques Charles, in studying the relationship between the volume of a gas and its temperature in 1787, discovered that the volume of a gas increases by 1/273 for each Celsius degree its temperature is increased. Extrapolating the data to a theoretical volume of zero, Charles reasoned that a temperature of −273.15°C is the lowest possible attainable temperature. He called this temperature *absolute zero*. The Kelvin temperature scale, named for the Scottish physicist William Kelvin (1824–1907), is based on the existence of absolute zero; the zero on the Kelvin scale is set exactly at absolute zero. Note that by international agreement the degree symbol and name are not used with the Kelvin scale. Thus, 150 K is read "150 Kelvin." The Kelvin scale is related to the Celsius scale as follows:

$$K = °C + 273$$

This expression is used in finding the *absolute temperature* when the Celsius temperature is known. Note that although there is an absolute zero to temperature, there is no known *upper* limit to temperature.

Temperature scale conversions:

$$°F = \left[\frac{9}{5}°C\right] + 32$$

$$°C = \frac{5}{9}(°F - 32)$$

$$K = °C + 273$$

Charles's studies led to the discovery that at a given pressure, the volume occupied by a gas is directly proportional to the Kelvin temperature of the gas. This relationship is known as *Charles's law*. Expressed as an equation this law is

$$V = kT$$

where

V = volume of the gas sample
T = absolute (Kelvin) temperature of the gas sample
k = a constant

Solving this expression for k we find that the ratio of the volume of a gas to its Kelvin temperature is a constant:

$$\frac{V}{T} = k$$

Thus, for a given gas sample, if the temperature is changed, this ratio must remain constant, so the volume must change to maintain the constant ratio. The ratio at a new temperature must be the same as the ratio at the original temperature, so

$$\frac{V_1}{T_1} = \frac{V_2}{T_2}$$

where

V_1 = original volume of sample of gas
T_1 = original Kelvin temperature
V_2 = new volume of the sample
T_2 = new Kelvin temperature of the sample

If we know the volume of a gas at one Kelvin temperature, we can find its volume at any other temperature by that equation. Conversely, the equation can be used to find Kelvin temperatures, given known volume.

COMBINED GAS LAW

Boyle's law and Charles's law may be combined into one expression as follows:

$$\frac{P_1 V_1}{T_1} = \frac{P_2 V_2}{T_2}$$

This expression is known as the *combined gas law*. The temperature and pressure at which the volume of a gas is measured are known as the *conditions of measurement*. When the volume of a gas is known at one set of conditions, its volume at a new set of conditions can be found with the combined gas law. Solving the equation for the new volume, we have

$$V_2 = V_1 \times \frac{P_1}{P_2} \times \frac{T_2}{T_1}$$

That equation tells us that the new volume of the gas equals the old volume multiplied by two correction factors, one for pressure and the other for temperature. Note the following very closely: If the new pressure is greater than the original pressure, the effect on the volume will be to make it smaller. Therefore, the pressure correction must be a fraction smaller than 1. Similarly, a new pressure less than the original will cause the gas to expand, thus making the correction greater than 1. Conversely, if the new temperature is higher than the original, the effect is to expand the gas. Thus the temperature correction must be greater than 1. If the new temperature is less than the original, the gas will contract, and the correction must then be less than 1. Always think through problems of this type before substituting into the combined gas equation. That will help you avoid careless mistakes. Let's go through an example.

EXAMPLE 1. A gas sample occupies 250 mL at 27°C and 780 Torr pressure. Find its volume at 0°C and 760 Torr pressure.

SOLUTION: First, convert the temperatures to the Kelvin scale by adding 273 (°C) to each:

$$T_1 = 27°\text{C} + 273 = 300 \text{ K}$$
$$T_2 = 0°\text{C} + 273 = 273 \text{ K}$$

Now, in the problem, temperature decreases, therefore volume decreases, therefore temperature correction must be less than 1.

Similarly, pressure decreases, therefore volume increases, therefore pressure correction must be greater than 1.

$$\text{Therefore: } V_2 = 250 \text{ mL} \times \frac{780 \text{ Torr}}{760 \text{ Torr}} \times \frac{273 \text{ K}}{300 \text{ K}}$$

Check this equation carefully with the combined gas equation.

Solving, we obtain

V_2 = 234 mL (or 230 mL to two significant figures)

EXPERIMENT 8: Hold your thumb on the end of a small hand bicycle pump, and pump the handle vigorously a few times to as much pressure as your thumb can hold. Feel the lower portion of the pump. Do you feel the increase in temperature? Repeat the pumping process, and this time, feel the upper part of the pump. Do you note any cooling? The ratio of P to T for any gas must remain constant. At the lower end, where the pressure is increased, the temperature must also rise, and at the upper end, where the pressure is lowered, the temperature also drops.

STANDARD CONDITIONS

We can see from the previous discussion that a statement of the volume of a gas without specifying the conditions under which it is measured is meaningless. All gases fill their containers completely. For example, if we have a mixture of oxygen, nitrogen, and carbon dioxide gases and put a sample of this mixture into a 1-ft^3 container, how much, by volume, of each gas will be present? The answer is 1 ft^3 of each, because each will fill the container. Thus we have a situation in which

$$1 \text{ ft}^3 \text{ of } O_2 + 1 \text{ ft}^3 \text{ of } N_2 + 1 \text{ ft}^3 \text{ of } CO_2 = 1 \text{ ft}^3 \text{ of mixture}$$

Temperature	Pressure
0°C	760 Torr
32°F	29.92 in.
273 K	1.000 atm = 101 kPa

Table 6.1 Standard Conditions

This answer sounds absurd, but it is absurd only because we are talking about volumes of gases without specifying the conditions.

To compare gas volumes, a set of *standard conditions* has been established. The standard temperature is the freezing point of water. The standard pressure is the average pressure at sea level. Table 6.1 summarizes standard conditions in various units. (1 atmosphere (atm) is defined to be the pressure at sea level)

LAW OF PARTIAL PRESSURES

John Dalton, in 1801, developed what came to be known as *Dalton's law of partial pressures,* which states that the total pressure of a mixture of two or more gases that do not chemically combine is the sum of the partial pressure of each. In a mixture of gases, each gas exerts the same pressure as if it occupied the volume alone. Because in a given gas mixture the temperature of the gases is the same and all fill the same container, the relative amount of each gas present is indicated by its partial pressure. For example, it is usually stated that air is composed of approximately 21% oxygen by volume and 79% nitrogen by volume. This actually means that the partial pressure of oxygen in air is 21% of the total air pressure, and the partial pressure of nitrogen in air is 79% of the total air pressure. Thus, we see that the volumetric composition of a gas mixture is indicated by the partial pressures of the constituents.

Insoluble gases are frequently collected by bubbling them into a container full of water to displace the water with the gas. In this process, the gas sample becomes saturated with water vapor. The total pressure of the collected gas is the sum of the pressure of the dry gas *plus* the pressure of the water vapor. The amount of dry gas collected can be found only after the vapor pressure of water at the temperature of collection is subtracted from the total gas pressure. Thus, Dalton's law of partial pressures enables us, in a sense, to dry a gas by arithmetic.

EXPERIMENT 9: Select a wad of steel wool large enough to stick in the bottom of a jar when wet. Thoroughly moisten the steel wool in vinegar (used to help clean the steel wool), press it into the bottom of the jar, and then invert the mouth of the jar into a basin of water so as to entrap the air in the jar. If possible, clamp the jar in place. You will observe the rusting taking place in the steel wool, and the rising of the water in the inverted jar. In rusting, the iron in the steel wool removes oxygen from the air, thus reducing the pressure of the entrapped air and permitting water from the basin to replace the oxygen. If the experiment is permitted to stand overnight, you will observe that the effect occurs only to a limited extent. When about one-fifth of the air has been replaced by water, the action ceases. This is evidence that air is made up of more than one gaseous substance and that the total pressure is the sum of the partial pressures of each. Table 6.2 gives the vapor pressure of water in Torr at various temperatures.

Now let's look at an example that incorporates all the relationships studied so far in this chapter.

EXAMPLE 2. A gas volume of 500 mL is collected, by the displacement of water, at a temperature of 77.0°F and a pressure of 748.8 Torr. Find the volume of the dry gas at standard conditions.

SOLUTION: To use Table 6.2 in this solution, we need the Celsius temperature of the gas.

$$77.0°\text{F} = \frac{9}{5}°\text{C} + 32$$

$$°\text{C} = \frac{5}{9}(77.0°\text{F} - 32) = 25.0°\text{C}$$

Temperature (°C)	Pressure (Torr)	Temperature (°C)	Pressure (Torr)	Temperature (°C)	Pressure (Torr)
0	4.6	22	19.8	40	55.3
5	6.5	23	21.1	45	71.9
10	9.2	24	22.4	50	92.5
15	12.8	25	23.8	60	149.4
16	13.6	26	25.2	70	233.7
17	14.5	27	26.7	80	355.1
18	15.5	28	28.4	90	525.8
19	16.5	29	30.0	100	760.0
20	17.5	30	31.8	200	11659.2
21	18.6	35	42.2	300	64432.8

Table 6.2 Vapor Pressure of Water

From Table 6.2 we find that the vapor pressure of water at 25 °C is 23.8 Torr. Therefore, the pressure of the dry gas is

$$748.8 \text{ Torr} - 23.8 \text{ Torr} = 725.0 \text{ Torr}$$

Standard conditions are 0 °C and 760 Torr pressure. Converting our temperatures to the absolute scale we have

$$T_1 = 25.0°\text{C} + 273 = 298.0 \text{ K}$$
$$T_2 = 0°\text{C} + 273 = 273.0 \text{ K}$$

Temperature decreases; therefore, correction factor is less than 1.

Pressure increases; therefore, correction factor is less than 1. Thus

$$V_2 = 500 \text{ mL} \times \frac{273.0 \text{ K}}{298.0 \text{ K}} \times \frac{725.0 \text{ Torr}}{760 \text{ Torr}} = 437 \text{ mL of dry gas}$$

PROBLEM SET NO. 6

1. Convert 30. °C to °F.

2. Convert 68 °F to °C.

3. Convert 95 °F to K.

4. A gas volume of 500. cm^3 at 770. Torr is compressed to 1540. Torr, at constant temperature. Find the new volume of the gas.

5. A gas volume of 350. mL at 47 °C is heated to 188 °C at constant pressure. Find the new volume of the gas.

6. A sample of gas at 20.0 °C and 750. Torr is heated to 252 °C while the volume is held constant. Find the new pressure of the gas.

7. The volumetric analysis of a sample of gas leaving a smokestack is as follows:

   10.8% CO_2 2.2% CO 4.5% O_2 82.5% N_2

   The pressure of the gas sample is 750. Torr. Find the partial pressure of each of the constituent gases.

8. A gas volume of 150. mL is collected at 22.0 °C and 740. Torr. Find the volume of the gas at standard conditions.

9. A gas occupies 330 mL at standard conditions. Find its volume at 86 °F and 30.40 in. of pressure.

10. A gas volume of 400. mL is collected by displacement of water from an inverted container so that the gas over the water is a mixture of the unknown gas and water vapor at 24 °C and 767.4 Torr. Find its volume at standard conditions when dry.

KINETIC-MOLECULAR THEORY

Some of the principal general properties of gases can be listed as follows:

1. Gases are compressible.

2. Gases fill any container.

3. Different gases mix completely.

4. Gases expand on heating.

5. Gases do not settle in their container.

The *kinetic-molecular theory* was developed as scientists attempted to find an explanation of these properties. According to this theory, gases consist of individual molecules that are relatively far apart with a lot of empty space between them. The particles are very small, so that the volume occupied by the gas molecules can be ignored because it is insignificant when compared with the volume of the container in which the gas is contained. This explains the ease with which gases can be compressed. The gas molecules are in constant, rapid motion. They move in straight lines until they collide with other molecules or with the inside walls of the container. This characteristic explains the filling of containers by gases and the mixing of gases. The moving gas molecules, in colliding with the walls of the container cause the pressure on the container. A given pressure is the result of the number of such collisions in a specified amount of time (unit time). Thus, gas pressure is increased by

- forcing more gas into the container—thus increasing the number of collisions per unit time;

- decreasing the volume of the gas by decreasing the size of the container—thus shortening the average distance between the molecules and thereby increasing the number of collisions per unit time;

- heating the gas in a closed container—thus increasing the speed of the molecules and thereby increasing the number of collisions per unit time.

The speed of the moving molecules is the result of the kinetic energy (energy of motion) they possess. This kinetic energy is increased by heating the gas and decreased by cooling it. The kinetic-molecular theory suggests that the collisions of the gas molecules with other molecules or with the walls of the container are perfectly elastic—that is, they take place without loss of energy either through friction or through any other means. If there were energy losses as a result of these collisions, a loss in kinetic energy would result, and the gas would ultimately settle in its container. The concept of elastic collisions explains the fact that gases do not settle.

IDEAL GASES

TRIVIA: The gas nitrous oxide, N_2O, commonly called laughing gas, was first discovered in 1772 but was used only for its entertainment purposes until an accident by one of its users in 1844 brought to light its anesthetic properties. It is used in dentistry to deaden the pain of tooth extraction.

An *ideal gas* is one whose molecules do not interact with one another and thus is one that follows the gas laws perfectly. Such as gas is nonexistent, because no known gas obeys the gas laws at all possible temperatures. What would happen if we cooled a gas to absolute zero? The gas would vanish!

$$V_2 = V_1 \times \frac{P_1}{P_2} \times \frac{0}{T_1} = 0$$

This, of course, does not happen in nature. There are two principal reasons why real gases do not behave as ideal gases:

1. The molecules of a real gas have mass, or weight, and the matter thus contained in them cannot be destroyed.

2. The molecules of a real gas occupy space, and thus can be compressed only so far. Once the limit of compression has been reached, neither increased pressure nor cooling can further reduce the volume of the gas.

In other words, a gas would behave as an ideal gas only if its molecules were true mathematical points—that is, if they possessed neither mass nor dimensions; however, at the ordinary temperatures and pressures used in industry or in the laboratory, molecules of real gases are so small, have so little mass, and are so widely separated by empty space, they behave almost as if they were mathematical points—that is, they follow the gas laws so closely that any deviations from these laws are insignificant. Nevertheless, it should be borne in mind that the gas laws are not strictly accurate, and results obtained from them are really only close approximations.

DENSITY OF GASES

The *absolute density* of a gas is the mass in grams of 1 mL of the gas at standard conditions. Compared with liquids and solids, gases have very low density. This is the result of the relatively large amount of empty space between gas molecules as suggested by the kinetic-molecular theory. Hydrogen has a density of 0.00009 g/mL. Oxygen has a density of 0.001429 g/mL. These figures are found experimentally by finding the actual mass of a known volume of a gas sample at known temperature and pressure conditions. The density is then computed with the aid of the combined gas law. Let's look at an example:

EXAMPLE 3. A 564.3-mL volume of chlorine gas at 27°C and 740. Torr has a mass of 1.607 g. Find the density of chlorine.

SOLUTION: First, the volume must be reduced to standard conditions.

$$V_2 = 564.3\text{ mL} \times \frac{273\text{ K}}{300\text{ K}} \times \frac{740.\text{ Torr}}{760\text{ Torr}} = 500.\text{ mL}$$

The mass of chlorine, of course, is not changed, because we are still dealing with the same amount of it. Density is the mass of

Gas	Formula	Absolute Density (g/mL)	Relative Density (air = 1)
acetylene	C_2H_2	0.001173	0.9073
air	—	0.0012929	1.00000
carbon dioxide	CO_2	0.0019769	1.5290
carbon monoxide	CO	0.0012504	0.9671
chlorine	Cl_2	0.003214	2.486
helium	He	0.00017847	0.13804
hydrogen	H_2	0.00008988	0.06952
nitrogen	N_2	0.00125055	0.96724
oxygen	O_2	0.001429	1.10527
sulfur dioxide	SO_2	0.002927	2.2639

Table 6.3 Density of Gases at Standard Conditions

1 mL of the gas at standard conditions. Because 500 mL of chlorine at standard conditions has a mass of 1.607 g, we can find the mass of 1 mL with a simple proportion:

$$\frac{x}{1 \text{ mL}} = \frac{1.607 \text{ g}}{500. \text{ mL}}$$

$$x = 0.003214 \text{ g/mL}$$

Therefore, the density of chlorine is 0.003214 g/mL.

Because the absolute densities of gases involve such small numbers, they are commonly compared with the density of air. Such a comparison, indicating the number of times a gas is more dense than air, is called the *relative density* of the gas. Table 6.3 gives the absolute and relative densities of several common gases.

LAW OF DIFFUSION

Gas molecules are small enough to diffuse (move through) through such porous materials as unglazed porcelain and natural rubber. The English scientist Thomas Graham in 1846 discovered that the rates of diffusion of gases are inversely proportional to the square root of their molar mass, when the gases are at the same temperature and pressure. This relationship is known as *Graham's law of diffusion*. Expressed as an equation, this law is

$$\frac{\text{Rate}_1}{\text{Rate}_2} = \sqrt{\frac{M_2}{M_1}}$$

where M is the molar mass of the gas.

This law is particularly useful in comparing rates of diffusion of different gases. Let's look at an example.

EXAMPLE 4. How much faster will hydrogen diffuse than helium?

SOLUTION: From the periodic table we find the molar mass of each gas:

helium: 4.0 u
hydrogen: 2.0 u

Therefore:

$$\frac{\text{Hydrogen rate}}{\text{Helium rate}} = \sqrt{\frac{4.0}{2.0}} = \sqrt{2.0} = 1.41$$

Thus, hydrogen diffuses about 1.4 times as fast as helium.

Graham's law of diffusion was put to use in the early work of preparing the original atomic bomb. Uranium metal was combined with fluorine to form the gas UF_6. Molecules of this gas containing the isotope of uranium of atomic weight 235 (U-235) diffused slightly more rapidly through a porous membrane than molecules of the gas containing uranium isotopes of atomic weight 238 (U-238). By repeated diffusions, U-235, the bomb material, was separated from the U-238.

GRAM MOLECULAR VOLUME

From the density of a given gas, we can find out what volume a mole of the gas will occupy at standard conditions through the use of a simple proportion. For example, in the case of hydrogen,

Density of hydrogen:	0.00008988 g/mL
Gram molecular mass:	2.016 g (the mass of 1 mol of H_2 = 2 × 1.008 g/mol)

Thus,

$$\frac{0.00008988\text{ g}}{1\text{ mL}} = \frac{2.016\text{ g}}{x}$$

$$x = 22{,}400\text{ mL (to 3 significant figures)} = 22.4\text{ L}$$

Similar calculations for the other gases yield approximately the same results. Thus, we may conclude that for any gas, the volume occupied at standard conditions by 1 mol of the gas is 22.4 L.

This is a significant fact, because now we can see that when we talk about a mole of a gas, we are referring to both a definite mass of a gas (its molecular mass) and a definite volume of it at standard conditions (22.4 L). This dual meaning of the mole is one of the most important concepts in chemistry.

AVOGADRO'S NUMBER

Avogadro's law states that equal volumes of gases contain equal numbers of molecules. Because at standard conditions 1 mol of any gas occupies the same gram molecular volume, 1 mol of every molecular substance must contain the same number of molecules. This number has been found by a variety of experimental methods to be 6.022×10^{23}. As you can see, this is a very large number. It is called *Avogadro's number*. Using this value, we can find the mass of a molecule of a substance. For example,

6.022×10^{23} molecules of O_2 have a mass of 32.0 g

Therefore, 1 molecule of O_2 has a mass of

$$\frac{32.0\text{ g/mol}}{6.022 \times 10^{23}\text{ molecules/mol}} = 5.31 \times 10^{-23}\text{ g per molecule}$$

This is

0.000 000 000 000 000 000 000 0531 g per molecule of O_2

Because a molecule of oxygen contains 2 atoms, each atom of oxygen must weigh half this amount, or 2.65×10^{-23} grams. As amazing as it may seem, numbers like this are very important in science.

FINDING MOLECULAR MASSES

To find the molecular mass of a substance, we simply weigh a known volume of the substance in the gaseous state at known temperature and pressure, reduce the volume to standard conditions, and then calculate the weight of 22.4 L at standard conditions using simple proportion. The direct weighing of gases is often quite difficult. Therefore, in actual practice, a solid or liquid substance that will decompose to produce the desired gas is weighed both before and after decomposition. The loss in weight is then the weight of the gas. Oxygen can be formed by heating solid compounds such as potassium chlorate, $KClO_3$. Carbon dioxide can be produced by heating calcium carbonate, $CaCO_3$. Sulfur dioxide can be formed by treating solid sodium sulfite, Na_2SO_3, with an acid.

Let's look at typical experimental results obtained in a laboratory to try to find the molecular weight of oxygen by heating $KClO_3$.

EXAMPLE 5. A 7.00-g portion of $KClO_3$ is heated and produces 1.55 L of oxygen gas at 27.0°C and 756 Torr. The residue from the heating has a mass of 5.00 g. Find the molecular mass of oxygen.

SOLUTION:
Mass of oxygen: 7.00 − 5.00 = 2.00 g
Volume of oxygen at standard conditions:

$$V_2 = 1.55 \text{ L} \times \frac{273 \text{ K}}{300. \text{ K}} \times \frac{756 \text{ Torr}}{760 \text{ Torr}} = 1.40 \text{ L}$$

Therefore, by proportion, the molecular weight of oxygen is

$$\frac{1.40 \text{ L}}{2.00 \text{ g}} = \frac{22.4 \text{ L}}{x}$$

$$x = \frac{22.4 \text{ L} \times 2.00 \text{ g}}{1.40 \text{ L}} = 32.0 \text{ g}$$

TRUE FORMULAS

We saw in the last chapter that we can compute the empirical formula of a substance if we know its percentage composition; however, we were unable to find the true formula of a molecular compound. The true formula gives the exact number of atoms of each element in a molecule of the substance. For molecular substances, the true formula must show

1. the ratio of the number of atoms of each element present and
2. the molecular mass of the substance (the sum of the atomic masses in the formula).

Therefore, in order to find the true formula of a substance, we must know both its percentage composition and its molecular mass. Let's look at an example to see how true formulas can be found.

EXAMPLE 6. A 500.-mL volume of a gaseous substance at standard conditions has a mass of 0.58 g. The substance contains 92.31% carbon and 7.69% hydrogen. Find its true formula.

SOLUTION: First, we find the molecular mass—that is, the mass of 22.4 L of the substance at standard conditions.

$$\frac{500. \text{ mL}}{0.58 \text{ g}} = \frac{22{,}400 \text{ mL}}{x}$$

$$x = \frac{22{,}400 \text{ mL} \times 0.58 \text{ g}}{500. \text{ mL}} = 26 \text{ g/mol}$$

The molecular weight of the substance is thus 26 g/mol. Next, we find the empirical formula:

$$\text{hydrogen:} \quad \frac{7.69}{1} = 7.69; \quad \frac{7.69}{7.69} = 1$$

$$\text{carbon:} \quad \frac{92.31}{12} = 7.69; \quad \frac{7.69}{7.69} = 1$$

Thus the empirical formula is C_1H_1, or simply CH.

The mass represented by this empirical formula is 12 g (C) + 1 g (H) = 13 g.

If we divide the true molecular mass by the formula mass of the empirical formula, we get the number with which we must multiply each of the subscripts in the empirical formula to give the true formula. Thus,

$$\frac{26}{13} = 2$$

Therefore, the true formula is C_2H_2.

Checking the mass, we see that the formula mass is now 24 + 2 = 26, which equals the molecular mass of the substance.

MASS-VOLUME RELATIONSHIPS

Because, for gases, the mole represents both a mass and a volume, we can apply this concept to equations to find out the volumes at standard conditions of gaseous reactants or products. Let's look at an example.

EXAMPLE 7. Oxygen is liberated from potassium chlorate by heat according to the equation

$$2\ KClO_3(s) = 2\ KCl(s) + 3\ O_2(g)$$

What volume of O_2 at standard conditions can be produced by the complete decomposition of 6.13 g of $KClO_3$?

SOLUTION: From the equation, we see that 2 mol of $KClO_3$ forms 3 mol of O_2, or, in other words, for every mole of $KClO_3$ decomposed, 3/2 mol of O_2 is formed. A mole of $KClO_3$ contains 122.6 g (sum of the atomic masses). Therefore, we are going to decompose 6.13/122.6 mol of $KClO_3$. The number of moles of O_2 produced by this decomposition will be

$$\frac{6.13\text{ g}}{1} \times \frac{1\text{ mol KClO}_3}{122.6\text{ g}} \times \frac{3\text{ mol O}_2}{2\text{ mol KClO}_3} = 0.075\text{ mol O}_2$$

At standard conditions, however, each mole of O_2 occupies 22.4 L. Therefore, the volume of O_2 produced at standard conditions must be

$$22.4\text{ L/mol} \times 0.075\text{ mol} = 1.68\text{ L of O}_2.$$

To find this volume at other than standard conditions, we can use the gas laws. *Note that all volumes found from equations apply only to standard conditions!*

VOLUME-VOLUME RELATIONSHIPS

Because, at standard conditions, a mole represents a definite volume of a gas, a ratio of moles of gases must indicate the same ratio of volumes of gases. Consider the following equation in which all the substances are gases:

$$N_2 + 3\ H_2 = 2\ NH_3$$

nitrogen		hydrogen		ammonia
1 mol	:	3 mol	:	2 mol
1 volume	:	3 volumes	:	2 volumes

Note that the molar ratio is identical to the volumetric ratio of the substances so long as all the volumes are measured *at the same conditions*.

Let's look at an example involving the volumes of gases in an equation.

EXAMPLE 8. A volume of 60 L of hydrogen, measured at room temperature and pressure, is to be used in the preparation of ammonia gas.

(a) What volume of nitrogen, measured at the same conditions, will be required for the reaction?

(b) What volume of ammonia, measured at the same conditions, can be prepared?

SOLUTION: The equation is

N_2	+	$3\ H_2$	=	$2\ NH_3$
1 vol	:	3 vol	:	2 vol

(a) The volume of nitrogen required is

$$\frac{x}{1} = \frac{60\text{ L}}{3}$$

$$x = 20\text{ L N}_2$$

(b) The volume of NH_3 that can be produced is

$$\frac{x}{2} = \frac{60\text{ L}}{3}$$

$$x = 40\text{ L NH}_3$$

THE IDEAL-GAS EQUATION

The gas laws (Boyle's law, Charles's law, and so on) all assume a fixed amount of gas; but often in chemistry, the number of moles of gas changes. How can we account for this factor? The answer is, by using the ideal-gas equation. If we combine all the gas laws, including Avogadro's law, we can derive the ideal-gas equation:

$$PV = nRT$$

where

P = pressure
V = volume
n = number of moles of gas
R = ideal-gas constant
T = Kelvin temperature

Any units may be chosen for the pressure and volume, but the value of the ideal constant will depend on the units chosen. Most commonly, the pressure is expressed in atmospheres (atm) and the volume in liters. The ideal-gas constant then becomes

$$R = \frac{0.0821 \text{ L} \cdot \text{atm}}{\text{K} \cdot \text{mol}}$$

EXAMPLE 9. An 88.0-g sample of CO_2 would occupy what volume at a pressure of 2.50 atm and a temperature of 27°C?

SOLUTION: First we convert the grams of carbon dioxide to moles:

$$\frac{88.0 \text{ g}}{1} \times \frac{1 \text{ mol}}{44.0 \text{ g}} = 2.0 \text{ mol}$$

Next, we convert the temperature to Kelvin:

$$\text{K} = 27°\text{C} + 273 = 300. \text{ K}$$

We solve the ideal-gas equation for the volume:

$$V = \frac{nRT}{P} = \frac{(2.0 \text{ mol})\left(0.0821 \frac{\text{L·atm}}{\text{K·mol}}\right)(300. \text{ K})}{2.5.0 \text{ atm}} = 19.7 \text{ L}$$

EQUIVALENT MASS OF AN ELEMENT

The *equivalent mass* of an element is defined as that amount of the element that combines with or displaces an atomic mass's worth (1.0 g) of hydrogen. It is also the amount of an element that combines with or displaces 8 parts by weight (8 g) of oxygen. Just as we have been referring to a mole of a substance, we are now ready to refer to an equivalent mass of a substance.

These two ideas—because they are ideas rather than mere defined quantities—form the basis of the quantitative applications of chemistry. We have seen how the idea of the mole is used with chemical equations. The coefficients of a balanced equation give the ratio of the number of moles of reactants and products involved in the reaction. Equivalents likewise refer to amounts of reactants and products. The rule is simply this: *one equivalent of any substance in nature reacts actually or theoretically with one equivalent of every other substance to produce one equivalent of each of the products involved.*

The concept of the equivalent is a powerful tool to use in prying open the secrets of chemistry; But just like any other tool, this one must be used skillfully and with understanding of its application. Let's focus our attention now on how to find the equivalent mass of an element.

The definitions of equivalent mass suggest experimental methods of measuring the equivalent mass of an element. The mass of an element that combines with or displaces either 1 g of hydrogen or 8 g of oxygen is the equivalent mass of the element in grams. Let's look at some examples.

EXAMPLE 10. Many metals displace hydrogen from acids. At standard conditions, 1 mol of hydrogen occupies 22.4 L; but a mole of hydrogen contains 2 atomic masses' worth of hydrogen, because the formula of hydrogen is H_2. Therefore, 11.2 L of hydrogen at standard conditions is 1 atomic mass's worth of hydrogen. Thus, that mass of a metal that would displace 11.2 L of hydrogen at standard conditions would be the equivalent mass of the metal.

A 1.5-g sample of zinc displaces 560. mL of hydrogen at 20°C and 748 Torr pressure. Find the equivalent mass of the metal.

SOLUTION: First we find the volume of hydrogen at standard conditions.

$$V_2 = 560.\text{ mL} \times \frac{273\text{ K}}{293\text{ K}} \times \frac{748\text{ Torr}}{760\text{ Torr}} = 513\text{ mL}$$

Then, by proportion:

$$\frac{1.5\text{ g}}{513\text{ mL}} = \frac{x}{11{,}200\text{ mL}}$$
$$x = 32.7\text{ g of Zn}$$

Thus, the equivalent mass of the zinc is expressed as 32.7 g/equiv.

EXAMPLE 11. Oxygen combines directly with most elements to form oxides. The analysis of an oxide to find its percentage composition gives a ratio of the mass of oxygen and the other element present. In fact, the percentage composition is the number of grams of each element present in a 100-g sample. Therefore, knowing the percentage composition, we can use a simple proportion to find the mass of an element that has combined with 8 g of oxygen.

Zinc oxide contains 80.34% zinc and 19.66% oxygen. Find the equivalent mass of zinc.

SOLUTION: By proportion:

$$\frac{80.34\text{ g}}{19.66\text{ g}} = \frac{x}{8\text{ g}}$$
$$x = 32.7\text{ g}$$

Thus, the equivalent weight of zinc is 32.7 g/equiv.

Using the techniques suggested in the preceding examples, we can find the equivalent mass of many elements. Table 6.4 gives both the atomic mass and the equivalent mass of some selected elements, together with the ratio of the atomic mass to the equivalent mass for each.

Look carefully at the ratio in the last column for each element. What is it? It is the oxidation number of the element. This is very important. We can establish the following rule:

$$\text{Oxidation number} = \frac{\text{Atomic Mass}}{\text{Equivalent Mass}}$$

$$\text{Equivalent Mass} = \frac{\text{Atomic Mass}}{\text{Oxidation number}}$$

You will recall that an element exhibits a valence other than zero only when it is in a compound. Furthermore, many elements are capable of exhibiting more than one valence. The particular oxidation number exhibited by an element depends on the compound under consideration. Because equivalent mass is related to oxidation number, the equivalent mass will vary according to the compound in

Element	Atomic Mass	Equivalent Mass	Atomic Mass / Equivalent Mass
hydrogen	1.0008	1.0008	1
sodium	22.990	22.990	1
potassium	39.098	39.098	1
silver	107.87	107.87	1
oxygen	15.999	8.000	2
magnesium	24.31	12.16	2
calcium	40.08	20.04	2
zinc	65.38	32.69	2
aluminum	26.98	8.99	3

Table 6.4 Equivalent Mass

which it is found. Elements that can have different oxidation numbers possess equivalent masses corresponding to each of their oxidation numbers. The equivalent mass of an element depends on the oxidation number of the element in a particular reaction.

Let's look at the reaction of iron with oxygen. Depending on the conditions under which the reaction takes place, iron forms two different oxides according to the following equations:

$$2\ Fe(s) + O_2(g) = 2\ FeO(s)$$
$$4\ Fe(s) + 3\ O_2(g) = 2\ Fe_2O_3(s)$$

In both of these oxides, oxygen has an oxidation number of -2.

In FeO, the oxidation number of iron is $+2$. In Fe_2O_3, the oxidation number of iron is $+3$.

Iron is exhibiting two different oxidation numbers, depending on the particular reaction under consideration. Therefore, iron has two different equivalent masses. In the first reaction, the equivalent mass of iron is half its atomic mass. In the second reaction, the equivalent mass of iron is one-third its atomic mass. The elements in Table 6.4 were selected because they normally have only one oxidation number and consequently have only one equivalent mass.

In the past few years, the use of equivalent mass and equivalents has declined dramatically. All the calculations that are commonly done using equivalents can be done using moles. Using moles makes the calculations somewhat more involved, however. Many industrial chemists still use the concepts of equivalents and equivalent masses because these simplify their calculations.

RULE OF DULONG AND PETIT

The quantity of heat absorbed by a substance in warming up—or given off by it in cooling—is specified in units known as calories. A *calorie* (cal) is that quantity of heat necessary to raise the temperature of 1 g of water 1 °C. Another unit of heat is the *joule* (J), with 1 cal = 4.184 J. The number of calories of heat necessary to raise the temperature of 1 g of any substance 1 °C is known as the *heat capacity* of the substance. The ratio of the heat capacity of a given substance to the heat capacity of water is known as the *specific heat* of the substance.

The French scientists Dulong and Petit studied the relationship between specific heat and atomic mass of elements. In the early 1800s they discovered the following relationship for elements in the solid state:

$$\text{Atomic mass} \times \text{Specific heat} = 6.2$$

This product is known as the *atomic heat* of the element. The value 6.2 is an average. Table 6.5 gives the atomic heat of several elements.

We can find the exact atomic mass of an element by using the rule of Dulong and Petit in conjunction with methods for finding the equivalent mass of an element. To do this we proceed in four steps:

1. Determine the exact equivalent mass.

2. Find the approximate atomic mass from the rule of Dulong and Petit.

3. Find the oxidation number by dividing the approximate atomic mass by the exact equivalent mass.

4. Multiply the equivalent mass by the oxidation number to obtain the exact atomic mass.

Element	Atomic Mass (g/mol)	Specific Heat	Atomic Heat
lithium	7	0.850	6.0
sodium	23	0.292	6.7
magnesium	24	0.245	5.9
phosphorus	31	0.181	5.6
potassium	39	0.180	7.0
calcium	40	0.155	6.2
iron	56	0.108	6.0
copper	64	0.092	5.9
zinc	65	0.092	6.0
silver	108	0.056	6.0
tin	119	0.054	6.4
gold	197	0.031	6.1

Table 6.5 Atomic Heat

Let's look at an example.

EXAMPLE 12. A compound contains 79.87% copper and 20.13% oxygen. The specific heat of copper is 0.092. Find the exact atomic mass of copper.

SOLUTION:
Step 1. The exact equivalent mass of copper (that mass of it combined with 8 g of oxygen) is

$$\frac{79.87 \text{ g}}{20.13 \text{ g}} \times \frac{x}{8 \text{ g}}$$

$$x = 31.74 \text{ g}$$

The equivalent mass of copper is 31.74 g.

Step 2. From the rule of Dulong and Petit:

$$\text{Approximate atomic mass of copper} = \frac{6.2}{\textit{specific heat}} = \frac{6.2}{0.092} = 67.4\ g$$

Step 3. The oxidation number must be a whole number. Dividing the approximate atomic mass by the equivalent mass we have:

$$\frac{67.4\ g}{31.74\ g} = 2.12$$

The oxidation number is the nearest whole number, which is 2.

Step 4. Multiplying the equivalent mass by the oxidation number gives us

$$31.74\ g/\text{equiv.} \times 2\ \text{equiv.} = 63.5\ g/\text{mol}$$

The exact atomic mass of copper is 63.5 g/mol.

SUMMARY

The properties of gases may be summarized in a number of laws in which the properties of pressure, volume, temperature, and amount are related to each other. These laws include those developed by Boyle, Charles, and others. A number of these laws are combined in the ideal-gas law. These laws allow us to predict what will happen if one or more variables are changed. Gas laws also allow us to interconvert the volume, moles, and mass of a gas. The kinetic-molecular theory of gases is one of the most successful models that has been developed in chemistry. It presents a model of the microscopic properties of gases that is consistent with the behavior of gases that we observe in the macroscopic world.

The gas laws and other relationships such as Avogadro's law may be used to determine the number of gas particles in a given gas sample. The gas-law relationships can also be used to calculate the molecular formula of a gas.

The Gas Laws:

Boyle's law:	$p = k\frac{1}{V}$ (keeping temperature and amount constant)
Charles's law:	$V = kT$ (keeping pressure and amount constant)
Combined gas law:	$\frac{P_1V_1}{T_1} = \frac{P_2V_2}{T_2}$
Graham's law of diffusion:	$\frac{\text{rate}_1}{\text{rate}_2} = \sqrt{\frac{M_2}{M_1}}$
Ideal-gas equation:	$PV = nRT$

PROBLEM SET NO. 7

1. A 555-mL sample of a gas at 22°C and 740. Torr pressure weighs 0.6465 g. Find the density of the gas.

2. How much faster will helium diffuse than chlorine?

3. Find the actual mass in grams of a single hydrogen atom.

4. A 0.24-g sample of a gas occupies 82.2 mL at 117 °C and 740. Torr pressure. Compute its molecular mass.

5. A gas consists of 85.72% carbon and 14.28% hydrogen. A 0.855-g sample of this gas occupies 523 mL at 29 °C and 733 Torr pressure. Find the true formula of the substance.

6. What volume of hydrogen, at standard conditions, can be produced by treating 10. g of zinc with hydrochloric acid, HCl?

7. What mass of calcium carbonate, $CaCO_3$, must be heated to produce 5 L of CO_2 at standard conditions?

8. What volume of pure O_2 will be required to burn 15 L of ethane, C_2H_6, to CO_2 and H_2O? (All gases are measured at standard conditions).

9. Compute the equivalent mass of phosphorus in the compound phosphine, PH_3.

CHAPTER 7

LIQUIDS AND SOLIDS

KEY TERMS

vapor pressure, boiling point, freezing point, amorphous solids, crystalline solids, heats of vaporization and fusion

If the volume of a gas is sufficiently reduced by compressing it or cooling it or both, the gas will condense to a liquid. Early scientific investigators discovered that a number of substances in the gaseous state at room temperature can condense to a liquid by pressure alone. Other gases resist liquefaction regardless of the pressure and condense only after the temperature has been reduced. These findings led to the idea that a critical temperature is involved in the liquefaction process. The *critical temperature* of a gaseous substance is the temperature above which it is impossible to liquefy the substance by pressure alone. The pressure required to liquefy a gas at its critical temperature is called the *critical pressure*. Table 7.1 gives the critical temperature and pressure of several common substances.

Substance	Critical Temperature (°C)	Critical Pressure (atm)
ammonia	132.4	111.3
argon	−122.4	48.0
carbon dioxide	31.0	72.9
carbon monoxide	−140.2	34.5
chlorine	144.0	76.1
ethyl alcohol	243.1	63.0
helium	−268.0	2.26
hydrogen	−239.9	12.8
nitrogen	−146.9	33.5
oxygen	−118.4	50.1
sulfur dioxide	157.5	77.9
water	374.2	218.3

Table 7.1 Critical Temperatures and Pressures

The more a gas is cooled below its critical temperature, the less pressure is required to liquefy it. When a substance is in the gaseous state at a temperature above its critical temperature, it is properly called a *gas*. A substance in the gaseous state at a temperature below its critical temperature is properly referred to as a *vapor*.

VAPOR PRESSURE

All liquids show a tendency to vaporize at room temperature and pressure. If a small sample of a liquid is placed in a dish and allowed to stand in contact with freely moving air, it will in time completely evaporate, which means that its molecules have escaped as a vapor. Liquids like gasoline and carbon tetrachloride evaporate much more rapidly than liquids like oil or mercury. The rate of evaporation is related to a property of liquids known as vapor pressure.

The *vapor pressure* of a liquid is the pressure exerted by a vapor when the liquid and its vapor are in dynamic equilibrium. To understand what vapor pressure is, think of a soda bottle partially filled with water with the cap on. The water starts to evaporate into the air in the bottle. The pressure of this air increases as the water vapor is added to it. Ultimately the water vapor will saturate the air in the bottle, and droplets of water will begin to condense back to a liquid on the upper walls of the bottle. The increase in the pressure of the air when it is saturated with water vapor is known as the vapor pressure of the water.

The vapor pressure of other liquids can be determined in the same fashion. For accurate measurement, the air to be used in the container in contact with the liquid initially has to be free of any molecules of the liquid vapor. The higher the vapor pressure, the more rapidly a liquid will evaporate into dry air. Another way to determine the vapor pressure of a liquid is to place the liquid into an "evacuated" flask (no air present) and allow the system to reach equilibrium. The gas pressure then measured is the vapor pressure of the liquid.

Because energy is required for a molecule of a liquid to escape as vapor from the liquid, the vapor pressure of a liquid is related to its temperature. Table 6.2 gives the vapor pressure of water at various temperatures. Note particularly the vapor pressure of water at 100°C. It is 760 Torr. That is standard atmospheric pressure, and 100°C is called the normal boiling point of water. The *boiling point* of a liquid is the temperature at which the vapor pressure of the liquid equals the pressure of the surrounding atmosphere. The normal boiling point of a liquid is the temperature at which its vapor pressure reaches standard atmospheric pressure, or 760 Torr.

Because atmospheric pressure changes slightly from day to day, the boiling point of fluctuates correspondingly. As the air pressure is reduced, the boiling point decreases. At high elevation, the boiling point of water is considerably reduced owing to the lower atmospheric pressure. For example, in Denver, Colorado, water boils at about 87°C, whereas at the top of Pike's Peak water boils at about 62°C. To easily compare boiling points of liquids, we usually examine the normal boiling points. Table 7.2 gives the normal boiling points of some common substances.

EVAPORATION

The phenomenon of evaporation is familiar to everyone. Like boiling, evaporation involves the change of a liquid to a vapor. Unlike boiling, evaporation takes place at any temperature.

Substance	Temperature (°C)	Substance	Temperature (°C)
benzene	80.1	hydrogen	−252.8
carbon dioxide	−78.5	oxygen	−183.0
chlorine	−34.1	sulfur dioxide	−10.0
ethyl alcohol	78.5	water	100.0

Table 7.2 Normal Boiling Points of Some Common Substances

To understand evaporation, we need to know a bit about energy. According to the kinetic-molecular theory, the molecules of a liquid are free to move around under the surface of a liquid. The motion is less rapid than in the case of gases, because the molecules of a liquid are virtually in contact with one another; but the molecules do move, because they possess kinetic energy (energy of motion) at any given temperature.

The amount of this energy possessed by each molecule of liquid is not uniform; some molecules possess considerably more energy than others in the same sample of a liquid. In general, most of the molecules have about the same amount of energy, but some possess appreciably more than average energy and some possess appreciably less. Furthermore, because the molecules are so close to one another, they collide frequently, and during such a collision energy is redistributed between the colliding molecules, causing a gain in kinetic energy in one of the molecules and a loss in kinetic energy in the other. This transfer of energy by collision can result in the formation of relatively high energy molecules.

The temperature of a liquid is a measure of the average energy of a sample of the liquid. If the temperature goes up, the average kinetic energy of the liquid goes up, because the particles are moving faster and have a higher kinetic energy. Similarly, a decrease in temperature results in a lowering of the average kinetic energy of the particles in the sample.

The escape of a liquid molecule in the evaporation process requires energy. Consequently, only those molecules of high energy can evaporate from a liquid. But the escape of these high-energy molecules results in a lowering of the average energy of the liquid sample. Thus, a cooling always accompanies evaporation.

EXPERIMENT 10: With pieces of cotton, spread water on the back of one hand and isopropyl (rubbing) alcohol on the back of the other, then wave your hands back and forth through the air. You will observe cooling from the evaporation of both liquids, but because the alcohol is more volatile (evaporates more easily), it causes more cooling.

Liquids evaporate because molecules at the surface having a higher kinetic energy escape into the air. The temperature of a liquid is a measure of its average kinetic energy. If the temperature rises, the average particle is moving faster, so the rate of evaporation increases.

Of course, as soon as the temperature of the liquid begins to drop below the temperature of the surrounding atmosphere, the atmosphere begins to warm the sample. This means that the atmosphere begins to add more energy to the liquid to replace the energy lost by evaporation. This additional energy, plus its accumulation by relatively few of the liquid molecules through molecular collisions, permits the evaporation of the liquid to continue until ultimately all the molecules of the liquid have gained enough energy to evaporate.

You have undoubtedly observed the cooling effect of liquids like water or alcohol evaporating from your skin. You perspire when you get hot, and the evaporating perspiration acts as a check to prevent your body from overheating. The temperature of the air is usually cooler by the seashore, or at the lake, or by a waterfall, because the evaporating water absorbs energy from the atmosphere and cools it. Many years ago, the dairy industry was confined to the northern states or to places of high elevation, because the then-common breeds of dairy cattle did not possess sweat glands and consequently could not survive in the high temperatures of the South. However, by selective breeding and crossbreeding of these earlier dairy cattle with cattle that do have sweat glands, scientists have been able to produce new varieties of dairy cattle that can live in more tropical climates. This development, of course, has bolstered the economy of warmer regions. These illustrations reemphasize the fact that natural phenomena follow definite laws of nature. Human progress is enhanced if we first learn these laws and then devise means of applying them to control our environment.

SURFACE TENSION

The molecules of a liquid are relatively close together. They are so close, in fact, that they exert appreciable attraction on one another. Inside the body of the liquid, where a given molecule is completely surrounded by other molecules, these attractive forces are equal in all directions. Thus they counteract one another, and no net unbalanced force remains on the molecule. However, on the surface of the liquid, a molecule is not completely surrounded by nearby molecules. The region above such a surface molecule is relatively vacant. Such a molecule is attracted by its neighbors on the surface and by the molecules below it. This results in a net unbalanced attractive force directed into the interior of the liquid. The effect is that the surface molecules form an encasing film on the liquid, or surface tension, that is relatively quite tough.

TRIVIA: The surface tension of water is so great that over a hundred paper clips can be added to a full glass of water without causing it to overflow.

You can readily observe the effects of surface tension. A tiny droplet of water remains intact and does not spread out to form a smooth film on a tabletop, because the surface film keeps the water from "seeking its own level." Drops of water falling free and passing through air are drawn into spheres by their surface film. A sphere, of course, is a shape containing a maximum volume within a minimum surface. You can float razor blades or needles on water, even though these objects are far denser than water, because the strong surface film supports them.

EXPERIMENT 11: Bend the end of a 4-in. length of wire into a single loop around a pencil. Remove the pencil, and then bend the loop so that its plane is perpendicular to the remaining length of wire. Fill a glass with water. Using the straight part of the wire as a handle, gently press the flat loop against the surface of the water. Note how the surface becomes indented before the loop finally breaks through into the water. This indentation is the result of the strong film on the water's surface. Then, slowly pull the loop up through the surface. Note how the water is pulled up above the level of the remaining surface.

The surface tension of a liquid can be changed by dissolving substances in the liquid. For example, soap greatly reduces the surface tension of water. This effect can be observed as follows.

EXPERIMENT 12: Fill a pan with water. Using the loop from Experiment 11, slowly ease a paper clip onto the surface of the water. The paper clip floats, owing to the surface tension of water, even though its density is greater than that of water. Dip the end of a toothpick into dishwashing detergent and then touch it to the water near the floating paper clip. The paper clip sinks.

The decrease in the horizontal forces of attraction in the water surface, caused by the dissolving detergent, results in the formation of net unbalanced forces in the surface directed away from the point of application of the detergent. In other words, the detergent breaks the surface tension of the water, allowing the paper clip to sink. This is one reason why soaps and detergents are used in washing dishes and clothes: they reduce water's surface tension and allow it to more efficiently "wet" the dishes and clothes.

In general, liquid molecules that have a higher attraction for one another (because of the presence of polar covalent bonds) have higher boiling points and higher surface tension. Water, because of its low molecular weight, should have a *much* lower boiling point than it actually does. Water's unusually high boiling point is due to the strong interaction (hydrogen bonding) among the water molecules.

CAPILLARY ACTION

Surface tension is related to another phenomenon associated with liquids. Between the molecules of a liquid there are attractive forces that can be called forces of *cohesion*. Similarly, there are attractive forces between the molecules of a liquid and the molecules of its container. These can be called forces of *adhesion*. If the forces of adhesion are greater than the forces of cohesion, the liquid will wet the container. If the reverse is true, the liquid will draw away from the container and not wet it. Figure 7.1 illustrates the difference in appearance of the surface of water and the surface of mercury in glass containers. Water wets glass. Mercury does not.

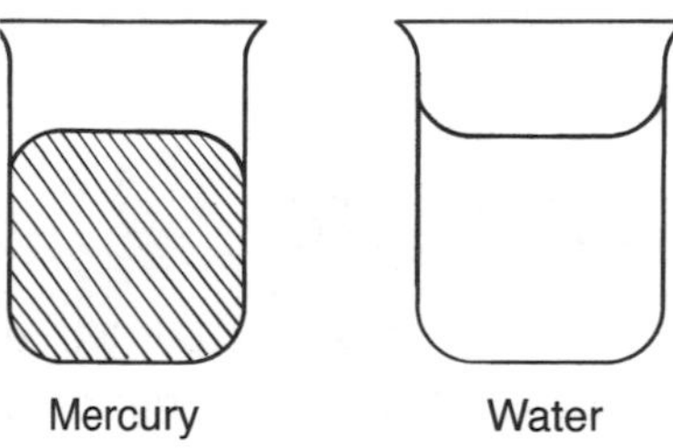

Figure 7.1 Surface Variation of Liquids with Different Ability to Wet Glass

If a liquid wets a given substance, the liquid will be drawn up into a small-diameter tube made of the substance. If the liquid does not wet a substance, a small-diameter tube made of the substance will depress the surface of the liquid. Figure 7.2 illustrates this effect with water and mercury in glass tubes. The rise or depression of a liquid surface in a small-diameter tube is known as *capillary action*. The amount of change in the level of the liquid in the tube is directly proportional to the surface tension of the liquid.

Capillary action has many applications. The subsurface water in fields is carried up to the roots of plants through tiny pores (capillary tubes) in the soil. A blotter works by capillary action. A towel consists of thousands of tiny capillary tubes that draw water from your

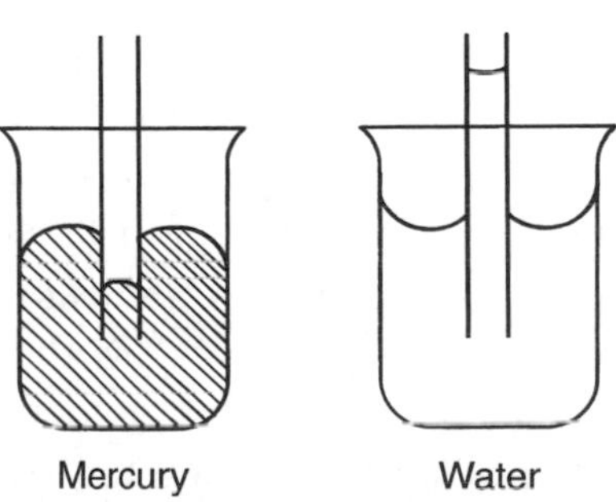

Figure 7.2 Capillary Action

skin to dry it. A sponge "drinks" water into its capillary tubes.

THE SOLID STATE

When liquids are sufficiently cooled, they congeal to the solid state. The temperature at which a substance solidifies from the liquid to the solid state is known as the *freezing point* of the substance. As in the case of the boiling point, the atmospheric pressure affects the freezing point of a substance. Some solids, like ice and bismuth, expand on cooling. Most solids shrink on cooling. An increase in pressure lowers the freezing point of solids that expand on cooling, which is why ice melts under pressure. For solids that shrink on cooling, such as iron, an increase in pressure raises the freezing point. In general, the effect of pressure on the freezing point is much less dramatic than its effect on the boiling point. In most situations, we can ignore the pressure effect.

In the solid state, the particles of a substance, whether these particles are molecules or ions, are not completely rigid. They are free to vibrate within definite spatial limits and thus possess some kinetic energy. Theoretically, all particles possess some energy until the temperature is reduced to absolute zero.

Amorphous Solids

Structurally, there are two classes of solids. One type is known as the *amorphous*, or *vitreous*, type. Amphorous solids have a completely random arrangement of the particles. This lack of definite geometric pattern results in specialized properties for this type. When these solids are broken with a hammer they exhibit a curved fracture surface. When heated they soften gradually and slowly transform to the liquid state with no clearly defined melting point. Glass is a typical example of this type of solid. There is some evidence to indicate that such solids are not true solids but in reality are highly rigid and extremely viscous liquids.

Crystalline Solids

The second type of solid is known as a *crystalline solid*. In this type, the particles are arranged in definite geometric patterns. The result is that, in contrast with amorphous solids, these solids exhibit definite physical properties. When struck with a hammer, crystalline solids break along definite planes. When heated, they melt at a specific temperature. Figure 7.3 shows the six basic configurations

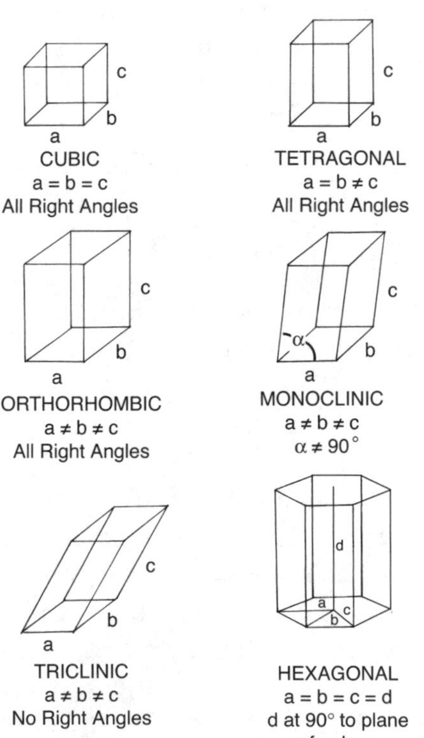

Figure 7.3 The Basic Crystal Configurations

of particles in crystals. These configurations are also known as crystal lattices. Examples of substances of the various types of crystal lattices are shown below.

- cubic lattice — table salt
- tetragonal lattice — sodium sulfide
- orthorhombic lattice — sulfur
- monoclinic lattice — sugar
- triclinic lattice — boric acid
- hexagonal lattice — quartz

ENERGY AND CHANGE OF STATE

Relatively large amounts of energy are involved when a substance changes from one physical state to another. When a substance changes from gas to liquid to solid, energy is released by the substance to its surroundings. A change from solid to liquid to gas involves the absorption of energy by the substance. The energy change at the liquid-to-gas transformation is known as the *heat of vaporization*. The heat of vaporization of a substance (liquid) is the number of calories required to change 1 g of the liquid to a gas at the transformation temperature. The heat of vaporization of water at 100°C is 540 cal/g.

The energy change at the solid-to-liquid transformation is called the *heat of fusion*. The heat of fusion of a substance is the number of calories required to melt 1 g of a solid to liquid at the transformation temperature. The heat of fusion of ice is 79.7 cal/g at 0°C.

gas → liquid → solid	Energy is released by the substance.
solid → liquid → gas	Energy is absorbed by the substance.

If we start with 1 g of ice and intend to boil it, the energy requirements are as follows:

To melt the ice	79.7 cal
To heat the water from 0°C to 100°C	100.0 cal
To vaporize the water at 100°C	540.0 cal
Total	719.7 cal

Notice that 619.7 cal, or about 86% of the total energy requirement, were involved with changes of state. Only about 14% of the energy absorbed went into raising the temperature of the water. Keep in mind that the energy absorbed or given off during changes in state does not cause a change in the temperature of the substance involved.

SUMMARY

Energy and pressure changes can cause substances to undergo changes of state. A gas can be condensed to a liquid by pressure only at temperatures below its critical temperature. Interactions among the molecules of a substance may affect its boiling and freezing points as well as other physical properties such as surface tension. Solids may be of two types: amorphous or crystalline.

PROBLEM SET NO. 8

1. Which of the substances in Table 7.1 are gases at room temperature (25°C)?

2. Using the data in Tables 7.1 and 7.2, compute the ratio of the absolute normal boiling point to the absolute critical temperature for each of the following:

   (a) carbon dioxide

   (b) chlorine

   (c) ethyl alcohol

   (d) hydrogen

   (e) oxygen

   (f) water

3. From your results in Problem 2, can you detect any relationship between these two properties of substances?

4. You are planning a trip by car to the top of Pike's Peak. Your car is equipped with a 180°F thermostat (which prevents the circulation of water through your radiator below that temperature). Prove by calculations why it would be wise to remove the thermostat before making the ascent.

5. How many calories would be required to completely vaporize 25 g of ice at 0°C?

CHAPTER 8

SOLUTIONS

KEY TERMS

solution, solvent, solute, molarity, normality, molality, percentage composition, solubility, electrolyte, nonelectrolyte, osmosis, colligative properties, colloid

A *solution* is a homogeneous mixture consisting of two components; a *solvent* that is the dissolving medium, and a *solute* that is the substance being dissolved. The solvent is normally present in a much larger amount than the solute(s). Solutions are mixtures, because an unlimited number of compositions involving a given solute and solvent are possible. In solutions, the solute is dispersed into molecules or ions, and the distribution of the solute is perfectly homogeneous throughout the solution. A tremendous amount of chemistry takes place in solution, so it is important for us to become familiar with both the terminology and the properties of solutions.

METHODS OF EXPRESSING CONCENTRATION

A *concentrated* solution contains a relatively large amount of solute per unit volume of solution. A *dilute* solution contains a relatively small amount of solute per unit volume of solution. (The words *strong* and *weak* should not be used when referring to the concentration of a solution, because these words have their own special meanings in chemistry.) The principal methods of expressing the concentrations of solutions are molarity, normality, molality, and percentage composition.

Molarity

JOKE: What do you call a tooth in a glass of water?—A one molar solution.

The *molarity* of a solution expresses the number of moles of solute per liter of solution.

$$\text{Molarity} = \frac{\textbf{Moles of solute}}{\textbf{Liters of solution}}$$

Molarity is abbreviated M. Its units are moles/liters. Thus, a solution that contains a half mole of solute per liter of solution is designated as 0.5 M:

$$\frac{0.5 \text{ mol}}{1.0 \text{ L}} = 0.5 \text{ M}$$

Normality

The *normality* of a solution expresses the number of equivalents of solute per liter of solution:

$$\text{Normality} = \frac{\textbf{Equivalents of solute}}{\textbf{Liters of solution}}$$

Thus, we must find the number of equivalents (equiv) before we can calculate normality. What is an equivalent? The answer to that question depends on the chemical reaction involved. In redox reactions, 1 equiv is the amount of an oxidizing agent that yields 1 mol of electrons (see Chapter 10). For acids, 1 equiv is the amount that yields 1 mol of H^+ ions. One equivalent of a base is the amount that yields 1 mol of OH^- ions.

In any of these cases, the first step in finding normality is to calculate the number of equivalents, using a quantity called the *equivalent mass*. The equivalent mass is the mass, stated in grams, of 1 equiv of the reactant. We can then use a simple proportion to tell us how many equivalents are present in a sample of a measured mass:

$$\text{No. of equivalents} = \frac{\text{Actual mass of substance}}{\text{Equivalent mass}}$$

For example, the equivalent mass of sulfuric acid is 49.0. (The molar mass of H_2SO_4 is 98. It yields 2 mol of H^+ ions. Therefore, its equivalent mass is 98 divided by 2.) Suppose we add 196 g of H_2SO_4 to enough water to make 1 L of solution. What is the normality?

First, we must find the number of equivalents: 196/49 = 4.0. Thus, we have 4.0 equiv of acid in our acid solution.

Since we made 1 L of solution, the normality is 4.0 equiv/1 L or 4.0. If we made 2 L of solution, the normality would be 4.0/2 = 2.0. Normality is abbreviated N. Thus, a solution containing 2.0 equiv of solute per liter of solution is designated as 2.0 N.

Often, it is easier to find normality by first calculating the molarity of the solution. The normality and molarity are related as follows:

$$\underset{(\text{equiv/L})}{\text{Normality}} = \underset{(\text{mol/L})}{\text{Molarity}} \times \text{Equiv/mol}$$

For example, suppose you want to calculate the normality of a 2.0 M sulfuric acid solution:

$$(2.0 \text{ mol } H_2SO_4/L) \times (2 \text{ equiv (mol } H^+)/ 1 \text{ mol } H_2SO_4) = 4.0 \text{ N}$$

NOTE: Normality is always equal to or greater than molarity. It is never less. Normality was once used extensively in laboratories and industry, but it has been largely replaced by molarity today.

Molality

The *molality* of a solution is the number of moles of solute per kilogram (1000 g) of solvent. Molality is abbreviated *m*. Its units are the number of moles of solute per kilograms of solvent. (Molarity and molality should not be confused. Molarity is defined in terms of the total volume of solution. Molality is defined in terms of a definite weight of solvent.) A solution that contains 0.50 mol of solute in 250 g of solvent would have a molality of

$$\frac{0.5 \text{ mol}}{1} \times \frac{1}{250\text{g}} \times \frac{1000 \text{ g}}{1 \text{ kg}} = 2.0 \text{ m}$$

Percentage Composition

Percentage composition can be expressed in terms of mass, volume, or mass/volume.

$$\% \text{ by mass (m/m)} = \frac{\text{Mass of solute}}{\text{Mass of solution}} \times 100$$

$$\% \text{ by volume (v/v)} = \frac{\text{Volume of solute}}{\text{Volume of solution}} \times 100$$

$$\underset{(\text{m/v})}{\% \text{ by mass/volume}} = \frac{\text{Mass of solute}}{\text{Volume of solution}} \times 100$$

Percentage by mass or mass/volume is usually used to refer to solids dissolved in liquids.

Percentage by volume is normally used in reference to gases in gases or liquids in liquids.

Ethyl alcohol solutions for human consumption often use the special concentration unit *proof*. The proof of such solutions is twice the v/v percentage. Thus, a 45% v/v solution would be 90 proof.

STANDARD SOLUTIONS

A *standard* solution is any solution of accurately known concentration. Standard solutions can be made up using three methods: weight per unit volume, "weight" dilution, and reaction. Let's look at these methods.

Weight per Unit Volume

In this method, a quantity of a pure chemical substance is accurately weighed and then dissolved in a quantity of the solvent. Then, additional solvent is added until the total volume of solution is accurately known. From the weight of solute, the volume of solution, and either the molecular mass or the equivalent mass of the solute, the molarity of the solution can be calculated. Let's look at an example.

EXAMPLE 1. A 21.2-g quantity of Na_2CO_3 is dissolved in enough water to make a final volume of 400 mL. Find the molarity of the solution.

SOLUTION:
The molecular mass of Na_2CO_3 is 106 g/mol.

Molarity is expressed as the number of moles per volume in liters.

$$\text{Number of moles} = \text{Actual mass} \div \text{Molecular mass}\left(\frac{\text{g}}{\text{g/mol}}\right)$$

Thus,

$$\text{Molarity} = \frac{\text{Actual mass/Molecular mass}}{\text{Volume in liters}}$$

$$= \frac{\text{Actual mass}}{\text{Molecular mass} \times \text{Volume in liters}}$$

$$\text{Molarity} = \frac{21.2\text{ g}}{106\text{ g/mol} \times 0.400\text{ L}} = 0.500\text{ M}$$

The net positive valence of Na_2CO_3 is $1 \times 2 = 2$.

The equivalent mass of Na_2CO_3 is

$$\frac{106\text{ g/mol}}{2\text{ equiv/mol}} = 53\text{ g/equiv}$$

The normality is 0.500 mol/L × 2 equiv/mol = 1.00 equiv/L (1.00 N)

Checking this result by working with the equivalent mass of Na_2CO_3, we have

$$\text{Normality} = \frac{\text{Actual mass}}{\text{Equivalent mass} \times \text{Volume in liters}}$$

$$= \frac{21.2\text{ g}}{53\text{ g/equiv} \times 0.400\text{ L}} = 1.0\text{ N}$$

Dilution

A definite volume of a more concentrated solution can be diluted with a definite amount of additional solvent to produce a more dilute solution of known concentration. Because the number of moles of solute does not change during dilution we can calculate the new concentration. From the following relationships:

Molarity × Volume in liter = Number of moles of solute

Molarity × Volume in milliliters = Number of millimoles of solute

We can see that when a solution is diluted, the product of the concentration and volume of the initial solution must be equal to the product of the concentration and volume of the diluted solution when the same system of units is used in both solutions. Expressed in an equation this relationship is

$$C_iV_i = C_fV_f$$

where

C_i is the concentration of the initial solution

V_i is the volume of the initial solution

C_f is the concentration of the final solution

V_f is the volume of the final solution

Let's work out an example of this type of problem.

EXAMPLE 2. How much water must be added to 50.0 mL of 1.2 M HCl solution to produce a 0.50 M HCl solution?

SOLUTION: We use the relationship

$$C_i \times V_i = C_f \times V_f$$

$$1.2 \text{ M} \times 50.0 \text{ mL} = 0.50 \text{ M} \times x$$

$$x = \frac{1.2 \text{ M} \times 50.0 \text{ mL}}{0.50 \text{ M}} = 120 \text{ mL}$$

Thus, 120 mL is the volume of the final solution. The amount of water to be added is then the difference between the volumes of the two solutions:

$$120 \text{ mL} - 50.0 \text{ mL} = 70 \text{ mL } H_2O$$

Reaction

It is a fundamental law of chemistry that a given number of equivalents of one substance react with precisely the same number of equivalents of any other substance. Therefore, if the concentration of a solution is unknown, its concentration can be found by measuring the volume of it that will react precisely with a definite mass of a pure solid substance or with a definite volume of a standard solution. The product of normality times volume in liters gives the number of equivalents of solute. Therefore, the following relationship holds true whenever solutions react:

$$N_1 \times V_1 = N_2 \times V_2$$

where

N_1 and V_1 are the normality and volume of one solution and

N_2 and V_2 are the normality and volume of the other.

When a solution is reacting with a solid substance, then the following is true:

$N_s \times V_s$ = number of equivalents of the solid substance

where N_s and V_s are the normality and volume of the solution involved. (Here the volume must be in liters!)

Let's study some examples of these types.

EXAMPLE 3. A 25.2-mL volume of 0.100 N HCl solution is required to neutralize 20.0 mL of an unknown base solution. Find the normality of the base solution. Find the concentration of the base solution.

SOLUTION: Applying the relationship:

$$N_1 \times V_1 = N_2 \times V_2$$

$$0.100 \text{ N} \times 25.2 \text{ mL} = x \times 20.0 \text{ mL}$$

$$x = \frac{0.100 \text{ N} \times 25.2 \text{ mL}}{20.0 \text{ mL}} = 0.126 \text{ N}$$

EXAMPLE 4: 48.5 mL of an unknown acid solution neutralizes 11.1 g of solid $Ca(OH)_2$. Compute the concentration of the acid solution.

SOLUTION: Applying the relationship, we obtain

$$N_s \times V_s \text{ (in liters)} = \text{number of equivalents of solid } Ca(OH)_2$$

$$x \times 0.0485 \text{ L} = \frac{11.1 \text{ g}}{1} \times \frac{1 \text{ mol}}{74.1 \text{ g}} \times \frac{2 \text{ equiv}}{1 \text{ mol}}$$

$$x = \frac{11.1 \text{ g}}{1} \times \frac{1 \text{ mol}}{74.1 \text{ g}} \times \frac{2 \text{ equiv}}{1 \text{ mol}} \div 0.0485 \text{ L}$$

$$= 6.18 \text{ equiv}$$

Notice that we initially had to determine the equivalent mass of $Ca(OH)_2$—its molecular mass, 74.1 g/equiv divided by its net positive valence, 2.

You can also work problems of this type using the molarity of the solution instead of the normality, if the balanced chemical equation can be written. Suppose, for example, that same 11.1 g of $Ca(OH)_2$ (s) was neutralized with 48.5 mL of a hydrochloric acid solution of unknown molarity. First, write the balanced chemical equation for the reaction:

$$2\ HCl(aq) + Ca(OH)_2(s) \rightarrow CaCl_2(aq) + 2\ H_2O(l)$$

Two moles of HCl are needed to neutralize 1 mol of $Ca(OH)_2$, so we can calculate the moles of $Ca(OH)_2$ using the mass and its molecular mass and then convert to moles of HCl and finally to molarity of the HCl solution:

$$\frac{11.1 \text{ g}}{1} \times \frac{1 \text{ mole } Ca(OH)_2}{74.1 \text{ g}}$$

$$\times \frac{2 \text{ mol HCl}}{1 \text{ mol } Ca(OH)_2} \times \frac{1}{0.0485 \text{ L}} = 6.18 \text{ M}$$

PROBLEM SET NO. 9

1. Find the molarity of the following solutions:
   (a) 0.02 mol NaOH in 80. mL of solution
   (b) 0.234 g of NaCl in 50 mL of solution
   (c) 222 g of $CaCl_2$ in 4 L of solution

2. Find the normality of the following solutions:
   (a) 0.2 moles of $CaCl_2$ in 200. mL of solution
   (b) 0.2 equivalents of $CaCl_2$ in 200. mL of solution
   (c) 2.76 g of K_2CO_3 in 400. mL of solution
   (d) 6.84 g of $Al_2(SO_4)_3$ in 250. mL of solution

3. Find the molality of the following solutions:
   (a) 4.0 g of NaOH in 400. g of water
   (b) 333 g of $CaCl_2$ in 6000. g of water

4. 6.0 g $MgSO_4$ is dissolved in 250. mL of solution.
   (a) Find the percentage by weight of the solute (specific gravity of water = 1).
   (b) Find the molarity of the solution.
   (c) Find the normality of the solution.

5. 10. mL of ethyl alcohol is dissolved in 40. mL of water. Find the percentage by volume of alcohol in the solution.

6. 34.0 g of $AgNO_3$ is dissolved in 750. mL of solution. Find the molarity of the solution.

7. 2.67 g of $AlCl_3$ is dissolved in 400. mL of solution. Find the normality of the solution.

8. How much water must be added to 35 mL of 0.8 M $NaNO_3$ solution to make a 0.5 M solution?

9. 24.0 mL of 0.1 N NaOH solution neutralize 25.0 mL of an unknown acid solution. Compute the concentrations of the acid solution.

10. 20.0 mL of $AgNO_3$ solution forms 0.285 g of AgCl from a salt–water solution. Compute the normality of the $AgNO_3$ solution. (Remember, the number of equivalents of a substance formed is equal to the number of equivalents of each of the substances required to form it.)

11. It requires 38.5 mL of an NaOH solution to neutralize 25.0 mL of a 0.100 M H_2SO_4 solution. Calculate the molarity of the sodium hydroxide solution.

SOLUBILITY

The *solubility* of a substance is the maximum amount of the substance that can be dissolved in a given amount of solvent at a specified temperature and pressure.

The factors that influence solubility are the nature of solute and solvent, temperature, and pressure. Let's consider each of these factors in terms of the common types of solutions: gas in a gas, gas in a liquid, liquid in a liquid, and solid in a liquid.

We discussed the gas in a gas type of solution earlier, in relation to Dalton's law of partial pressures and to gases in general. All gases are soluble in one another and mix in all proportions. Molecules of one gas diffuse into the void between the molecules of the other and ultimately form a homogeneous system.

In the case of solubility of a gas in a liquid, the temperature and pressure have important effects. When a gas is brought in contact with a liquid, some of the molecules of the gas enter the liquid and dissolve in it. It is difficult to predict the solubility of a given gas in a given solvent. Considering the polar nature (see Chapter 4) of each is probably the best indicator—that is, polar gases have greater solubility in polar solvents, and polar solvents more readily dissolve polar gases. Gases like ammonia and hydrogen chloride are extremely soluble in water because they react with the water molecules, whereas hydrogen and oxygen are only very slightly soluble in water because they are nonpolar and do not react. The effect of temperature in such cases is quite definite. Just as an increase in temperature speeds up the rate of evaporation of a liquid, so too an increase in temperature increases the rate at which dissolved gas molecules are expelled from a liquid. The rule is: *the solubility of any gas in a liquid solvent decreases as the temperature rises, and becomes zero at the boiling point of the solvent.*

EXPERIMENT 13: Select two small bottles of any carbonated soft drink, one ice-cold and one warm. Open the bottles, and holding the bottles over a sink, with thumbs held over the tops, *gently* shake them. Which drink foams up more? Carbon dioxide gas has been dissolved under pressure in both, but CO_2 is less soluble in the warm sample, and therefore it escapes more easily from the warm drink—in the form of foam.

The effect of pressure on gas solubility is also quite definite. An increase in pressure on the gas will speed up the rate at which gas mole-

> Temperature and pressure affect the solubility of a gas in a liquid. Solubility decreases as the temperature rises and becomes zero at the boiling point of the solvent. The greater the pressure of the gas, the greater its solubility in a liquid.

cules enter the liquid. The relationship between pressure and the solubility of a gas in a liquid is stated in *Henry's law* thus: *For gases that do not react chemically with the solvent, the mass of gas that dissolves in a liquid at a given temperature is proportional to the partial pressure of the gas over the solution.* In other words, the greater the partial pressure, the greater the solubility.

In the case of liquids dissolving in liquids there are three possibilities:

1. The two liquids are completely miscible (capable of being mixed).
2. The two liquids are partially miscible.
3. The two liquids are immiscible (not capable of being mixed).

The chemical structure of the substances determines to a great extent whether two liquids will dissolve in each other. Water and alcohol are soluble in each other in all proportions. Alcohol and carbon tetrachloride are soluble in each other in all proportions, but water and carbon tetrachloride are practically insoluble in each other. The behavior of this last pair of liquids is better understood if we consider the structure of each substance.

The water molecule is a bent molecule, as shown in Figure 8.1. Oxygen has a higher

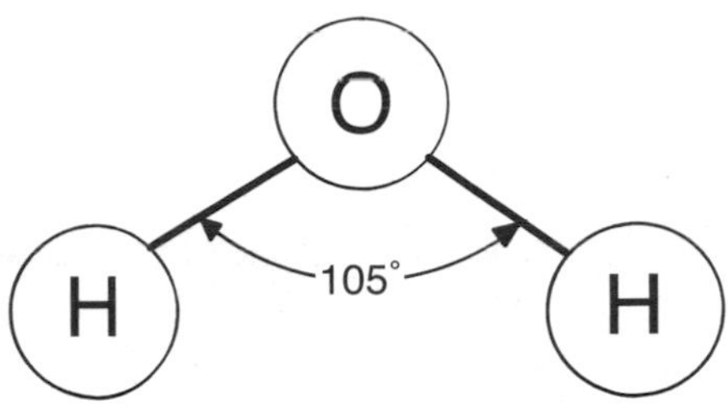

Figure 8.1 The Water Molecule

electronegativity than hydrogen, and so the bonding pairs of electrons are drawn closer to the oxygen. Therefore, the oxygen end of the molecule acquires a partial negative electrical character, and the hydrogen end acquires a partial positive electrical charge. Thus, the water molecule is highly polar (see Chapter 4) and possesses a high degree of electrical activity. In contrast, carbon tetrachloride is made up of highly symmetrical molecules. In these molecules, the carbon atom is centrally located, and the four chlorine atoms surround the carbon, forming the apexes of a tetrahedron, as shown in Figure 8.2. Such a molecule has very little polarity and electrically is virtually inert. The great difference in polarity

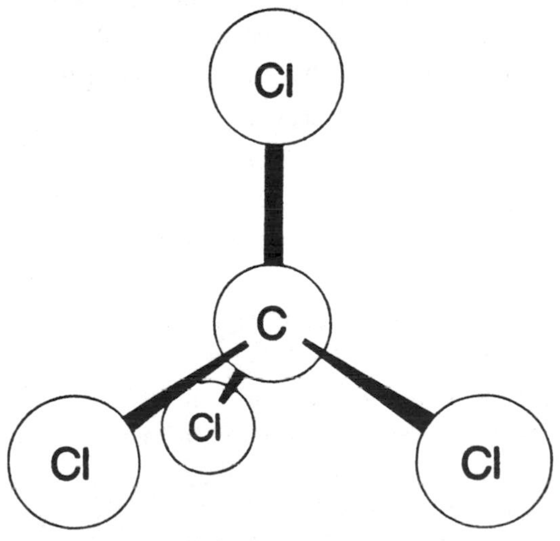

Figure 8.2 Carbon Tetrachloride

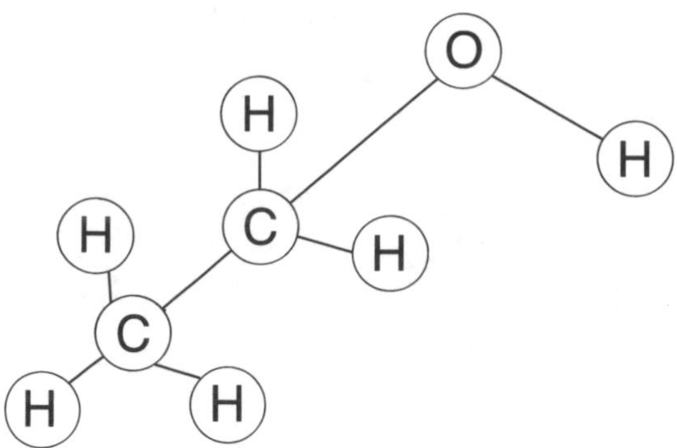

Figure 8.3 Ethyl Alcohol

between molecules of carbon tetrachloride and water accounts for the immiscibility of these two liquids.

The molecules of ethyl alcohol have a structure that produces a degree of polarity intermediate between those of water and carbon tetrachloride, as illustrated in Figure 8.3. One end of the alcohol molecule has carbon atoms surrounded by hydrogen atoms and is structurally akin to the molecules of carbon tetrachloride. The other end of the molecule of alcohol is similar to the water molecule, and the lack of symmetry produces a moderate amount of polarity. The similarities are sufficient to enable alcohol to dissolve in each of the other two liquids. The rule based on this principle is *like dissolves like.* "Like" refers to the degree of polarity and the similarity of structural features of the molecules concerned.

A change in temperature changes the solubility of liquids partially miscible in one another, but the direction and magnitude of such a change in solubility follows no general rule. Change in pressure has a negligible effect on the solubility of liquids in one another.

The most commonly encountered type of solution is a solid in a liquid. If a small amount of salt is added to a glass of water, the solution is said to be *unsaturated* because more of the solute can be dissolved in the amount of solvent that is present. If salt continues to be added, a point is ultimately reached at which no further salt can be dissolved, and the solution is said to be *saturated*. The concentration of a solute in a saturated solution is the solubility of the solute at the particular temperature.

With some salts, like photographic hypo (sodium thiosulfate, $Na_2S_2O_3$), it is possible to form an unstable supersaturated solution, which holds in solution more solute than can theoretically be held by the solvent at that particular temperature. Such a solution will revert to a saturated solution by precipitating the excess solute as crystals if the supersaturated solution is shaken, or comes in contact with dust particles, or is "seeded" with a small crystal of the solute.

As a rule, the solubility of a solid in a liquid increases as the temperature rises. There are, however, important exceptions. Some salts decrease in solubility at increased temperature. Such salts form scales in boilers and leave deposits in kettles and steam irons.

The *law of partition* states that if a solid is soluble in two liquid solvents that, in turn, are immiscible in each other, the solid will distribute itself between the solvents in quantities proportional to its solubility in each solvent. This phenomenon has application in the extraction of a solute from one solvent by another. For example, iodine is about 650 times more soluble in carbon tetrachloride than in water. Each time a sample of water containing dissolved iodine is shaken with carbon tetrachloride, the dissolved iodine will distribute itself between these two solvents as follows: 650 parts by mass of iodine in carbon tetrachloride to 1 part by mass of iodine in water. Successive treatments of the water solution of iodine with fresh quantities of carbon tetrachloride will rapidly and effectively remove all appreciable traces of iodine from the water.

EFFECTS OF NONELECTROLYTE SOLUTES ON PROPERTIES OF SOLVENTS

Solutes can be divided into two classes:

1. Those that when dissolved in water produce a solution that conducts an electric current. Such solutes are called *electrolytes.* Acids, bases, and salts belong to this class.

2. Those that when dissolved in water produce a solution not capable of conducting an electric current. Such solutes are called *nonelectrolytes.* Let's look at the effect of nonelectrolytes on the vapor pressure, boiling point, freezing point, and osmotic pressure of solvents.

Lowering of Vapor Pressure

If a solute less volatile than the solvent is dissolved in the solvent, the vapor pressure of the solution will be lower than the vapor pressure of the solvent. You will recall that the pressure of the vapor of a liquid is the result of the escape of the vapor from the surface of the liquid into the air in contact with the liquid. You also know that a solute distributes itself uniformly throughout the solvent and thus takes up a portion of the surface of the solution. This has the effect of decreasing the number of molecules of solvent in contact with the air above it and consequently reduces the rate at which these molecules escape as vapor. This reduction, in turn, lowers the vapor pressure.

The extent to which nonvolatile nonelectrolytes lower the vapor pressure of their solvents depends on the concentration of the nonvolatile nonelectrolyte in the solution. *Raoult's law* states this effect: *Equimolar quantities of different nonvolatile solutes, when added to equal masses of the same solvent, lower the vapor pressure the same amount, and the ratio of the amount of lowering to the vapor pressure of the pure solvent equals the ratio of the number of moles of solute to the number of moles of solution.* In mathematical form, Raoult's law looks like this:

$$\frac{P-p}{P} = \frac{n}{N+n}$$

where

P = vapor pressure of solvent

p = vapor pressure of solution

N = number of moles of solvent

n = number of moles of solute

In dilute solutions, where the number of moles of solvent is considerably larger than the number of moles of solute, the quantity $(N + n)$ in Raoult's law may be changed simply to (N) without introducing too much error. This greatly simplifies calculations, and it modifies Raoult's law to read:

$$\frac{P-p}{P} = \frac{n}{N}$$

Of course, the number of moles of any substance equals its actual mass divided by its molecular mass. Thus, in the preceding expression,

$$n = \frac{w}{m} \text{ and } N = \frac{W}{M}$$

where

w and W are the actual masses of solute and solvent respectively, and

m and M are the molecular masses of solute and solvent, respectively.

Substituting these terms into the preceding expression, and solving for m (the molecular mass of the solute) we obtain:

$$m = \frac{(w) \times (M) \times (P)}{(W) \times (P - p)}$$

Thus, through the application of studies into the nature of vapor pressure we arrive at a method for finding the molecular mass of solids of unknown composition. We simply find a suitable solvent for the solid, measure the actual masses of the two substances forming the solution, measure the vapor pressure of the solution, and then calculate the unknown molecular mass. This process illustrates the familiar fact that investigation into one facet of the behavior of nature frequently leads to information revealing the secrets of nature in other areas.

Elevation of Boiling Point

That a solution has a lower vapor pressure than the pure solvent neatly explains another phenomenon associated with solutions: the boiling point of a solution is higher than the boiling point of the pure solvent. Examine Figure 8.4 carefully. Here, the vapor pressure

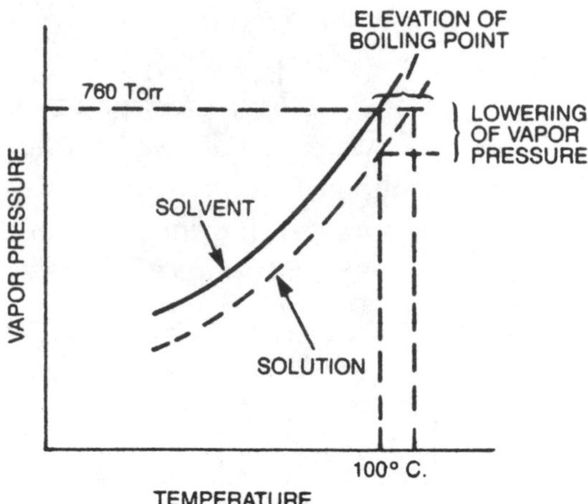

Figure 8.4 Effect of Solute on Vapor Pressure of a Liquid

of a solvent has been plotted versus temperature and is shown by the solid line. Such a curve would be seen if we plotted the data for water from Table 6.2. The dotted curve shows how the vapor pressure curve would be depressed if a solute were added to the solvent. Now, a liquid boils when its vapor pressure equals the atmospheric pressure around it. The horizontal broken line in Figure 8.4 represents standard atmospheric pressure, 760 Torr. Notice that the vapor pressure curve of the solution reaches the atmospheric pressure line at a higher temperature than does the vapor pressure curve of the solvent. Thus, the solution has a higher boiling point.

Investigations into this phenomenon have revealed that the amount of elevation of the boiling point for a given solution is directly proportional to the molality of the solution. Equimolal solutions of various solutes in the same solvent cause the same elevation in boiling point. For water, 1.0 m solutions boil at 0.52°C higher than pure water at standard pressure. For water solutions, then, the following applies:

$$B - b = 0.52°\text{C} \times m$$

where

B is the boiling point of the solution

b is the boiling point of water

m is the molality of the solution

NOTE: The constant 0.52°C/m applies only to solutions in which water is the solvent. Each other solvent has its own constant, which is in each case the number of degrees its boiling point is elevated when a nonvolatile solute is added to it to a concentration of 1 m.

This relationship provides another means of finding the molecular weight of a solute. Note the following expression:

Molality = Number of moles solute per 1000 g solvent

Thus:

$$\text{Molality} = \frac{\text{Actual mass of solute/1000 g of solvent}}{\text{Molecular mass of solute}}$$

If we let

w = actual mass of solute per 1000 g solvent, and m = molecular mass of solute, and we substitute this into the equation above, we get

$$B - b = \frac{0.52°\text{C}\ w}{m}$$

Solving for m, we obtain the expression

$$m = \frac{0.52\ w}{B - b}$$

Let's work through a problem of this type.

EXAMPLE 5. A 42.75-g portion of a substance is dissolved in 250. g of water. The solution boils at 100.26°C at standard pressure. Find the molecular mass of the solute.

SOLUTION:
First, we find how many grams of the substance would be contained in 1000 g of water:

$$w = 42.75\ \text{g} \times \frac{1000\ \text{g}}{250.\ \text{g}} = 171\ \text{g}$$

Next, we find the change in boiling point:

$$B - b = 100.26°\text{C} - 100°\text{C} = 0.26°\text{C}$$

Substituting these values in the equation $m = \frac{0.52\ w}{B - b}$, we obtain

$$m = \frac{0.52°\text{C}/m \times 171\ \text{g}}{0.26°\text{C}} = 342\ \text{g}/m$$

Thus, the molecular mass of the substance is 342 g/mol.

Lowering of Freezing Point

A solution freezes at a temperature below the freezing point of its solvent. This phenomenon is also related to vapor pressure. Examine Figure 8.5 carefully. You know that water boils at 100°C and freezes at 0°C at standard pressure. Also the boiling point decreases as the pressure drops, and the freezing point rises slowly as the pressure decreases. Thus, as the pressure decreases, the range of temperature between the freezing and boiling points of water becomes increasingly smaller. Curve AB in Figure 8.5 shows how the boiling point changes as the pressure decreases, and curve AC shows the same for the freezing point.

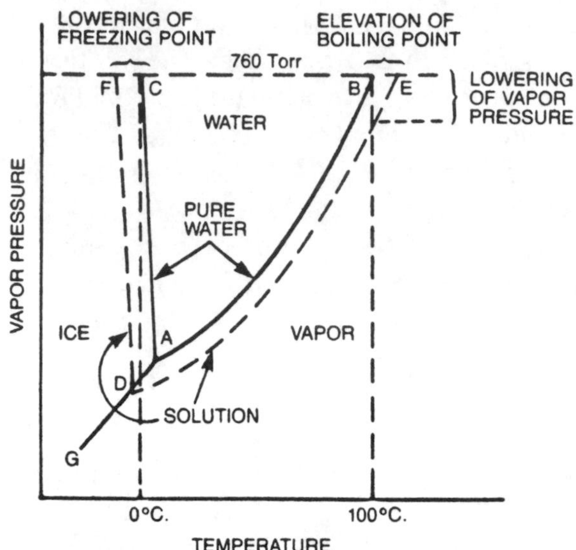

Figure 8.5 Temperature–Pressure Effects of Solutes on the Ice–Water–Vapor System

The curves meet at point A, which is called the *triple point* because here ice, water, and vapor are all present at the same temperature and pressure. Curve AG shows the temperatures at which water vapor freezes directly to ice at pressures below the triple point.

As in the case of boiling-point elevation, the amount of freezing-point depression is directly proportional to the molality of the solution. For water, solutions of nonelectrolytes of unit molality freeze at $-1.86\,°C$. For water solutions, then, the following applies:

$$F - f = 1.86\ m$$

where

F = the freezing point of water

f = the freezing point of the solution

m = the molality of the solution

The constant $1.86\,°C/m$ applies only to water. Every other solvent has its own particular constant, which is the number of degrees a solution of unit molality freezes below the freezing point of the solvent.

Again, we have a method of finding the molecular mass of the solute. The derivation of an expression giving us the molecular mass of the solute from freezing-point data proceeds exactly as in the case of boiling-point elevation.

The expression becomes

$$m = \frac{1.86\,°C/m \times w}{F - f}$$

Let's work through a problem of this type.

EXAMPLE 6. A 14.25-g sample of a substance is dissolved in 125 g of water. The solution freezes at $-9.30\,°C$. Find the molecular mass of the solute.

SOLUTION:

$$w = 14.25\text{ g} \times \frac{1000\text{ g}}{125\text{ g}}$$

$$= 114\text{ g per 1000 g of solvent}$$

$$F - f = 0.00\,°C - (-9.30\,°C) = 9.30\,°C$$

$$m = \frac{1.86\,°C/m \times 114\text{ g}}{9.30\,°C} = 22.8\text{ g}/m$$

Thus, the molecular mass of the solute is 22.8 g/mol.

OSMOTIC PRESSURE

Imagine that a container is divided into two parts by a semipermeable membrane—a thin sheet of a substance whose pores will allow the passage of solvent but not solute. One side of the container is filled with a pure solvent and the other side with a solution. A solvent always passes through the semipermeable membrane from the pure solvent side to the solution side. This process is called *osmosis.* Enough pressure can be applied to the solution side to just stop osmosis. The amount of pressure that must be applied to stop osmosis is called the *osmotic pressure.* The higher the concentration of the solution, the higher the osmotic pressure. In fact, if additional pressure is applied so that the osmotic pressure is exceeded, the flow reverses and solvent moves from the solution to the solvent. This process is called *reverse osmosis* and is used commercially in the purification and desalination of water.

Quantitatively, the osmotic pressure can be related to the molarity of the solution by the equation

$$\pi = MRT$$

where π is the osmotic pressure in atmospheres, M is the molarity of the solution, R is the ideal-gas constant, and T is the Celsius temperature.

COLLIGATIVE PROPERTIES

Suppose you make a sucrose solution by mixing some table sugar with water. Some of the properties of that solution are related to the solute—the sweet taste, for example; however, there are a number of properties that depend solely on the number of solute particles. These are called *colligative properties,* solution properties that depend on the *number* of solute particles but not their type.

We have already looked at the four primary colligative properties with respect to nonelectrolytes: vapor-pressure lowering, boiling-point elevation, freezing-point depression, and osmotic pressure. These colligative properties also apply to electrolyte solutions, but in those cases you must calculate the molality of particles and not just the solute. For example, if an NaCl solution is 1.0 *m*, it is 2.0 *m* with respect to ions, because for every mole of NaCl put into solution, 2 mol of ions (1 mol of Na^+ and 1 mol of Cl^-) are formed. After the concentration of particles is calculated, the problems are solved in the same way as for nonelectrolytes.

COLLOIDS

In solutions, the solute particle size is small, around 1 nanometer (1×10^{-9}m). In suspensions like dirty water—in which the particles are not dissolved—the solute particles are large (greater than 1000 nm) and will settle out when the solvent is left standing. A *colloid* is an intermediate between a true dissolved particle and a suspended solid that will settle out of solution. Colloids have solute particle sizes between those of solutions and suspensions, 1–1000 nm. The solutes do not settle out upon standing, but because they are larger than solution solute particles, they reflect light that can be seen through the colloid. If you shine an intense light beam through a solution, you do not see the beam as it passes through the solution; but if you shine a light beam through a colloid, the beam becomes visible. Perhaps you have driven in a fog and have seen your car's headlight beams reflected in the fog. That fog was a colloid of a liquid (tiny water droplets) in a gas (the air). This property of a colloid of reflecting a light beam and making it visible is called the *Tyndall effect,* and it is the principal way of distinguishing a colloid from a true solution.

APPLICATIONS OF THE EFFECTS OF SOLUTES ON SOLVENTS

Interesting and useful applications of the lowering of the vapor pressure of solvents by solutes are common. For example, if a dish containing pure water and one containing a solution are placed side by side inside a tightly covered container, each liquid will begin to emit vapor into the air in the container. Because the vapor pressure of the solution is less than the vapor pressure of the water, the air in the container will become saturated with vapor, relative to the solution, first. Thus, any additional vapor emitted by the water will condense back to liquid—but in the solution. Consequently, the air never gets a chance to become saturated with vapor relative to the water. The result is that all the water eventually finds its way into the solution. This process is known as *isothermal distillation*.

EXPERIMENT 14: Select two 1-oz glasses. Fill one half full of water. Place the same volume of saturated sugar water in the other. With a small piece of adhesive tape, mark the level of the liquid in each glass. Place the two glasses inside an airtight container, preferably made of glass or clear plastic so you can observe any changes. Cover the container and set it aside in a location free from severe temperature changes. You will observe the sugar solution increase in volume at the expense of the water.

This phenomenon is the principle of operation of the various drying agents used in the cellar or in the closets of homes in moist climates. Solid chemicals such as calcium chloride or potassium carbonate, which are very soluble in water, become moist in damp air. A saturated solution forms on their surface. The vapor pressure of this solution is less than the vapor pressure of the water in the moist air, and so more moisture is absorbed by the solution. Ultimately the entire solid dissolves in the moisture it absorbs, but even so, the solution continues to absorb moisture until the vapor pressure of the solution equals the pressure of the water in the atmosphere. This phenomenon is known as *deliquescence*, and deliquescent chemicals used as drying agents are called *desiccants*.

Applications of the lowering of freezing points include adding alcohol or ethylene glycol (permanent antifreeze) to a car radiator to lower the freezing point of the water, and salting roads in the winter to prevent ice from forming on their surface.

SUMMARY

A solution consists of a solute dissolved in a solvent. A concentrated solution contains more solute per unit volume than does a dilute solution. A standard solution is one of accurately known concentration. The solubility of a substance is the maximum amount of the substance that can be dissolved in a given amount of a solvent at a specified temperature and pressure.

For gases dissolved in liquids, as the temperature increases, the solubility decreases; as the pressure increases, the solubility increases.

For liquids dissolved in liquids, like dissolves like. For solids dissolved in liquids, as the temperature increases, the solubility usually increases.

Nonvolatile nonelectrolytes alter the properties of their solvents as follows:

1. They lower the vapor pressure.
2. They elevate the boiling point.
3. They lower the freezing point.
4. They increase the osmotic pressure.

Osmosis is the movement of solvent molecules through a semipermeable membrane from a pure solvent to a solution. The pressure that it takes to just stop this osmosis is the solution's osmotic pressure. By the application of pressure exceeding the osmotic pressure, solvent can be forced from the solution to the pure solvent side by a process called reverse osmosis.

Colligative properties are properties of solutions that do not depend on the type of solute particles but only on their number. Colloids have solute particle sizes intermediate between those of true solutions and those of suspensions. They are distinguished from true solutions by the Tyndall effect. Deliquescent solids dissolve in the moisture they absorb from the atmosphere.

PROBLEM SET NO. 10

1. The vapor pressure of water at 20.°C is 17.5 Torr, and that of a solution of 23 g of glycerin in 500. g water is 17.34 Torr. Find the approximate molecular weight of the glycerin.

2. A 12.4-g sample of ethylene glycol, $C_2H_6O_2$, is dissolved in 200. g water. At what temperature will this solution boil at standard pressure?

3. A 15.5-g portion of a solute is dissolved in 100. g of water. At standard pressure, the solution boils at 101.3°C. Find the molecular weight of the solute.

4. An 11.5-g quantity of a solute is dissolved in 100. g of water. The solution freezes at −2.325°C. Find the molecular mass of the solute.

5. Six quarts of ethylene glycol is mixed with 12 qt of water in the radiator of a car. The specific gravity of the glycol is 1.26. Its molecular mass is 62 g/mol. The specific gravity of water is 1.00. Find the Fahrenheit temperature at which this solution will freeze.

CHAPTER 9

SOLUTIONS OF ELECTROLYTES

KEY TERMS

acid, base, salts, equilibrium, Le Châtelier's principle, law of mass action, pH, common-ion effect, hydrolysis, solubility product principle

Electrolytes are solutes that when dissolved in water produce a solution that conducts an electric current. The behavior of solutions of electrolytes is remarkably different from the behavior of solutions of nonelectrolytes. Their chemistry is both interesting and important.

EXPERIMENT 15: In this experiment, you will first construct a conductivity apparatus. The materials you will need for it are a 9-V battery, a wide rubber band, a 100-ohm resistor, and a light-emitting diode (LED). Place the wide rubber band around the 9-V battery so that the band covers the terminal. Insert the wire from one end of the resistor under the rubber band so that it is in good contact with one of the terminals. The rubber band will hold it in place. Insert a wire of the LED under the rubber band so that it is in good contact with the other terminal. Briefly touch the free ends of the resistor and LED together. The LED should light. This is your conductivity apparatus.

Now, to test a substance for conductivity, place the free ends of the resistor and LED in the substance. If the substance is a conductor, the LED will light. The brighter the light, the better the conductor.

EXPERIMENT 16: Prepare water solutions of each the following substances: sugar, alcohol, table salt, baking soda, washing soda, and lye. Using the conductivity apparatus prepared in the previous experiment, check the conductivity of each of the solutions by lifting the container up to the wires of the conductivity apparatus. The LED will light if current passes between the immersed wires. Clean the wires with pure water and dry with a cloth after testing each solution. Also check on the conductivity of pure water, pure alcohol, dry sugar, and dry table salt. Which of the solutes are electrolytes? Must the solvent be present to have conductivity?

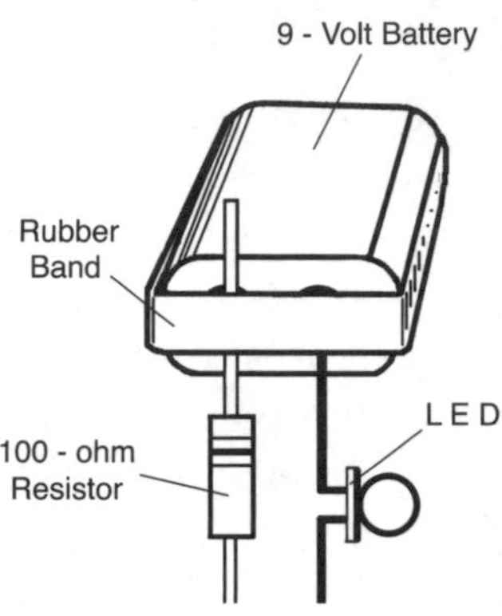

Figure 9.1 Conductivity Apparatus

ABNORMAL BEHAVIOR OF ELECTROLYTES

The following are two major differences between electrolytes and nonelectrolytes:

1. Solutions of electrolytes conduct electricity; solutions of nonelectrolytes do not.

2. Electrolytes alter the properties of solvents in the same way nonelectrolytes do, but to a much greater degree.

Table 9.1 gives the freezing points of water solutions. (Remember that the freezing point of pure water is 0°C.) In Table 9.1 the first three substances are nonelectrolytes. The others are electrolytes. Note the differences in the freezing-point depression of water caused by the two types of solutes. The nonelectrolytes average 1.86°C per mole of solute. HCl, KCl, and NaCl depress the freezing point almost twice that amount, whereas Na_2SO_4, $CaCl_2$, and $NiCl_2$ depress the freezing point more than twice (and nearly three times) as much as nonelectrolytes. Boiling-point elevation data and vapor-pressure lowering data for these same solutes would show the same degree of dissimilarity. Such abnormal behavior requires explanation.

The theory of ionization, first proposed by Svante Arrhenius in 1884 and then modified by subsequent investigation, adequately explains the behavior of dilute solutions of electrolytes. According to this theory, electrolytes dissociate into positively and negatively charged ions in solution. This process is called *ionization*. These charged ions are free to migrate through the solution and thus are responsible for the conductivity of solutions of electrolytes.

Solute	Molarity	Freezing Point (°C)
glycerin	0.1	−0.187
ethyl alcohol	0.1	−0.183
sugar	0.1	−0.188
HCl	0.1	−0.352
KCl	0.1	−0.345
NaCl	0.1	−0.348
Na_2SO_4	0.1	−0.434
$CaCl_2$	0.1	−0.494
$NiCl_2$	0.1	−0.538

Table 9.1 Freezing Points of Several Aqueous Solutions

The following equations show the dissociation of some electrolytes into their ions.

$$HCl \rightarrow H^+ + Cl^-$$

$$NaCl \rightarrow Na^+ + Cl^-$$

$$Na_2SO_4 \rightarrow 2\ Na^+ + SO_4^{2-}$$

$$CaCl_2 \rightarrow Ca^{2+} + 2\ Cl^-$$

Note that the charge on each ion is the same as the oxidation number of the atom or radical.

In solution, electrolytes dissociate into positively and negatively charged ions. Because the charged ions are free to move around, solutions of electrolytes conduct electricity.

Electrolytes are divided into three types of substances:

1. *Acids*—substances that ionize in solution to produce hydrogen ions (H^+)

2. *Bases*—substances that ionize in solution to produce hydroxide ions (OH^-)

3. *Salts*—substances that ionize in solution but produce neither hydrogen nor hydroxide ions

The ionization equation of an electrolyte furnishes a strong clue about the abnormal behavior of solutions of electrolytes. Notice that in the equation

$$KCl \rightarrow K^+ + Cl^-$$

for each mole of KCl that ionizes, 1 mol of potassium ions and 1 mol of chloride ions form. Thus, there are 2 mol of particles in solution for each mole of solute dissolved, or twice Avogadro's number of particles in solution. A nonelectrolyte, which separates into molecules in solution, produces only 1 mol of particles in solution for each mole of solute dissolved. These facts lead directly to the idea that the alteration of properties of solvents by solutes depends not merely on the concentration of the solute but, more precisely, on the total number of particles in solution. Because a salt like KCl produces twice as many particles in solution as substances like glycerin or sugar do, we might expect KCl to lower the freezing point of water twice as much as an equimolar solution of a nonelectrolyte. The data in Table 9.1 are in general agreement with that idea. Actually, KCl lowers the freezing point of water about 1.85 times as much as the nonelectrolytes do. Similarly, $NiCl_2$, which ionizes as

$$NiCl_2 \rightarrow Ni^{2+} + 2\ Cl^-$$

and which thus produces 3 moles of particles per mole of solute, lowers the freezing point of water almost three times (2.9 actually) as much as nonelectrolytes do.

Why doesn't KCl lower the freezing point of water precisely twice as much as a nonelectrolyte? At first, investigators though that perhaps electrolytes do not ionize completely, that is, 100%; however, subsequent investigation proved that idea untrue. All the electrolytes in Table 9.1 essentially ionize 100%. The answer lies, rather, in that the ions possess electrical charges and are free to move about in the solution. Opposite charges attract one another, so as the ions move about they occasionally come close enough to an oppositely charged ion to be attracted to it and be held momentarily by it. This has the effect of producing a single particle from two particles.

Water molecules are polar, and it is the polarity of water that causes ionic solutes to dissociate. The polar water molecules literally surround an ion, partially reduce the intensity of its attractive force, and float the ion off into the solution. When ions momentarily recombine, the water molecules quickly pull them apart. Nevertheless, at any one instant, a percentage of the ions of an electrolyte will be momentarily held together. The number so held depends on the concentration of the solution and the nature of the ions present. The number will be larger in relatively more concentrated solutions, larger if the ions themselves are of relatively large size, and larger if the charge on the ions is greater than 1 unit.

Let's look at the math involved as a result of these recombinations. Suppose that a solution initially contains 100 K^+ ions and 100 Cl^- ions. Let's suppose further that 10% of these ions are recombined at any instant. The total number of particles in solution will then be

90 K^+ ions

90 Cl^- ions

+ 10 KCl particles

or 190 total particles

Thus, the original 200 particles are reduced in number by 5%. This explanation, based on the theory of ionization, explains why the colligative properties are not quite as predicted.

CONCENTRATION OF IONS IN SOLUTION

When an electrolyte that ionizes 100% is dissolved in water, the concentration of the ions present depends on the formula of the electrolyte dissolved. For example, if 0.1 mol of NaCl is dissolved in a liter of water, the concentration of each ion will be 0.1 M. A glance at the ionization equation of NaCl shows us why.

$$NaCl \rightarrow Na^+ + Cl^-$$

The equation tells us that for each mole of NaCl dissolved, 1 mol of each ion is formed. Now, let's look at the ionization equation of a salt like aluminum sulfate, $Al_2(SO_4)_3$:

$$Al_2(SO_4)_3 \rightarrow 2\ Al^{3+} + 3\ SO_4^{2-}$$

Here, for each mole of salt dissolved, 2 mol of aluminum cation and 3 mol of sulfate anion are released into the solution. Thus, if 0.1 mol of aluminum sulfate is dissolved in a liter of water, the concentration of Al^{3+} will be 0.2 M, and the concentration of SO_4^{2-} will be 0.3 M.

STRONG AND WEAK ELECTROLYTES

Ions possess electrical charges, and as a result, solutions containing ions are able to carry electrical currents (see Figure 9.2). The conductivity of solutions of electrolytes has been thoroughly studied. Conductivity is, by definition, the reciprocal of electrical resistance. The unit of conductivity is the *mho*, pronounced "reciprocal ohm." The conductivities of many solutions have been accurately measured. We can assume that the total amount of electricity carried by a solution of an electrolyte is the sum of the electricity carried by each of the ions present. Table 9.2 gives the conductivity of 0.1 M solutions of the electrolytes indicated.

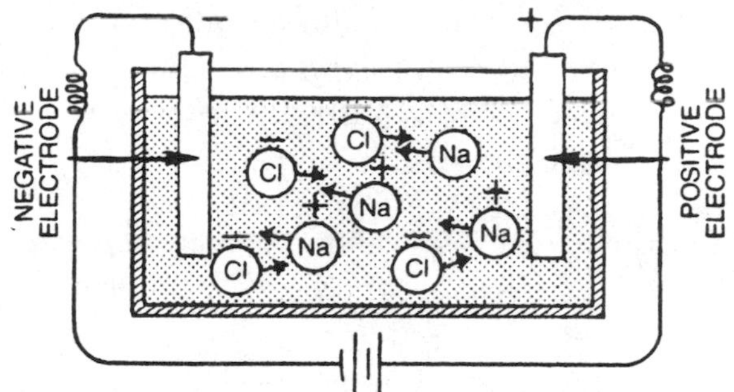

Figure 9.2 Electrical Conduction of Solutions of Electrolytes

Electrolyte	Molarity	Conductivity (mho)
acetic acid, $HC_2H_3O_2$	0.1	4.67
hydrochloric acid, HCl	0.1	350.6
nitric acid, HNO_3	0.1	346.4
sodium acetate, $NaC_2H_3O_2$	0.1	61.9
sodium chloride, NaCl	0.1	92.0
sodium nitrate, $NaNO_3$	0.1	87.4

Table 9.2 Conductivity of Several Aqueous Solutions

Let's apply some arithmetic operations to these data to learn something about the nature of these electrolytes. In the following expressions C indicates conductivity, and the symbol C_{HCl} means "the conductivity of HCl."

$$C_{HCl} \rightarrow C_{H^+} + C_{Cl^-} = 350.6 \text{ mho}$$
$$C_{NaNO_3} \rightarrow C_{Na^+} + C_{NO_3^-} = 87.4 \text{ mho}$$

Adding these two equations, we get

$$C_{HCl} + C_{NaNO_3} \rightarrow C_{H^+} + C_{NO_3^-} + C_{Na^+} + C_{Cl^-} = 438.0 \text{ mho}$$

Notice that the conductivity of the solution is fundamentally based on the ability of ions to carry electricity through the solution.

Now, suppose that we subtract from the preceding sum the conductivity of NaCl solution, which has the effect of removing the influence of Na^+ and Cl^- from both sides of the equation:

$$C_{HCl} + C_{NaNO_3} \rightarrow C_{H^+} + C_{NO_3^-} + C_{Na^+} + C_{Cl^-} = 438.0 \text{ mho}$$
$$C_{NaCl} \rightarrow C_{Na^+} + C_{Cl^-} = 92.0 \text{ mho}$$
$$C_{HNO_3} \rightarrow C_{H^+} + C_{NO_3^-} = 346.0 \text{ mho}$$

This gives us a theoretical figure for the conductivity of nitric acid, HNO_3, solution, which is in excellent agreement with the value in Table 9.2. Let's repeat this process to compute the conductivity of acetic acid, $HC_2H_3O_2$.

$$C_{HCl} \rightarrow C_{H^+} + C_{Cl^-} = 350.6 \text{ mho}$$
$$C_{NaC_2H_3O_2} \rightarrow C_{Na^+} + C_{C_2H_3O_2^-} = 61.9 \text{ mho}$$

Adding, we get

$$C_{HCl} + C_{NaC_2H_3O_2} \rightarrow C_{H^+} + C_{C_2H_3O_2^-} + C_{Na^+} + C_{Cl^-} = 412.5 \text{ mho}$$

Subtracting,

$$C_{NaCl} \rightarrow C_{Na^+} + C_{Cl^-} = 92.0 \text{ mho}$$

we get

$$C_{HC_2H_3O_2} \rightarrow C_{H^+} + C_{C_2H_3O_2^-} = 320.5 \text{ mho}$$

This time our computed value for the conductivity of an electrolyte shows no agreement whatsoever with the value in Table 9.2. The value for acetic acid in the table is only about 1.4% of the computed value. Acetic acid and some other electrolytes conduct a current of electricity only feebly. These facts lead to the idea that there are two different classes of electrolytes: one known as *strong electrolytes*, and the other known as *weak electrolytes*.

A strong electrolyte is 100% ionized. A weak electrolyte is ionized only to a slight extent in solution. *Always remember that the terms "strong" and "weak" refer to the degree of ionization of electrolytes, and never to their concentration.*

EXPERIMENT 17: Using the device you constructed for Experiment 15, compare the conductivity of a solution of table salt and a solution of baking soda with the conductivity of vinegar (acetic acid) and ammonia water (ammonium hydroxide). Which are strong and which are weak electrolytes?

The strong electrolytes are the following:

1. The strong acids: hydrochloric acid, HCl; sulfuric acid, H_2SO_4; and nitric acid, HNO_3.

(A few less common acids such as HBr and HI are also strong.)

2. The strong bases: sodium hydroxide, NaOH; potassium hydroxide, KOH; magnesium hydroxide, $Mg(OH)_2$; and calcium hydroxide, $Ca(OH)_2$. (The latter two are only slightly soluble in water.) In general, the hydroxides of the metals in groups 1A and 2A of the periodic table are strong electrolytes.

3. Practically all salts.

REACTIONS OF ELECTROLYTES

The chemistry and properties of ions in solution are completely independent of the source of the ions. For example, Cu^{2+} ions are blue in water solution whether they come from $CuCl_2$, $CuSO_4$, or $Cu(NO_3)_2$. Likewise, they will form a precipitate of $Cu(OH)_2$ in alkaline solutions regardless of their salt of origin. (A *precipitate* is an insoluble solid that forms in a solution. It is so named because it precipitates, or falls, to the bottom of the reaction vessel. In chemical equations, products that precipitate are sometimes followed by a downward-facing arrow (↓). Solutions of electrolytes are, in reality, solutions of the ions of the electrolyte. Therefore, the chemical reactions of electrolytes are in fact the reactions of the free ions in solution.

Ions in solution react with one another only under the following conditions:

1. if they can combine to form weak electrolytes

2. if they can combine to form relatively insoluble substances

3. if they oxidize or reduce one another or other molecules present in the solution

The reactions of the third type will be discussed in the next two chapters. Let's concentrate for now on the first two types of ionic reactions.

You now know all the strong electrolytes. Any electrolyte not included in the preceding list of strong electrolytes may be considered weak. Incidentally, water is one of the most important weak electrolytes. At room temperature it ionizes 0.00001%. As to the relatively insoluble substances, Table 9.3 will help you decide whether a given electrolyte is soluble.

Sample Ionic Reactions

In the following cases, solutions of the two electrolytes indicated are to be mixed. Let's see how we can tell what reaction, if any, will take place.

1. $NaCl + KNO_3$

   Each species ionizes:

   $NaCl \rightarrow Na^+ + Cl^-$

   $KNO_3 \rightarrow NO_3^- + K^+$

   Examination of the ions present in the mixed solutions reveals that the potential products, KCl and $NaNO_3$, are both salts (strong electrolytes). Both are soluble. Therefore, no reaction takes place in solution in this case.

2. $NaCl + AgNO_3$

   Each species ionizes:

   $NaCl \rightarrow Na^+ + Cl^-$

   $AgNO_3 \rightarrow NO_3^- + Ag^+$

Ionic Substances	General Rules
Na^+	All sodium salts are *soluble.*
K^+	All potassium salts are *soluble.*
NH_4^+	All ammonium salts are *soluble.*
Ag^+	All silver salts, except $AgNO_3$, are *insoluble.*
$C_2H_3O_2^-$	All acetates, except those of silver, mercury, and bismuth, are *soluble.*
NO_3^-	All nitrates are *soluble.*
Cl^-	All chlorides, except AgCl, CuCl, $PbCl_2$, and Hg_2Cl_2, are *soluble.*
SO_4^{2-}	All sulfates except those of Ba^{2+}, Pb^{2+}, Ca^{2+}, Hg^+, and Ag^+ are *soluble.*
OH^-	All hydroxides except those of Na^+, K^+, NH_4^+, Ba^{2+}, Ca^{2+}, and Sr^{2+} are *insoluble.*
CO_3^{2-}	All carbonates except those of Na^+, K^+, and NH_4^+ are *insoluble.*
S^{2-}	All sulfides except those of Na^+, K^+, NH_4^+, Mg^{2+}, Ca^{2+}, Ba^{2+}, and Sr^{2+} are *insoluble.*

Table 9.3 Solubility of Electrolytes

Examining the ions, we see that the potential products are $NaNO_3$ and AgCl. Both are salts (strong electrolytes), but AgCl is not soluble, so a reaction takes place in this case. The equation for the reaction may be written in the following three forms:

(a) Formula form:

$$NaCl(aq) + AgNO_3(aq) \rightarrow AgCl(s) + NaNO_3(aq)$$

The arrow indicates the AgCl precipitates out and that the remaining liquid is a solution of $NaNO_3$.

(b) Ionic form:

$$(Na^+, Cl^-) + (Ag^+, NO_3^-) \rightarrow AgCl(s) + Na^+ + NO_3^-$$

This form indicates that a solid precipitates from two soluble electrolytes and shows the ions left in solution.

(c) Net ionic form:

$$Ag^+ + Cl^- = AgCl(s)$$

This form shows the heart of the reaction, indicating that the free ions from any source will undergo the reaction.

3. $Na_2CO_3 + HCl$

Each species ionizes:

$$Na_2CO_3 \rightarrow 2\ Na^+ + CO_3^{2-}$$

$$2\ HCl \rightarrow 2\ Cl^- + 2\ H^+$$

An examination of the ions indicates that one of the products is the weak acid, carbonic acid, H_2CO_3. Therefore, a reaction takes place. (Each of the substances in the second equation was given a coefficient of 2 to balance the charges in the two equations.) This reaction can then be written:

(a) Formula form:

$$Na_2CO_3(aq) + 2\ HCl(aq) \rightarrow H_2CO_3(aq) + 2NaCl(aq)$$

The H_2CO_3 remains in solution. (*Note*: Carbonic acid, H_2CO_3, is really a solution of the gas CO_2 in water. When CO_2

dissolves in water, it forms molecules of H_2CO_3 thus:

$$CO_2(g) + H_2O(l) \rightarrow H_2CO_3(aq)$$

The gas CO_2 is not highly soluble in water, so if the amount of H_2CO_3 formed in this reaction is in excess of the amount that would be present in a saturated solution of CO_2, the H_2CO_3 will decompose, and CO_2 will bubble out (effervesce) of the solution until all the excess CO_2 has been emitted. (This same discussion applies to another weak acid, sulfurous acid, H_2SO_3, which is a solution of sulfur dioxide, SO_2, in water.)

(b) Ionic form:

$$(2\ Na^+, CO_3^{2-}) + 2\ (H^+, Cl^-) \rightarrow H_2CO_3(aq) + 2\ Na^+ + 2\ Cl^-$$

(c) Net ionic form:

$$2\ H^+ + CO_3^{2-} \rightarrow H_2CO_3(aq)$$

4. NaOH + HCl

Each species ionizes: $NaOH \rightarrow Na^+ + OH^-$

$$HCl \rightarrow Cl^- + H^+$$

Examining the ions, we see that one of the products is the weak electrolyte water, H_2O, so a reaction takes place:

(a) Formula form:

$$NaOH(aq) + HCl(aq) \rightarrow H_2O(l) + NaCl(aq)$$

(b) Ionic form:

$$(Na^+, OH^-) + (H^+, Cl^-) \rightarrow H_2O(l) + Na^+ + Cl^-$$

(c) Net ionic form:

$$H^+ + OH^- \rightarrow H_2O(l)$$

This reaction between an acid and a base, forming water, is known as *neutralization*. The net ionic form of the reaction tells us that any acid, strong or weak, will neutralize any base, strong or weak.

Chemical Equilibrium

Weak electrolytes are only slightly, or partially, ionized. The conductivity data for acetic acid in Table 9.2 and the arithmetic treatment of it led us to that conclusion. What accounts for the fact that every 0.1 M solution of acetic acid at room temperature is ionized to the same extent?

The explanation of the phenomenon of partial ionization can be found if we think about things we already know. When any electrolyte, strong or weak, is added to water, the polar water molecules begin to dissociate the electrolyte into its ions. Thus, a concentration of ions begins to build up in solution. The ions, of course, possess opposite electrical charges and thus can attract one another. In the case of strong electrolytes, water molecules are sufficiently polar to prevent any permanent recombination of these ions; but in the case of weak electrolytes, water is less effective, and recombination of ions begins to take place as soon as any appreciable concentration of ions is present in the solution. Thus, in the case of weak electrolytes, two processes take place simultaneously in opposite directions, *dissociation* and *recombination*. Initially, the dissociation takes place at a faster rate than the recombination, but eventually, as the concentration of ions builds up, the rate of recombination catches up to the rate of dissociation. After that point, both processes continue to proceed at the same rate. The apparent effect of this is no change, because the two processes nullify each other. When two opposing processes take place simultaneously at the same rate, a state of *equilibrium* exists.

We have already studied several other equilibrium situations. When the air in contact with a liquid is saturated with the liquid's vapor, the rate of evaporation equals the rate of condensation. When a solution is saturated with a solute, the rate of dissolution is just equal to the rate of precipitation from solution. In each of these cases, both processes continue, but at the same rate, creating the illusion of static conditions.

The *position* of equilibrium is the extent to which one of the processes progresses before the opposite process catches up with it. For example, a 0.01 M solution of acetic acid at room temperature is ionized about 4%. This means that if 0.01 mol of pure acetic acid is dissolved in a liter of water, about 4% of the solute will have ionized by the time equilibrium is reached. It also means that if 0.01 mol of hydrogen ion and 0.01 mol of acetate ion are added to a liter of water from different sources, the two ions will have combined to an extent of about 96% by the time equilibrium is reached.

The position of equilibrium is not fixed. It depends on the nature of the substances involved, temperature, pressure (when gases are involved), and concentration. For example, each different weak electrolyte has its own degree of ionization at a given temperature, pressure, and concentration. An increase in temperature speeds up a process that absorbs energy and slows one that gives off energy. Thus, an increase in temperature will shift the position of equilibrium in the direction of the process absorbing energy. In many reactions involving gases, the number of moles of gas on each side of the equation is not necessarily the same. A difference in number of moles of gas is, of course, a difference in volume. An increase in pressure always shifts the position of equilibrium in the direction of the reaction producing the smaller volume. *Le Châtelier's principle* sums up the effects of changes in any of the factors influencing the position of equilibrium. It states: *A system in equilibrium, when subjected to a stress resulting from a change in temperature, pressure, or concentration, and causing the equilibrium to be upset, will adjust its position of equilibrium to relieve the stress, and reestablish equilibrium.*

LAW OF MASS ACTION

The effect of concentration on equilibrium is expressed in the *law of mass action* (also called the *law of chemical equilibrium*). This law states: *The velocity of a reaction is proportional to the product of the molar concentrations of the reacting substances, taken to proper powers, the powers being the coefficients of the reactants in the balanced equation for the reaction.* Consider the following hypothetical reaction:

$$mA + nB \rightleftharpoons pC + qD$$

Let's assume that an equilibrium will be established between the substances on the left and those on the right. The double arrows in the equation indicate that this is an equilibrium system. This means that we are thus dealing with two reactions, one proceeding to the right and the other proceeding to the left. According to the law of mass action, the rate of the reaction proceeding to the right is

$$r_1 = k_1 \times [A]^m \times [B]^n$$

where

r_1 is the reaction velocity or rate,

[A] is the molar concentration of substance A,

[B] is the molar concentration of substance B,

k_1 is the constant of proportionality.

Note that sequence brackets are used to indicate molarity.

Similarly, by the law of mass action, the velocity or rate of the reaction proceeding to the left is

$$r_2 = k_2 \times [C]^p \times [D]^q$$

Now, at equilibrium, the two rates will be equal. Thus, if

$$r_2 = r_1$$

then

$$k_2 \times [C]^p \times [D]^q = k_1 \times [A]^m \times [B]^n$$

Rearranging, we obtain,

$$\frac{[C]^p \times [D]^q}{[A]^m \times [B]^n} = \frac{k_1}{k_2} = K$$

This equation is known as the generalized law of mass action, or reaction quotient, or equilibrium expression. It applies to all equilibrium situations. Notice that the numerator contains the molar concentrations of the substances (to proper powers) that usually appear on the right side in a balanced chemical equation. This is an adopted convention. The constant *K* is a constant for a given temperature. Let's now examine some of the applications of the equilibrium expression to weak electrolyte systems.

> Equilibrium means balance. In a solution at equilibrium, the reaction to the left is proceeding at exactly the same rates as the reaction to the right.

Ionization Constant

If we know the percentage ionization of a weak electrolyte and its concentration, we can compute *K* in the equilibrium expression for the electrolyte. In this case, *K* is known as the *ionization constant.*

EXAMPLE 1. A 0.1 M solution of acetic acid, $HC_2H_3O_2$, is 1.32% ionized at equilibrium. Find its ionization constant.

SOLUTION: Acetic acid ionizes:

$$HC_2H_3O_2 \rightleftharpoons H^+ + C_2H_3O_2^-$$

The equilibrium expression for this reaction is

$$\frac{[H^+] \times [C_2H_3O_2^-]}{[HC_2H_3O_2]} = K$$

We are given that 1.32% of the acid is in ionic form.

Therefore, $[H^+] = 0.1 \text{ M} \times 0.0132 - 0.00132 \text{ M}$

$[C_2H_3O_2^-]$ = the same as $[H^+]$, because the ions are formed in equal amounts

$[HC_2H_3O_2] = 0.1 - 0.00132 = 0.09868 \text{ M}$

Substituting into the equilibrium expression, we obtain

$$K = \frac{[0.00132 \text{ M}]^2}{[0.09868 \text{ M}]} = 1.77 \times 10^{-5}$$

This constant applies to all equilibrium solutions of acetic acid at room temperature.

NOTE: In finding the concentration of $HC_2H_3O_2$, we subtracted a very small number from a relatively larger one. If we had ignored this subtraction, our answer would not have changed very much at all. Therefore, in future applications we will ignore the effect of that part of the weak electrolyte that has ionized. We will assume, instead, that the concentration of electrolyte remaining un-ionized is the same as the original concentration of the electrolyte.

Once we know the ionization constant for a weak electrolyte, we are then able to calculate additional properties of other solutions of this electrolyte.

EXAMPLE 2. The ionization constant of acetic acid is 1.8×10^{-5}. Find $[H^+]$ in a 0.01 M solution of this acid.

SOLUTION: First, we write the ionization equation:

$$HC_2H_3O_2 \rightleftharpoons H^+ + C_2H_3O_2^-$$

Next, we write the equilibrium expression:

$$\frac{[H^+] \times [C_2H_3O_2^-]}{[HC_2H_3O_2]} = 1.8 \times 10^{-5}$$

Now, let $[H^+] = x$.

Then, $[C_2H_3O_2^-] = x$, because both are formed in the same amount.

$$[HC_2H_3O_2] = 0.01 - [H^+] \approx 0.01$$

Substituting values, we get

$$\frac{x^2}{0.01} = 1.8 \times 10^{-5}$$

$$x^2 = 1.8 \times 10^{-7}$$

$$x = 4.2 \times 10^{-4}\ \text{M} = 0.00042\ \text{M}$$

EXAMPLE 3. What is the percentage of ionization of the acid in Example 2?

SOLUTION: If the 0.01 M acetic acid were 100% ionized, the concentration of H^+ would be 0.01 M. Thus, to find the percentage ionization of the acid, we divide the actual H^+ concentration by 0.01 and multiply the result by 100, thus:

$$\frac{4.2 \times 10^{-4}}{1 \times 10^{-2}} \times 100 = 4.2\%$$

EXPERIMENT 18: Check the conductivity of pure vinegar in the conductivity apparatus made in Experiment 15. Then, add small quantities of water to the vinegar, checking the conductivity after each addition. You will observe the gradual brightening of the LED, because the percentage ionization of a weak electrolyte increases with dilution. Repeat the procedure with ammonia water.

TRIVIA: Hydrogen sulfide is more toxic than hydrogen cyanide, but our sense of smell can detect minute quantities of hydrogen sulfide, giving us ample warning of its presence.

The pH Scale

The concentration of hydrogen ions in any aqueous solution is important, because the hydrogen ion is responsible for all acid properties. Water is a very weak electrolyte, which means that it ionizes to a slight extent as follows:

$$H_2O \rightleftharpoons H^+ + OH^-$$

Actually, H^+ ions (bare protons) cannot exist in water. They covalently bond with water molecules to form H_3O^+, called the *hydronium ion*. The acidic properties of water are ascribed to H^+ or H_3O^+. In this text we will write H^+, keeping in mind that the symbol represents H_3O^+. Note also that the dissociation of water can be written

$$H_2O(l) + H_2O\ (l) \rightleftharpoons H_3O^+(aq) + OH^-(aq)$$

This reaction of water with itself is called *self-ionization*. Thus, any time water is present, some of its ions are also present. Water is 0.00001% ionized at room temperature. Therefore, the concentration of each of the ions in pure water at room temperature is 1.0×10^{-7} M.

As you may have observed, the concentration of hydrogen ions in pure water and in dilute solutions of weak acids is usually a very small number. To avoid the use of such numbers, scientists have devised a scale for indicating the concentration of hydrogen ions, known as the *pH scale*. It is defined as follows: *the pH is the negative logarithm of the molar concentration of the hydrogen ion.* In mathematical form,

$$\text{pH} = -\log\ [H^+]$$

The pOH is defined in similar terms. Mathematically,

$$\text{pOH} = -\log\ [OH^-]$$

The pH of pure water can now be computed:

$$\text{pH} = -\log\ [1.0 \times 10^{-7}] = 7.0$$

Because, in pure water, $[OH^-] = [H^+]$, the pOH is also 7.0. From the law of mass action, it can be shown that the following important relationship holds true whenever water is present:

$$[H^+] \times [OH^-] = K_w = 1.0 \times 10^{-14}$$

K_w is the dissociation constant for the self-ionization of water.

Therefore, for any system containing water

$$\text{pH} + \text{pOH} = 14$$

We have seen that the pH of pure water is 7. If a solution has a pH of less than 7, the concentration of hydrogen ion is greater than it is in pure water. Therefore, such solutions exhibit acidic properties. Similarly, if the pH is greater than 7, the solution is basic:

$$\text{pH} < 7.0 = \text{acid}$$

$$\text{pH} > 7.0 = \text{base}$$

Table 9.4 gives the common logarithms (base = 10) of numbers from 1.0 to 9.9. It will be of value in finding the pH of solutions in working problems; however, most people now use scientific calculators. Just be sure that you take the negative of log $[H^+]$.

Let's now study a few examples involving the calculation of pH.

EXAMPLE 4. A 0.1 M solution of acetic acid has a hydrogen ion concentration of 1.3×10^{-3} M. Find the pH of the solution.

SOLUTION:

$$\text{pH} = -\log\ [1.3 \times 10^{-3}]$$

$$= -(2.89)$$

$$\text{pH} = 2.89$$

The pH of 2.89 in this example is less than 7. This value indicates that the solution is definitely acidic.

EXAMPLE 5. The ionization constant of ammonium hydroxide, NH_4OH, is 1.8×10^{-5}. For a 0.01 M solution of this substance, find the

(a) concentration of OH^-

(b) pH

(c) percentage ionization

SOLUTION:

(a) The ionization equation is

$$NH_4OH \rightleftharpoons NH_4^+ + OH^-$$

The equilibrium expression, then, is

$$\frac{[NH_4^+] \times [OH^-]}{[NH_4OH]} = 1.8 \times 10^{-5}$$

N	log	N	log	N	log	N	Log	N	Log
1.0	0.000	3.0	0.477	5.0	0.699	7.0	0.845	9.0	0.954
1.1	0.041	3.1	0.491	5.1	0.708	7.1	0.851	9.1	0.959
1.2	0.079	3.2	0.505	5.2	0.716	7.2	0.857	9.2	0.964
1.3	0.114	3.3	0.519	5.3	0.724	7.3	0.863	9.3	0.968
1.4	0.146	3.4	0.531	5.4	0.732	7.4	0.869	9.4	0.973
1.5	0.176	3.5	0.544	5.5	0.740	7.5	0.875	9.5	0.978
1.6	0.204	3.6	0.556	5.6	0.748	7.6	0.881	9.6	0.982
1.7	0.230	3.7	0.568	5.7	0.756	7.7	0.886	9.7	0.987
1.8	0.255	3.8	0.580	5.8	0.763	7.8	0.892	9.8	0.991
1.9	0.279	3.9	0.591	5.9	0.771	7.9	0.898	9.9	0.996
2.0	0.301	4.0	0.602	6.0	0.778	8.0	0.903		
2.1	0.322	4.1	0.613	6.1	0.785	8.1	0.908		
2.2	0.342	4.2	0.623	6.2	0.792	8.2	0.914		
2.3	0.362	4.3	0.633	6.3	0.799	8.3	0.919		
2.4	0.380	4.4	0.643	6.4	0.806	8.4	0.924		
2.5	0.398	4.5	0.653	6.5	0.813	8.5	0.929		
2.6	0.415	4.6	0.663	6.6	0.820	8.6	0.934		
2.7	0.431	4.7	0.672	6.7	0.826	8.7	0.940		
2.8	0.447	4.8	0.681	6.8	0.833	8.8	0.944		
2.9	0.462	4.9	0.690	6.9	0.839	8.9	0.949		

Table 9.4 Common Logarithms

Let $[OH^-] = x$.

Then,

$[NH_4^-] = x$ (both are formed in the same amount)

$$[NH_4OH] = 0.01 - x \approx 0.01$$

Substituting into the equilibrium expression, we obtain

$$\frac{x^2}{[0.01]} = 1.8 \times 10^{-5}$$

$$x^2 = 1.8 \times 10^{-7}$$

Thus, $[OH^-] = x = 4.2 \times 10^{-4}$ M

(b) $pOH = -\log [OH^-]$

$$= -\log [4.2 \times 10^{-4}] = 3.38$$

$$pH = 14 - pOH = 14 - 3.38 = 10.62$$

Because the pH is greater than 7, the solution is basic.

(c) Percentage ionizaion $= \dfrac{[OH^-]}{[NH_4OH]} \times 100$

$$= \frac{4.2 \times 10^{-4}}{1 \times 10^{-2}} \times 100$$

$$= 4.2\%$$

COMMON-ION EFFECT

An interesting phenomenon occurs when an ion common to the solution is added to a solution of a weak electrolyte. Consider the ionization of ammonium hydroxide: $NH_4OH \rightleftharpoons NH_4^+ + OH^-$. Suppose that more ammonium ion was added to a solution of NH_4OH by dissolving some NH_4Cl salt in the solution. According to Le Châtelier's principle, the increase in concentration of ammonium ion would upset the equilibrium, and the position of equilibrium would shift toward the left to relieve the stress. This, then, should have the effect of decreasing the concentration of hydroxide ions in solution. In other words, the addition of the salt would have the effect of partially neutralizing the base! Let's check this with a numerical example.

EXAMPLE 6. To a 0.01 M solution of NH_4OH, sufficient NH_4Cl is added to increase the concentration of NH_4^+ to 0.03 M. Find:

(a) the concentration of OH^- in the solution

(b) the pH of the solution

SOLUTION: The ionization equation is

$$NH_4OH \rightarrow NH_4^+ + OH^-$$

The equilibrium expression is

$$\frac{[NH_4^+] \times [OH^-]}{[NH_4OH]} = 1.8 \times 10^{-5}$$

(a) Let: $[OH^-] = x$

$$[NH_4^+] = 0.03 \text{ M}$$

$$[NH_4OH] = 0.01 - x \approx 0.01$$

Substituting into the equilibrium expression, we get

$$\frac{[0.03] \times x}{[0.01]} = 1.8 \times 10^{-5}$$

$$[OH^-] = x = 6.0 \times 10^{-6} \text{ M}$$

In Example 5 we saw that the OH^- concentration in a 0.01 M solution of NH_4OH was 4.2×10^{-4}, so the effect of adding the common ion was indeed to lower the concentration of hydroxide ion.

(b) $pOH = -\log [6.0 \times 10^{-6}] = 5.22$

$pH = 14 - pOH = 14 - 5.22 = 8.78$

The addition of the ammonium ion changed the pH from 10.62 to 8.78, a value much closer to 7. Thus, the salt reduced the pH of the solution.

This example demonstrated the *common-ion effect*, which states that *the addition of an ion in common with an ion of the solute represses the dissociation of the solute, thereby reducing the solubility of the solute.*

HYDROLYSIS

When an acid and a base neutralize each other, the products formed are a salt and water. For this type of reaction, four different types of salts are possible:

1. a salt of a strong acid and a strong base (e.g., NaCl from NaOH and HCl)

2. a salt of a strong acid and a weak base (e.g., NH_4Cl from HCl and NH_4OH)

3. a salt of a weak acid and a strong base (e.g., Na_2CO_3 from H_2CO_3 and NaOH)

4. a salt of a weak acid and a weak base (e.g., NH_4CN from HCN and NH_4OH)

The reaction of a substance with water is *hydrolysis*. Some salts are composed of ions of weak electrolytes that may react with the ions of the water solvent to form molecules of weak electrolytes in solution. Let's investigate each type of salt to see what possible hydrolysis reactions might occur.

CASE 1: Salt of a strong base and a strong acid

NaCl ionizes:

$$NaCl \rightarrow Na^+ + Cl^-$$

Its solvent, H_2O, ionizes:

$$H_2O \rightleftharpoons OH^- + H^+$$

Examination of the ions formed in solution (cation of one, anion of the other and vice versa) shows that the potential products in each case, NaOH and HCl, are both strong electrolytes, both will ionize 100%, and thus neither will form molecules in solution. Therefore, this class of salt does not hydrolyze in solution.

CASE 2: Salt of a weak base and a strong acid

NH_4Cl ionizes:

$$NH_4Cl \rightarrow NH_4^+ + Cl^-$$

Its solvent, H_2O, ionizes:

$$H_2O \rightleftharpoons OH^- + H^+$$

Examination of the ions present quickly shows that the weak electrolyte NH_4OH will form in this solution. Note that this reaction will remove hydroxide ions from solution and bind them up in the NH_4OH molecules. Hydrogen ions will thus be left in excess in the solution, and the solution will become acidic. Therefore, this class of salt hydrolyzes to produce acidic solutions.

CASE 3: Salt of a strong base and a weak acid

Na_2CO_3 ionizes:

$$Na_2CO_3 \rightarrow 2\ Na^+ + CO_3^{2-}$$

Its solvent, H_2O, ionizes:

$$2\ H_2O \rightleftharpoons 2\ OH^- + 2\ H^+$$

Examination of the ions present quickly shows that the weak electrolyte H_2CO_3, carbonic acid, will form in this solution. Note that this reaction will remove hydrogen ions from the solution and bind them up in molecules of carbonic acid. Hydroxide ions will thus be left in excess in solution, and the solution will become basic. Therefore, this class of salt hydrolyzes to produce basic solutions.

CASE 4: Salt of a weak base and a weak acid.

NH_4CN, ammonium cyanide, ionizes:

$$NH_4CN \rightarrow NH_4^+ + CN^-$$

Its solvent, H_2O, ionizes:

$$H_2O \rightleftharpoons OH^- + H^+$$

Examination of the ions present shows that two weak electrolytes, ammonium hydroxide, NH_4OH, and hydrocyanic acid, HCN, will form in this solution. Most of the ions will be removed from this solution and it will thus be a very poor conductor of electricity. The acidity or alkalinity of this solution depends on which of the two weak electrolytes formed is weaker. In this case, the ionization constant of NH_4OH is 1.8×10^{-5}, whereas the ionization constant of HCN is 4×10^{-10}. This comparison of ionization constants shows that HCN is a much weaker electrolyte than NH_4OH, and that HCN will be ionized to a much lesser extent. Therefore, more hydrogen ions will be removed from this solution than hydroxide ions, and the solution will become very slightly basic. Salts of this class hydrolyze extensively, producing solutions that are either slightly acidic or slightly basic, depending on the particular weak electrolytes formed in the solution.

EXPERIMENT 19: Prepare aqueous solutions of each of the following: table salt (NaCl), baking soda ($NaHCO_3$), and washing soda (Na_2CO_3). To each of these salt solutions add a few drops of the purple-cabbage solution prepared in Experiment 5. In which two solutions is hydrolysis taking place? Compare them for degree of hydrolysis. You will observe that the washing soda is considerably more alkaline.

A salt will hydrolyze under two other conditions:

1. if an insoluble hydroxide forms

2. if a volatile gas forms

Let's look briefly at these possibilities.

Magnesium chloride, $MgCl_2$, ionizes:

$$MgCl_2 \rightarrow Mg^{2+} + 2\ Cl^-$$

Its solvent, H_2O, ionizes:

$$2\ H_2O \rightleftharpoons 2\ OH^- + 2\ H^+$$

Examination of the ions present shows that $Mg(OH)_2$, a strong base but an insoluble one, will form in this solution. This reaction will remove hydroxide ions from solution, leave hydrogen ions behind in excess, and thus produce an acid solution.

We saw earlier that NaCl does not hydrolyze in solution; however, if a solution of NaCl is vigorously boiled, the volatile gas HCl will be expelled from the solution. Thus, hydrogen ions will be removed from the solution, and the solution will become basic as a result of the buildup of hydroxide ions.

Hydrolysis is an extremely important phenomenon that has many applications in industry. For example, the water in the boilers of factories or ships is made alkaline by adding sodium carbonate to it rather than sodium hydroxide, in most cases. A second example: if acid is spilled on the floor, a solution of sodium carbonate or sodium bicarbonate effectively neutralizes the acid.

SOLUBILITY PRODUCT

We have a tendency to estimate the solubility of a substance by eye. For example, if we add some solute to water, stir the solution, and the solute disappears, we say that the solute was soluble. If, on the other hand, the solute does not disappear, we say that it is insoluble. Actually, when we use the word "insoluble" in chemistry, we mean "slightly soluble," because all electrolytes are soluble to some extent in water; however, many of them are so slightly soluble that we cannot detect by eye any dissolving taking place. These slightly soluble substances play an important part in analytical chemistry, a field in which we measure the composition of a substance or a mixture.

A saturated solution of an electrolyte has some special properties. As you know, in a saturated solution a state of equilibrium exists between the ions in solution and the ions in the crystals of the solute. All the rules of equilibrium thus apply to saturated solutions.

EXPERIMENT 20: Taste both table salt and baking soda (NaCl and $NaHCO_3$). Prepare a saturated solution of table salt. Add a quarter teaspoon of baking soda. Allow the precipitate to settle. After carefully pouring off the excess liquid, filter the solution through a handkerchief to collect the solid precipitate. Permit the solid in the handkerchief to dry. Taste it. What is it? You are observing the effect of a common ion, the sodium ion, on a saturated solution.

Furthermore, a law similar to the law of mass action applies to saturated solutions of slightly soluble electrolytes. This law is the *solubility product principle*, and it states: *At saturation, the product of the molar concentrations of the ions of the electrolyte, raised to proper powers, is a constant.* The powers are the coefficients in the ionization equation of the electrolyte. The constant is constant at a particular temperature. Thus, for the hypothetical electrolyte

$$A_mB_n = mA + nB$$

Its solubility product expression is

$$[A]^m \times [B]^n = K_{sp}$$

where, again, brackets indicate the molar concentration of the substances, and K_{sp} represents the solubility product constant. We can find the solubility product constant of a given electrolyte if we know its solubility in water.

EXAMPLE 7. The solubility of calcium fluoride, CaF_2, is 0.0016 g/100. mL of water. Find the solubility product constant of CaF_2.

SOLUTION: The molecular weight of CaF_2 is 78 g/mol. Its solubility is 0.016 g/L.

Calculating the solubility of CaF_2 in moles per liter gives us

$$\frac{0.016\text{ g}}{1\,L} \times \frac{1\text{ mol}}{78\text{ g}} = 2.05 \times 10^{-4}\text{ M}$$

CaF_2 ionizes:

$$CaF_2 \rightleftharpoons Ca^{2+} + 2\,F^-$$

The solubility product expression, therefore, is

$$[Ca^{2+}] \times [F^-]^2 = K_{sp}$$

To find the constant we proceed in two steps:

1. Find the molar concentration of each of the ions.

2. Substitute these concentrations into the expression and solve.

The ionization equation tells us that for each mole of CaF_2 that dissolves, 1 mol of calcium ion and 2 mol of fluoride ion are released into the solution. Thus, $[Ca^{2+}]$ in solution = 2.05×10^{-4} M, and $[F^-]$ in solution = $2 \times 2.05 \times 10^{-4} = 4.1 \times 10^{-4}$ M. Substituting into the equilibrium expression, we have

$$[2.05 \times 10^{-4}] \times [4.1 \times 10^{-4}]^2 = K_{sp}$$

$$[2.05 \times 10^{-4}] \times [16.8 \times 10^{-8}] = K_{sp}$$

$$K_{sp} = 3.4 \times 10^{-11}$$

The solubility product constant for slightly soluble substances, like the ionization constant for weak electrolytes, is a property of the substance. It applies to all saturated solutions of the substance. If we know the solubility product constant of a substance, we can use the solubility product expression to find the concentration of the ions in a saturated solution.

EXAMPLE 8. The solubility product constant of silver chloride, AgCl, is 1.6×10^{-10}. Find the concentration of the ions in a saturated solution of this salt.

SOLUTION: AgCl ionizes:

$$AgCl \rightleftharpoons Ag^+ + Cl^-$$

Its solubility product expression is $[Ag^+] \times [Cl^-] = 1.6 \times 10^{-10}$.

Let $x = [Ag^+]$. Then, $x = [Cl^-]$ also, because the two ions are formed in equal amounts according to the ionization equation.

Substituting into the solubility product expression, we obtain

$$x^2 = 1.6 \times 10^{-10}$$

$$[Ag^+] = [Cl^-] = x = 1.3 \times 10^{-5}\text{ M}$$

The following example illustrates the solubility product principle and the behavior of solutions of ions of slightly soluble substances. Study it carefully.

EXAMPLE 9. A solution contains chloride ion, Cl^-, and chromate ion, CrO_4^{2-}. The Cl^- concentration is 0.001 M, and the CrO_4^{2-} concentration is 0.005 M. Silver ion, in the form of a solution of silver nitrate, is to be added slowly to this solution. The K_{sp} of AgCl is 1.6×10^{-10}. The K_{sp} of Ag_2CrO_4 is 9.0×10^{-12}.

(a) At what silver ion concentration will AgCl begin to precipitate?

(b) At what silver ion concentration will Ag_2CrO_4 begin to precipitate?

(c) Which salt will precipitate first?

SOLUTION:

(a) AgCl ionizes:

$$AgCl \rightleftharpoons Ag^+ + Cl^-$$

Its solubility product expression is $[Ag^+] \times [Cl^-] = 1.6 \times 10^{-10}$.

Let

$$[Ag^+] = x$$
$$[Cl^-] = 0.001 \text{ M}$$

Substituting those values into the solubility product expression, we have

$$x \times [0.001] = 1.6 \times 10^{-10}$$
$$[Ag^+] = x = 1.6 \times 10^{-7} \text{ M}$$

Thus, the silver ion concentration necessary to begin the precipitation of AgCl is 1.6×10^{-7} M.

(b) Ag_2CrO_4 ionizes:

$$Ag_2CrO_4 \rightleftharpoons 2\,Ag^+ + CrO_4^{2-}.$$

Its solubility product expression is $[Ag^+]^2 \times [CrO_4^{2-}] = 9.0 \times 10^{-12}$.

Let

$$[Ag^+]^2 = x$$
$$[CrO_4^{2-}] = 0.005 \text{ M}$$

Substituting those values into the solubility product expression, we have

$$x^2 \times [0.005] = 9.0 \times 10^{-12}$$
$$x^2 = \frac{9.0 \times 10^{-12}}{5.0 \times 10^{-3}} = 1.8 \times 10^{-9}$$
$$[Ag^+] = x = 4.2 \times 10^{-5} \text{ M}$$

Thus, the silver ion concentration necessary to begin the precipitation of Ag_2CrO_4 is 4.2×10^{-5} M.

(c) The results of parts (a) and (b) show that it takes less silver ion to cause AgCl to begin to precipitate than it takes to cause $AgCrO^4$ to begin to precipitate as silver ion is added to this solution. Therefore, AgCl will begin to precipitate first from the solution. Under actual conditions, the white salt AgCl will continue to precipitate as silver ion is added until the amount of chloride ion left in solution is negligible. At that point the concentration of silver ion will have increased sufficiently in the solution so that the red salt Ag_2CrO_4 will begin to precipitate. The first appearance of the red salt indicates that the silver has precipitated all the chloride. Thus, we have an important method of analyzing solutions for their chloride content.

We mention one more application of Le Châtelier's principle. An electrolyte that is only slightly soluble in water can be put into solution if its ions can be made to form molecules of weak electrolytes. Consider the following situation: Calcium carbonate, $CaCO_3$, is only slightly soluble in water. It ionizes: $CaCO_3 \rightleftharpoons Ca^{2+} + CO_3^{2-}$. If an acid is added to the saturated $CaCO_3$ solution, the hydrogen ions from the acid react with the carbonate ions to form molecules of carbonic acid, H_2CO_3. This upsets the equilibrium between the ions of $CaCO_3$ in solution and its ions in the solid crystals, and drives the reaction to the right, according to Le Châtelier's principle. The result is that the entire solid dissolves in the acidified solution. This phenomenon makes acid solutions good solvents for all salts except salts of strong acids.

SUMMARY

Electrolytes dissociate into ions in solution. Acids ionize to produce hydronium ions. Alkalis ionize to produce hydroxide ions. Salts ionize but produce neither of these two ions. The chemistry of ions in solutions is independent of the source of the ions.

A strong electrolyte is 100% ionized. A weak electrolyte is only partially ionized.

Ions in solution react if

1. they can form a weak electrolyte;
2. they can form an insoluble substance;
3. they oxidize or reduce one another.

A state of equilibrium exists when two opposing processes take place simultaneously at the same rate. Le Châtelier's principle states that a system in equilibrium when subjected to a stress resulting from a change in temperature, pressure, or concentration that causes the equilibrium to be upset will adjust its position of equilibrium to relieve the stress and reestablish equilibrium.

The law of mass action states that the velocity of a reaction is proportional to the product of the molar concentrations of the reacting substances, taken to proper powers, the powers being the coefficients of the reactants in the balanced equation for the reaction.

The pH of a solution is the negative logarithm of the molar concentration of the hydrogen ion. Neutral solutions have a pH of 7. Acidic solutions have a pH of less than 7. Basic solutions have a pH of greater than 7.

The common-ion effect means that the addition of an ion in common with an ion of a solute represses the dissociation of the solute. Hydrolysis is the reaction between the ions of an electrolyte and the ions of water in solution.

The solubility product principle states that at saturation, the product of the molar concentrations of the ions of the electrolyte, raised to proper powers, is a constant. The powers are the coefficients in the ionization equation of the electrolyte. Substances only slightly soluble in water can be caused to dissolve if weak electrolytes can be formed in their saturated solution.

PROBLEM SET NO. 11

1. A 44.4-g of $CaCl_2$ is dissolved in 200. g of water. Assuming that the salt remains 100% ionized, find the temperature at which this solution will freeze.

2. What is the concentration of each of the ions present in each of the following solutions?

   (a) 0.05 M $MgCl_2$

   (b) 0.25 M H_2SO_4

   (c) 0.1 M $Fe_2(SO_4)_3$

3. Which of the following pairs of electrolytes will react with each other in solution? Write a balanced equation for each reaction that takes place.

   (a) $K_2CO_3 + CaCl_2$

   (b) $HNO_3 + Na_2SO_3$

   (c) $H_2SO_4 + AgNO_3$

   (d) $HNO_3 + Na_2SO_4$

4. A 0.1 M solution of nitrous acid, HNO_2, is 6.3% ionized at room temperature. Find its ionization constant.

5. The ionization constant of acetic acid, $HC_2H_3O_2$, is 1.8×10^{-5}. Find $[H^+]$ in a 0.1 M solution of this acid.

6. Find the percentage ionization of the solution in problem 5.

7. The ionization constant of HCN is 4×10^{-10}. Find the pH of a 0.1 M solution of HCN.

8. The ionization constant of NH_4OH is 1.8×10^{-5}. Find the pH of a 0.1 M solution of NH_4OH.

9. The ionization constant of acetic acid is 1.8×10^{-5}. To a 0.05 M solution of this acid sufficient sodium acetate is added to raise the acetate ion concentration to 0.01 M. Find the pH of the resulting solution.

10. With the aid of ionization equations, show how the following salts will, or will not, hydrolyze, and indicate whether their solutions will be acidic, basic, or neutral.

    (a) KNO_3 (d) Na_2SO_3

    (b) KNO_2 (e) $FeCl_3$

    (c) NH_4NO_3

11. The solubility of $Mg(OH)_2$, magnesium hydroxide, is 0.0009 g/100 mL. Find its solubility product constant.

12. The solubility product constant of silver acetate, $AgC_2H_3O_2$, is 2×10^{-3}. Find the concentration of each of the ions in a saturated solution of this salt.

13. The solubility product constant of magnesium carbonate, $MgCO_3$, is 1×10^{-5}. A solution has a carbonate ion concentration of 0.05 M. What concentration of magnesium ion will be required to begin the precipitation of $MgCO_3$?

14. Which of the following salts will dissolve in acid? (Use ionization equations to verify your answer.)

    (a) $MgCO_3$

    (b) AgCl

    (c) $Cu(OH)_2$

    (d) $BaSO_4$

15. The solubility product constant of calcium hydroxide, $Ca(OH)_2$, is 8×10^{-6}. Find the pH of a saturated solution of this substance.

CHAPTER 10

OXIDATION–REDUCTION

KEY TERMS

oxidation, reduction, redox reaction, oxidizing agent, reducing agent

Structurally, all chemical changes fall into one of two categories: those that involve a change in oxidation numbers and those that do not.

None of the reactions in the last chapter involved a change in oxidation number. The motivating force behind them was the force of attraction between oppositely charged ions. More specifically, such reactions occur only when the forces of electrical attraction are strong enough to overcome the nullifying influence of the polar water molecules. The polarity of water is responsible for all dissociations. The strong electrical attraction between the ions of weak electrolytes and substances of low solubility is responsible for the formation of those compounds.

Let's now turn our attention to reactions that involve changes in oxidation numbers.

DEFINITIONS

Consider the following reaction:

$$2\ Cu(s) + O_2(g) \rightarrow 2\ CuO(s)$$

When copper is heated in air, the black oxide forms on the surface of the metal. Now let's write the oxidation number of each element under its symbol wherever it appears in the equation. (See Chapter 4 for the rules for assigning oxidation numbers.)

$$\overset{0}{2\ Cu} + \overset{0}{O_2} \rightarrow \overset{+2,-2}{2\ CuO}$$

As a result of the reaction, the oxidation number of copper increased from 0 to +2, and the oxidation number of oxygen decreased from 0 to −2. A substance whose oxidation number has increased has been oxidized. A substance whose oxidation number has decreased has been reduced. In general,

> *Oxidation* is a process involving an increase in oxidation number.

> *Reduction* is a process involving a decrease in oxidation number.

Bearing in mind what we already know about the structure of matter and its relation to valence number, let's examine this reaction more closely. How can the oxidation number of an element increase? The answer is simple. The oxidation number of an element can increase only through the loss of electrons. Similarly, the oxidation number of an element can decrease only through the gain of electrons. Therefore, we can redefine our terms:

> *Oxidation* is a process involving the loss of electrons.

> *Reduction* is a process involving the gain of electrons.

In the example reaction, the oxygen atoms gained electrons, and the copper atoms lost electrons. Such a reaction is often called a *redox reaction*, a shortened form of "oxidation–reduction." The substance that gains electrons in a redox reaction is the *oxidizing agent*. The substance that loses electrons is the *reducing agent*. In the example, the free, elemental oxygen acted as the oxidizing agent; the free, elemental copper acted as the reducing agent.

By now, you may have realized that oxidation–reduction reactions occur because of a flow of electrons from the reducing agent to the oxidizing agent. This is important, because a flow of electrons is, by definition, a current of electricity. The causes and effects of electrical currents will be explored more fully in the next chapter.

One other facet of redox reactions should be noted at this time. Both oxidation and reduction must occur simultaneously. One cannot happen without the other. Oxidation provides the electrons that combine with the other chemical species undergoing reduction. In fact, the law of conservation of matter requires that *the total number of electrons gained in a redox reaction must equal the total number of electrons lost*. Let us check this in the balanced reaction under consideration. Each copper atom loses 2 electrons, for a total of 4. Each oxygen atom gains 2 electrons, for a total of 4. Thus, the number of electrons gained equals the number lost. This fact provides us with a powerful tool for balancing oxidation–reduction reactions.

JOKE: What did one atom tell another?
–I think I lost an electron.
–Are you sure?
–Yes, I'm positive.

BALANCING REDOX EQUATIONS

Two methods are commonly used for balancing redox equation: the ion–electron method and the oxidation number–change method. Let's look first at the ion–electron method.

To balance a redox reaction, remember this: the total number of electrons gained must equal the total number of electrons lost.

In the ion–electron method, the unbalanced redox equation is rewritten as two *half-reactions*, the oxidation and the reduction. After each equation is balanced, they are combined. Consider the following redox equation:

$$HCl(aq) + MnO_2(s) \rightarrow MnCl_2(aq) + H_2O(l) + Cl_2(g)$$

First, we write the oxidation numbers of all elements involved in the redox reaction:

$$\overset{+1,-1}{HCl}(aq) + \overset{+4,-2}{MnO_2}(s) \rightarrow \overset{+2,-1}{MnCl_2}(aq) + \overset{+1,-2}{H_2O}(l) + \overset{0}{Cl_2}(g)$$

Then, we write two separate reactions involving the elements whose oxidation numbers have changed. If that element is part of a covalently bonded molecule, we use the entire molecule. If it exists in solution (where many redox reactions occur), we then show only the atom or ion. In the preceding example the oxidation numbers of chlorine (Cl) and manganese (Mn) changed so the two half-reactions are

$$Cl^- \rightarrow Cl_2$$
$$MnO_2 \rightarrow Mn^{2+}$$

NOTE: We are now writing ionic charges, not oxidation numbers. The oxidation numbers are used only to break the overall reaction into half-reactions.

Next, let's balance by inspection all atoms with the exception of oxygen and hydrogen. We cannot change subscripts, only add coefficients:

$$2\ Cl^- \rightarrow Cl_2$$
$$MnO_2 \rightarrow Mn^{2+}$$

Now, we balance the oxygen atoms. How we do this depends upon whether the reaction occurs in acidic or in basic solution. If the reaction occurs in acid solution, we add to the side deficient in oxygen the same number of water molecules as oxygen atoms needed. If the reaction occurs in basic solution, we add 2 OH^- ions for every oxygen atom needed to the side deficient in oxygen. To the other side of the equation we add half as many water molecules as OH^- ions used. Since our example involves hydrochloric acid, the reaction is occurring in acid solution, so we add water molecules:

$$2\ Cl^- \rightarrow Cl_2$$
$$MnO_2 \rightarrow Mn^{2+} + 2\ H_2O$$

Next, we balance the hydrogen atoms. If the reaction is in an acid solution, we add to the side needing hydrogen as many H^+ cations as hydrogen atoms needed. If in basic solution, we add to the side needing hydrogen as many H_2O molecules as hydrogens added. Then we add to the other side of the equation the same number of OH^- ions. Because the example reaction is acidic, we balance it with hydrogen ions:

$$2\ Cl^- \rightarrow Cl_2$$
$$4\ H^+ + MnO_2 \rightarrow Mn^{2+} + 2\ H_2O$$

Now, we balance the ionic charge of each half-reaction by adding electrons to the appropriate side of each:

$$2\ Cl^- \rightarrow Cl_2 + 2\ e^-$$
$$2\ e^- + 4\ H^+ + MnO_2 \rightarrow Mn^{2+} + 2\ H_2O$$

The electrons should be on opposite sides of the two half-reactions.

Next, we balance electron gain with electron loss in the two half-reactions. The electrons that are gained in the reduction half-reaction are the same electrons lost in the oxidation half-reaction, so the number of electrons gained and lost must be the same. If needed, we adjust the numbers using an appropriate multiplier for the entire half-reaction. In this case, the number of electrons lost is equal to the number gained, so a multiplier is not needed.

Now, add the two half-reactions together and cancel any chemical species appearing on both sides. The electrons should always cancel:

$$2\ Cl^- + 2\ e^- + 4\ H^+ + MnO_2$$
$$\rightarrow Cl_2 + 2\ e^- + Mn^{2+} + 2\ H_2O$$

Finally, we convert the equation back to the molecular form by returning to the equation those species whose oxidation number did not change:

$$4\ HCl(aq) + MnO_2(s)$$
$$\rightarrow Cl_2(g) + Mn(Cl)_2(aq) + 2\ H_2O(l)$$

Now let's balance the same equation using the oxidation number–change method. Again, we assign oxidation numbers:

$$\overset{+1,-1}{HCl}(aq) + \overset{+4,-2}{MnO_2}(s)$$
$$\rightarrow \overset{+2,-1}{MnCl_2}(aq) + \overset{+1,-2}{H_2O}(l) + \overset{0}{Cl_2}(g)$$

The changes in oxidation numbers for manganese and chlorine atoms are apparent. Notice that the oxidation number for manganese changed from +4 to +2. This means that manganese has gained 2 electrons. The oxidation number of chlorine changed from −1 to 0, because chlorine lost an electron. Since the number of electrons lost must equal the number of electrons gained, 2 atoms of chlorine must react for each atom of manganese that reacts. Thus, tentatively the equation becomes

$$2\ HCl(aq) + MnO_2(s) \rightarrow MnCl_2(aq) + H_2O(l) + Cl_2(g)$$

which balances the electrons; however, we need 2 more chlorine atoms to balance the $MnCl_2$. Consequently, we change the coefficient of HCl to 4, and of H_2O to 2:

$$4\ HCl(aq) + MnO_2(s) \rightarrow MnCl_2(aq) + 2H_2O(l) + Cl_2(g)$$

Now, the equation is balanced.

Let's now use the ion–electron method to balance a redox equation that involves ions:

$$Fe^{2+} + MnO_4^- + H^+ \rightarrow Mn^{2+} + Fe^{3+} + H_2O$$

We first break the reaction into two half-reactions:

$$Fe^{2+} \rightarrow Fe^{3+}$$
$$MnO_4^- \rightarrow Mn^{2+}$$

Since both iron and manganese are balanced, we will balance the oxygens and then the hydrogens using the rules for acidic solutions:

$$Fe^{2+} \rightarrow Fe^{3+}$$
$$8\ H^+ + MnO_4^- \rightarrow Mn^{2+} + 4\ H_2O$$

Balancing the electrons, we obtain

$$Fe^{2+} \rightarrow Fe^{3+} + e^-$$
$$5\ e^- + 8\ H^+ + MnO_4^- \rightarrow Mn^{2+} + 4\ H_2O$$

We now balance electron loss with electron gain:

$$[Fe^{2+} \rightarrow Fe^{3+} + e^-] \times 5$$
$$5\ e^- + 8\ H^+ + MnO_4^- \rightarrow Mn^{2+} + 4\ H_2O$$

Finally, we add the equations together, canceling the 5 electrons that appear on both sides, and obtain the following balanced equation:

$$5\ Fe^{2+} + 8\ H^+ + MnO_4^- \rightarrow 5\ Fe^{3+} + Mn^{2+} + 4\ H_2O$$

In balanced ionic equations, not only must the atoms balance, but also the net charge on each side of the equation must be the same. Examining our equation, we see that a net charge of +17 appears on both sides of the equation; therefore, the equation is balanced.

SUMMARY

Oxidation is a process involving an increase in oxidation number or a loss of electrons. Reduction is a process involving a decrease in oxidation number or a gain of electrons. An oxidizing agent gains electrons. A reducing agent loses electrons. A flow of electrons accompanies oxidation–reduction reactions. The total number of electrons gained must equal the total number of electrons lost in an oxidation–reduction reaction. An ionic redox reaction is balanced only when both the number of atoms and the net charges are equal on both sides of the equation.

PROBLEM SET NO. 12

1. Which of the following are oxidation–reduction reactions?

   (a) $2\ Fe^{2+} + Cl_2 \rightarrow 2\ Fe^{3+} + 2\ Cl^-$

   (b) $Cu^{2+} + H_2S \rightarrow CuS + 2\ H^+$

   (c) $2\ SO_2 + O_2 \rightarrow 2\ SO_3$

2. Balance the following equation:

$$H_2S + I_2 \rightarrow S + I^- + H^+$$

   (a) Which substance is the oxidizing agent?

   (b) Which substance is reduced?

3. Balance the following skeleton equation:

$$Cu^{2+} + I^- \rightarrow I_2 + Cu^+$$

   (a) Which substance is the reducing agent?

   (b) Which substance is oxidized?

   (c) Why would the addition of a coefficient of 2 in front of I^- *not* balance this equation?

4. Balance the following equation:

$$MnO_4^- + Sn^{2+} + H^+ \rightarrow Mn^{2+} + Sn^{4+} + H_2O$$

   (a) Which substance is the oxidizing agent?

   (b) Which substance is oxidized?

5. Balance the following equation:

$$Cr_2O_7^{2-} + Fe^{2+} + H^+ \rightarrow Cr^{3+} + Fe^{3+} + H_2O$$

   (a) Which substance is the reducing agent?

   (b) Which substance is reduced?

CHAPTER 11 ELECTROCHEMISTRY

KEY TERMS

electrolyte, electromotive force (emf), electrode, electrolysis, cathode, anode, activity series of metals, reduction potential, polarization, corrosion

Matter has an electrical nature. The atoms that make up all matter consist of nuclei carrying positive electrical charges surrounded by one or more electrons possessing a negative electrical charge. Furthermore, a relatively large class of substances, called electrolytes, consist of charged particles called ions. An *electrolyte* is a substance that conducts an electrical current in the molten state or when dissolved in water. In the last chapter we looked at oxidation–reduction reactions, in which a flow of electrons occurs. The *force of electrostatic attraction is* responsible for all ionic reactions of a non–oxidation–reduction type (Chapter 9).

Redox reactions also involve electrical changes, but they are driven by a different force, the *electromotive force*. This electromotive force is a measure of the electrical driving force necessary to complete an electrochemical reaction. That force produces a current of electricity that oxidizes one substance and reduces the other. As we investigate this force we can ask and answer two questions:

1. What is the effect of an electric current on solutions of electrolytes?
2. What electrical effects accompany oxidation–reduction reactions?

ELECTRICAL UNITS

It is necessary to define the following units of measure before we proceed:

Coulomb. A coulomb (C) is a unit that measures the quantity of electricity. It is the amount of electricity that will deposit 0.001118 g of silver from a solution of silver nitrate onto an *electrode* (a solid piece of conducting metal) at standard conditions, which are 0°C and 1 atm.

Ampere. An ampere (A) is a unit that measures the *rate of flow* of electricity. It is that current that will deposit 0.001118 g of silver in 1 second. In other words, an ampere is a current of 1 coulomb per second.

Ohm. An ohm (Ω)is a unit that measures electrical *resistance.*

Volt. A volt (V) is a unit of *electromotive force.* It is the difference in electrical potential (electrical potential energy) required to cause a current of 1 A to pass through a resistance of 1 Ω.

ELECTROLYSIS

The passage of an electric current through a solution of an electrolyte is known as *electrolysis*. Electrolysis decomposes the electrolyte. Figure 11.1 shows a typical electrolysis setup. It consists of a battery with its two terminals connected to electrodes. The electrodes are immersed in a solution of the electrolyte. Often,

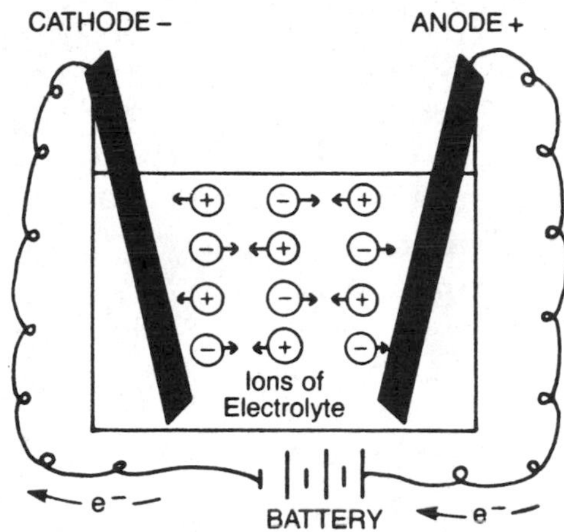

Figure 11.1 Electrolysis Cell

the electrodes are made of a solid conducting material, like platinum, that does not react with the substances in solution.

In solution, electrolytes dissociate into ions. Positively charged ions are called *cations*. Negatively charged ions are *anions*. The battery has the effect of pulling electrons out of one of the metallic electrodes and pushing them onto the other. This causes the first electrode, with the deficiency of electrons, to become positively charged. The other electrode, with the excess of electrons, becomes negatively charged. The negative electrode attracts the positive cations. This electrode is called the *cathode*. The positive electrode attracts the negative anions. This electrode is called the *anode*.

At the electrodes, chemical reactions take place. The positive cations combine with electrons from the negative cathode and become neutral atoms. At the anode, the negative anions yield their electrons to the positive anode, also becoming neutral atoms. Thus, at both electrodes, ions are changed to free atoms. Since a change in charge, and consequently a change in oxidation number, takes place, all these reactions are of the oxidation–reduction type. *Oxidation takes place at the anode. Reduction takes place at the cathode.*

As an example, let's start with a solution of hydrochloric acid, HCl. In solution, the HCl ionizes:

$$HCl \rightarrow H^+ + Cl^-$$

In the electrolysis cell, the H^+ ions migrate toward the cathode, and the Cl^- ions migrate toward the anode. At the cathode, each hydrogen ion picks up an electron from the negatively charged cathode and becomes a hydrogen atom. Pairs of hydrogen atoms then unite to form molecules of hydrogen gas, H_2, which bubbles out of solution: The equation is

$$2\ H^+ + 2\ e^- \rightarrow H_2$$

Each chloride ion loses its electron to the anode, and pairs of chlorine atoms unite to form chlorine gas, Cl_2, which bubbles out of solution. The equation is

$$2\ Cl^- \rightarrow Cl_2 + 2\ e^-$$

The net effect is the decomposition of HCl into hydrogen and chlorine gases. The overall equation is

$$2\ HCl(aq) \rightarrow H_2(g) + Cl_2(g)$$

Note that during electrolysis the solution remains electrically neutral. Molecules of chlorine and hydrogen gas form in equal number. The electrolysis continues until the electrolyte is decomposed completely.

In an electrolysis cell, chemical reactions take place at the electrodes. The positive cations combine with electrons from the negative cathode and become neutral atoms. At the anode, the negative anions yield their electrons to the positive anode, also becoming neutral atoms. Since a change in charge, and consequently a change in oxidation number, takes place, all these reactions are of the oxidation–reduction type.

One other feature of this electrolysis is that water molecules may also be oxidized at the anode, along with the chloride ions. This introduces another possible anode reaction into the system:

$$2\ H_2O(l) \rightarrow O_2(g) + 4\ H^+ + 4\ e^-$$

This anode reaction requires more electrical energy than does the oxidation of the chloride ion. Thus, when large numbers of chloride ions are present, the chloride ions are selectively oxidized, and the second reaction does not occur. However, as the concentration of chloride ions decreases near the end of electrolysis, some oxygen forms and contaminates the chlorine gas.

Now, consider the electrolysis of a solution of copper sulfate, $CuSO_4$. In solution, $CuSO_4$ is ionized:

$$CuSO_4(aq) \rightarrow Cu^{2+} + SO_4^{2-}$$

The solvent, water, is also ionized, so its ions are also present in the solution. The following reactions are possible:

Cathode:	*Anode:*
$Cu^{2+} + 2\ e^- \rightarrow Cu$	$SO_4^{2-} \rightarrow 2\ e^- + SO_4$
$H^+ + e^- \rightarrow H$	$4\ OH^- \rightarrow 2\ H_2O + O_2 + 4\ e^-$

The reaction that occurs at each electrode depends on which requires the smaller amount of energy. At the cathode, it takes less energy to reduce the copper ions than to reduce hydrogen ions. Therefore, the copper deposits onto the cathode. Similarly, it takes less energy at the anode to oxidize the hydroxide ions than the sulfate ions. Therefore, oxygen gas forms at the anode. We can write the overall reaction for this electrolysis as

$$2\ CuSO_4(aq) + 2\ H_2O(l) \rightarrow 2\ Cu(s) + O_2(g) + 2\ H_2SO_4(aq)$$

Faraday's Laws

Michael Faraday, while studying electrolysis, calculated a predictable, quantitative relationship between the amounts of elements formed and the amount of electricity used. In 1833, he formulated his discoveries into two laws that bear his name:

1. The mass of a given element liberated at an electrode during electrolysis is directly proportional to the quantity of electricity that passes through the solution.

2. When the same quantity of electricity passes through solutions of different electrolytes, the weights of the substances liberated at the electrodes are directly proportional to their equivalent masses (Chapter 6).

The quantity of electricity required to deposit one equivalent mass of any element is 96,500 coulombs. That quantity of electricity is known as a *faraday* (F). The following relationships are useful in solving electrolysis problems .

$$\text{Number of coulombs} = \text{Number of amperes} \times \text{Time in seconds}$$

$$\text{Number of faradays} = \frac{\text{Number of coulombs}}{96{,}500}$$

Let's work an example of a problem that applies Faraday's laws.

EXAMPLE 1: A current of 20 A passes through a solution of $CuSO_4$ for 2 hours. What weight of copper will be deposited?

SOLUTION: The atomic mass of copper is 63.55. The oxidation number of copper in $CuSO_4$ is +2. Therefore, the equivalent mass of copper is (63.55 g/mol)/(2 equiv/mol) = 31.78 g/equiv. The time in seconds is 2 h × 60 min/h × 60 s/min = 7200 s. The number of coulombs is 20 A × 7200 s = 144,000 C. The number of F is 144,000/96,500 = 1.492 F. The mass of copper deposited is 31.78 g/equiv × 1.492 F = 47.42 g.

NOTE: Each of these steps can be combined, and a single formula can be developed:

$$\text{Mass of deposit} = \frac{\text{Equivalent mass} \times \text{amperes} \times \text{Times of seconds}}{96{,}500}$$

Applications

Practical applications of electrolysis fall into three categories: analytical chemistry, production chemistry, and electroplating. Analytical methods help chemists identify unknown substances or measure the amounts of known substances. The value of electrolysis in performing such analyses arises from two of its characteristics. First, charged ions in solution carry electricity as they move to the electrodes. The more ions present, the more conductive the solution. Second, in many reactions of ions in solution, the total number of ions decreases as the reaction progresses. Determining the conductivity of a solution or the number of ions present makes identifications and measurements of unknowns possible.

For example, if a solution of sulfuric acid is added to a solution of barium hydroxide, the products formed are water (a weak electrolyte) and barium sulfate (an insoluble substance). The reaction in ionic form is

$$(2\ H^+, SO_4^{2-}) + (Ba^{2+}, 2\ OH^-) \rightarrow BaSO_4(s) + 2\ H_2O(l)$$

In reactions like this, when equivalent amounts of the two electrolytes react, the number of ions in solution falls to nearly zero. As the acid is added to the base the conductivity of the solution diminishes. It reaches a minimum at the equivalence point (the point at which one substance has been reacted with an equivalent amount of another substance) and increases again as excess acid is added. From the volume of standard acid required to reach the point of minimum conductivity with a measured volume of unknown base, the concentration of the unknown base can be computed. Other analytical methods are based on finding the weight of an element liberated by a known current in a measured amount of time. The equivalent mass of the unknown metal then is computed, which in turn can lead to its identification.

The most important economic application of electrolysis is production chemistry. Many valuable elements are produced commercially from electrolysis cells, such as hydrogen, oxygen, sodium, potassium, magnesium, calcium, and aluminum. Many metals, such as copper, zinc, and silver, can be purified by electrolysis as the final step in their production. In addition, many important chemical compounds are obtained as by-products of electrolysis.

In electroplating, the anode is a block of the metal to be plated, and the electrolyte is a salt of that same metal. The object to be plated is made the cathode and immersed in the electrolyte solution. The anode goes into solution, and pure metal is deposited on the cathode as the electroplating process proceeds. For example, in copper plating, copper is used as the anode, and copper sulfate solution is the electrolyte. The possible anode reactions, then, are

$$SO_4^{2-} \rightarrow 2\ e^- + SO_4$$
$$4\ OH^- \rightarrow 4\ e- + 2\ H_2O + O_2$$
$$Cu \rightarrow 2\ e^- + Cu^{2+}$$

Because the last reaction requires the least energy, it is the only one to occur. Copper ions are replaced in the solution as fast as they are deposited on the cathode, so the concentration of copper ions in the electrolyte remains constant. The mass of metal plating deposited on the cathode equals the mass of metal that dissolves from the anode.

ACTIVITY SERIES OF METALS

Let's look now at the electrical effects accompanying the chemical change, beginning with the behavior of metals toward an acid like

Occurrence	Metal	Reactivity to H_2O	Reactivity to HCl, dil. H_2SO_4	Reactivity to HNO_3, conc. H_2SO_4
Never found free in nature	K-Potassium Ca-Calcium Na-Sodium Mg-Magnesium	React with cold water to give H_2	Explosive action	Extremely explosive
	Al-Aluminum Mn-Manganese Zn-Zinc Cr-Chromium Fe-Iron	Much less active, hot metal gives H_2 with steam	Liberate hydrogen gas	Liberate gas OTHER THAN HYDROGEN
Rarely found free in nature	Fe-Iron Co-Cobalt Ni-Nickel Sn-Tin Pb-Lead	Very poor activity with steam	Slow action Very slow action	NO, NO_2, or NH_3 from HNO_3 SO_2 or H_2S from conc. H_2SO_4
	H-HYDROGEN			
Often found free in nature	Sb-Antimony Bi-Bismuth Cu-Copper Hg-Mercury Ag-Silver	Inactive with water, no gas of any kind liberated	Inactive with these acids	Action decreases as we go down list
Found free in nature	Pt-Platinum Au-Gold			Inactive

Table 11.1 Activity Series of Metals

HCl. Some metals, when added to a water solution of this acid, will dissolve in the acid and liberate hydrogen gas from the solution. The rate of the reaction varies. For example, sodium metal reacts explosively if dropped in the acid solution. Zinc causes a vigorous but quiet evolution of hydrogen gas. Lead liberates hydrogen only very slowly, and at temperatures well above room temperature

Some other metals, including copper, silver, and gold, are different. The acid does not attack them, and they do not liberate hydrogen from the solution.

Table 11.1, *the activity series of metals,* lists metals in their order of reactivity with water and some acids—with the most reactive metal at the top. The table reveals another important principle: *a metal higher on the list will replace the ions of metals lower on the list from solution; and it will liberate the lower metal in the free state.* For example, if iron nails are dropped into a solution of copper sulfate, $CuSO_4$, the iron (higher on the list) will dissolve, and copper metal (lower on the list) will precipitate from the solution. The reaction is

$$Fe(s) + Cu^{2+} \rightarrow Fe^{2+} + Cu(s)$$

EXPERIMENT 21: From a hobby shop or chemical supply house obtain 1 oz (28 g) of copper sulfate crystals. Dissolve about 3 g of these crystals in a small glass of water. Thoroughly clean a nail with sandpaper. Place the nail in the copper sulfate solution. Let it stand overnight. You will observe that some copper plates onto the surface of the nail and some falls to the bottom of the glass.

In contrast, if a strip of copper metal is placed in a solution of iron (II) (ferrous) sulfate, $FeSO_4$, no reaction takes place. Thus, the activity series of metals allows us to predict whether a particular reaction will take place. By the way, this precipitation of less active metals by more active ones is a process frequently used in extracting metals from their ores.

All reactions based on the activity series of metals are oxidation–reduction reactions (Chapter 10). To find out what makes this type of reaction take place, let's analyze the following reaction:

$$Fe(s) + Cu^{2+} \rightarrow Fe^{2+} + Cu(s)$$

Think of this reaction as occurring in two steps:

$$Fe \rightarrow Fe^{2+} + 2\ e^{-}$$
$$Cu^{2+} + 2\ e^{-} \rightarrow Cu$$

The two parts are half-reactions (Chapter 10). The first half-reaction shows iron metal giving up 2 electrons and becoming iron(II) (ferrous) ions in solution. The second half-reaction shows copper ions combining with 2 electrons to become atoms of metallic copper.

NOTE: As soon as the full reaction begins, the system contains *both metals and both ions at the same time.* Yet, the reaction continues in only one direction; that is, the iron dissolves, and the copper precipitates. This happens because *electrons continue to flow from the iron metal to the copper ions.* No electrons flow from the copper metal to either of the ions, nor do any electrons flow from the iron metal to the iron(II) (ferrous) ions. In summary, iron loses electrons more easily than copper, and it loses them specifically to the copper ions.

For electrons to flow, thereby creating an electric current, an electromotive force (emf) must push the electrons along their path. An emf is a difference in electrical potential. An emf exists between all oppositely charged bodies and between a charged body and a neutral one. Thus, there is an emf between each metal and each of the ions in the example reaction; however, since the electrons actually flow from the iron metal to the copper ions, the largest emf must be between these two particles.

The emfs associated with a large number of oxidation–reduction half-reactions have been measured. Some of them are listed in Table 11.2.

Notice that in each of the half-reactions, electrons are *being accepted*. Each emf listed is a *reduction potential* (that is, for the reduction reaction in which an element gains electrons). All reactions are shown in terms of the reduction reaction with the standard hydrogen electrode as a reference point and assigned an emf of 0.00 volts. The more positive the value of the voltage associated with the half-reaction (emf), the more readily the reaction occurs.

> The activity series of metals lists metals in their order of reactivity with water and some acids—with the most reactive metal at the top. A metal higher on the list will replace the ions of metals lower on the list from solution; and it will liberate the lower metal in the free state.

Table 11.2 Half-Reaction (Electrode) Reduction Potentials

Reduction Half-Reaction	Voltage (emf)
$F_2 + 2\,e^- \rightarrow 2\,F^-$	+2.87
$SO_4 + 2e^- \rightarrow SO_4^{2-}$	+1.90
$BrO_4^- + 2\,H^+ + 2\,e^- \rightarrow BrO_3^- + H_2O$	+1.74
$PbO_2 + 4\,H^+ + SO_4^{2-} + 2\,e^- \rightarrow PbSO_4 + 2\,H_2O$	+1.70
$IO_4^- + 2\,H^+ + 2\,e^- \rightarrow IO_3^- + H_2O$	+1.65
$MnO_4^- + 8\,H^+ + 5\,e^- \rightarrow Mn^{2+} + 4\,H_2O$	+1.491
$PbO_2 + 4\,H^+ + 2\,e^- \rightarrow Pb^{2+} + 2\,H_2O$	+1.455
$Ce^{4+} + 1\,e^- \rightarrow Ce^{3+}$	+1.4430
$Au^{3+} + 3\,e^- \rightarrow Au$	+1.42
$Cl_2 + 2\,e^- \rightarrow 2\,Cl^-$	+1.359
$Cr_2O_7^{2-} + 14\,H^+ + 6\,e^- \rightarrow 2\,Cr^{3+} + 7\,H_2O$	+1.33

Table 11.2 *(Continued)*

Reduction Half-Reaction	Voltage (emf)
$O_2 + 4\,H^+ + 4\,e^- \rightarrow 2\,H_2O$	+1.229
$Br_2(l) + 2\,e^- \rightarrow 2\,Br^-$	+1.087
$VO_2^+ + 2\,H^+ + 1\,e^- \rightarrow VO^{2+} + H_2O$	+0.9994
$AuCl_4^- + 3\,e^- \rightarrow Au + 4\,Cl^-$	+0.99
$N_2O_4 + 2\,e^- \rightarrow 2\,NO_2^-$	+0.88
$HNO_2 + 7\,H^+ + 6e^- \rightarrow NH_4^+ + 2\,H_2O$	+0.86
$Ag^+ + 1\,e^- \rightarrow Ag$	+0.7994
$Hg_2^{2+} + 2\,e^- \rightarrow Hg_2$	+0.7961
$Fe^{3+} + 1\,e^- \rightarrow Fe^{2+}$	+0.771
$4\,MnO_2 + 4\,NH_4^+ + Zn^{2+} + 4\,e^- \rightarrow 2\,Mn_2O_3 + Zn(NH_3)_4^{2+} + 2\,H_2O$	+0.738
$ClO_2^- + H_2O + 2e^- \rightarrow ClO^- + 2\,OH^-$	+0.66
$BrO_3^- + 3\,H_2O + 6\,e^- \rightarrow Br^- + 6\,OH^-$	+0.61
$ClO_4^- + 4\,H_2O + 8\,e^- \rightarrow Cl^- + 8\,OH^-$	+0.56
$I_2(s) + 2\,e^- \rightarrow 2\,I^-$	+0.5355
$Cu^+ + 1\,e^- \rightarrow Cu$	+0.521
$NiO_2 + 2\,H_2O + 2\,e^- \rightarrow Ni(OH)_2 + 2\,OH^-$	+0.49
$2\,BrO^- + 2\,H_2O + 2e^- \rightarrow Br_2 + 4\,OH^-$	+0.45
$O_2 + H_2O + 4\,e^- \rightarrow 4\,OH^- -$	+0.401
$Cu^{2+} + 2\,e^- \rightarrow Cu$	+0.337
$HAsO_2 + 3\,H^+ + 3\,e^- \rightarrow As(s) + 2\,H_2O$	+0.2475
$Cu^{2+} + 1e^- \rightarrow Cu^+$	+0.153
$Sn^{4+} + 2\,e^- \rightarrow Sn^{2+}$	+0.15
$2\,H^+ + 2\,e^- \rightarrow H_2$	+0.0000

Table 11.2 ***(Continued)***

Reduction Half-Reaction	Voltage (emf)
$Fe^{3+} + 3\ e^- \rightarrow Fe$	−0.036
$Pb^{2+} + 2\ e^- \rightarrow Pb(s)$	−0.1263
$Sn^{2+} + 2\ e^- \rightarrow Sn$	−0.1364
$Ni^{2+} + 2\ e^- \rightarrow Ni$	−0.23
$Co^{2+} + 2\ e^- \rightarrow Co$	−0.28
$PbSO_4 + 2\ e^- \rightarrow Pb + SO_4^{2-}$	−0.356
$Tl^+ + 1\ e^- \rightarrow Tl(s)$	−0.3363
$Cd^{2+} + 2\ e^- \rightarrow Cd(s)$	−0.4026
$Cr^{3+} + 1\ e^- \rightarrow Cr^{2+}$	−0.41
$Fe^{2+} + 2\ e^- \rightarrow Fe$	−0.4402
$[Au(CN)_2]^- + 1\ e^- \rightarrow Au + 2\ CN^-$	−0.50
$2\ SO_3^{2-} + 3\ H_2O + 4\ e^- \rightarrow S_2O_3^{2-} + 6\ OH^-$	−0.58
$Cr^{3+} + 3\ e^- \rightarrow Cr$	−0.74
$Cd(OH)_2 + 2\ e^- \rightarrow Cd + 2\ OH^-$	−0.761
$Zn^{2+} + 2\ e^- \rightarrow Zn$	−0.7628
$2\ H_2O + 2\ e^- \rightarrow H_2 + 2\ OH^-$	−0.8277
$Fe(OH)_2 + 2\ e^- \rightarrow Fe + 2\ OH^-$	−0.877
$Cr^{2+} + 2\ e^- \rightarrow Cr(s)$	−0.91
$Mn^{2+} + 2\ e^- \rightarrow Mn$	−1.185
$Al^{3+} + 3\ e^- \rightarrow Al$	−1.66
$Mg^{2+} + 2\ e^- \rightarrow Mg$	−2.37
$Na^+ + e^- \rightarrow Na$	−2.711
$Nb_2O_5(s) + 10\ H^+ + 10\ e^- \rightarrow 2\ Nb(s) + 5\ H_2O(l)$	−2.711
$Ca^{2+} + 2\ e^- \rightarrow Ca$	−2.87
$K^+ + e^- \rightarrow K$	−2.924
$Li^+ + 1\ e^- \rightarrow Li$	−3.045

As you saw in Chapter 10, an oxidation–reduction reaction is a combination of two half-reactions. In the full reaction, the half-reaction with the more positive emf will proceed from left to right as written in Table 11.2. It is called the reducing half-reaction, and the substance losing the electrons is the reducing agent in the full reaction. The half-reaction with the more negative emf potential will run in the opposite direction, or proceed from right to left as written in Table 11.2. It is called the oxidizing half-reaction, and the substance combining with the electrons is the oxidizing agent in the full reaction. Substances on the right side of the equations in Table 11.2 are reducing agents in order of increasing emf. Substances on the left side are oxidizing agents in order of decreasing emf. The strength of the oxidizing agent *increases* as the emf value becomes more positive, and the strength of the reducing agent *increases* as the emf value becomes more negative. Thus, elemental lithium is the most powerful reducing agent listed, and gaseous fluorine, F_2, is the most powerful oxidizing agent listed.

The total emf driving the full reaction to take place is the algebraic difference between the emfs of the two half-reactions. Let's see how this works for the iron–copper system.

From Table 11.2:

$$Fe^{2+} + 2\ e^- \rightarrow Fe \quad (-0.440\ V)$$
$$Cu^{2+} + 2\ e^- \rightarrow Cu \quad (+0.337\ V)$$

The copper half-reaction has the more positive emf; therefore, it proceeds left to right as written. Since the iron half-reaction has a more negative emf, it proceeds in the opposite direction, from right to left. As a rule, we first write the two half-reactions in the direction they actually proceed. A reversal of a reduction half-reaction makes a negative value positive:

$Fe \rightarrow Fe^{2+} + 2\ e^-$
(Oxidation half-reaction, V = +0.440)

$Cu^{2+} + 2\ e^- \rightarrow Cu$
(Reduction half-reaction, V = +0.337)

Adding:

$Fe + Cu^{2+} \rightarrow Fe^{2+} + Cu$ (Full reaction)

The electrons cancel out in the addition. The total emf for this reaction is

$$+0.440\ V + (+0.337)\ V = 0.777\ V$$

The size of the total emf indicates the vigor of the reaction. Substances close together in Table 11.2 are powered by only a small total emf; consequently, they react only slowly and mildly. Substances far apart in Table 11.2 are driven by a larger total emf; they react more rapidly and vigorously.

TRIVIA: Silver-plated objects often can be cleaned of their tarnish by boiling them in an aluminum pan with a little detergent added. A redox reaction occurs in which the silver sulfide is reduced to silver metal.

BATTERIES

The flow of electrons in oxidation–reduction reactions presents an interesting possibility. If the flow of electrons can be harnessed in some way and caused to pass through wires, we can obtain useful electrical energy from chemical action. The result is the handy device commonly known as the battery.

Voltaic Cells

Figure 11.2 illustrates one method of harnessing the electricity of a chemical change. The arrangement is known as a *voltaic cell*, sometimes called a *galvanic cell*.

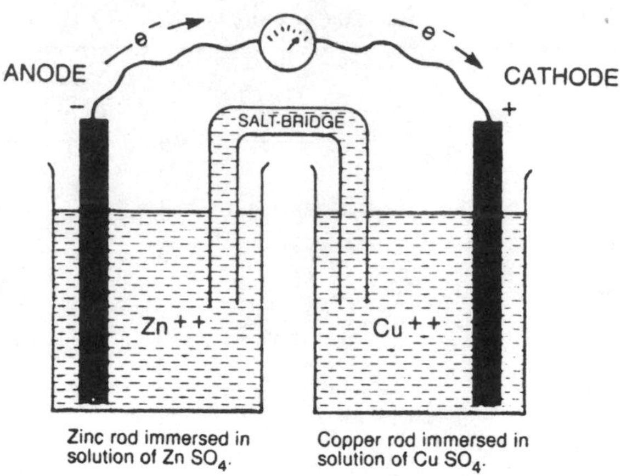

Figure 11.2 Voltaic Cell

The basis of a voltaic cell is solutions of two different ions. Strips of metal of the same element as the positive ions of the solution are immersed into each solution, forming the electrodes. A salt bridge, which is a concentrated solution of a third electrolyte that does not react with either of the two original solutions, is placed between the two solutions. Charged ions from the solution in the salt bridge migrate into the solutions, keeping them electrically neutral. A wire would not work in place of the bridge, for it would not donate charged ions to the solutions. Finally, when a wire connects the two strips of metal, a current flows through the circuit. The emf or voltage of the cell depends on the substances used.

In Figure 11.2, the two half-reactions are

$$Zn^{2+} + 2\ e^- \rightarrow Zn \quad (-0.763\ V)$$
$$Cu^{2+} + 2\ e^- \rightarrow Cu \quad (+0.337\ V)$$

Because the copper half-reaction is more positive, it proceeds as written. The zinc half-reaction proceeds in the opposite direction: zinc atoms in the zinc rod go into solution as ions. The electrons the zinc atoms give up move from the zinc rod, through the external circuit, to the copper rod. There they combine with copper ions and become copper atoms. The copper atoms are deposited on the rod and form a copper plate. The total reaction is

$$\begin{array}{ll} Zn \rightarrow Zn^{2+} + 2\,e^- & (+0.763\text{ V}) \\ \underline{Cu^{2+} + 2\,e^- \rightarrow Cu} & (+0.337\text{ V}) \\ Zn + Cu^{2+} \rightarrow Zn^{2+} + Cu & \end{array}$$

The total voltage (emf) is

$$+0.763\text{ V} + +0.337\text{ V} = 1.100\text{ V}$$

Many other combinations of half-reactions from Table 11.2 can be arranged into voltaic cells of the type illustrated in Figure 11.2.

Dry Cells

The common dry cell is another device that generates electrical energy from chemical action. Figure 11.3 shows a cross section of a dry cell. The dry cell is misnamed, because it actually contains water. Ions migrate through the water in the paste, and a dry cell will not work without water.

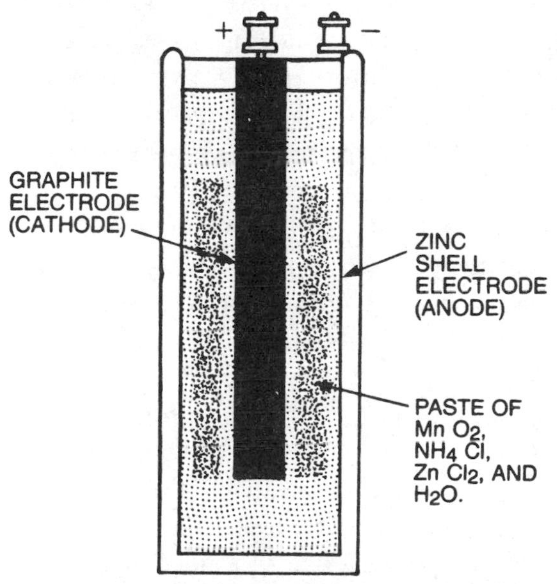

Figure 11.3 Dry Cell

The half-reactions in the dry cell may be written as follows:

$$Zn^{2+} + 2\,e^- \rightarrow Zn \qquad (-0.763\text{ V})$$

$$4\,MnO_2 + 4\,NH_4^+ + Zn^{2+} + 4\,e^- \rightarrow 2\,Mn_2O_3 + Zn(NH_3)_4^{2+} + 2\,H_2O \qquad (+0.738\text{ V})$$

The total emf of the dry cell is

$$+0.763\text{ V} + (+0.738\text{ V}) = 1.5\text{ V}$$

The second half-reaction, which proceeds from left to right, actually takes place in steps as follows:

Step 1. The electrons from the zinc shell are absorbed by the ammonium ions in the paste:

$$2\,NH_4^+ + 2\,e^- \rightarrow 2\,NH_3 + H_2$$

Step 2. The hydrogen gas thus formed accumulates around the graphite electrode. Because the electrode potential of hydrogen is intermediate between the emfs of the two half-reactions of the dry cell (see Table 11.2), it tends to set up its own voltaic action within the cell. This reaction greatly reduces the voltage of the dry cell. The phenomenon is known as *polarization*. To offset this tendency, manganese dioxide, MnO_2, is included in the paste. The MnO_2 oxidizes the hydrogen gas as it is formed, acting as a *depolarizer*:

$$2\,MnO_2 + H_2 \rightarrow Mn_2O_3 + H_2O$$

Step 3. The ammonia gas formed in step 1 reacts with the zinc ions from the zinc shell:

$$Zn^{2+} + 4\,NH_3 \rightarrow Zn(NH_3)_4^{2+}$$

This third step keeps the concentration of zinc ions in the paste constant and equal to the amount originally added to the paste in the form of zinc chloride, $ZnCl_2$.

The depolarization effect of MnO_2 happens rather slowly. If a dry cell is used to produce a great amount of electric current in a short time, hydrogen gas accumulates faster than the MnO_2 can react with. Consequently, the cell polarizes and its voltage diminishes. If the cell sits idle for a while, the continuing action of the MnO_2 in oxidizing the hydrogen gas restores its voltage.

Lead Storage Cells

Another important device for producing electric current from chemical action is the lead storage cell. It consists of two lead gratings immersed in a solution of sulfuric acid, H_2SO_4. One grating is impregnated with spongy lead to provide a large surface area for reaction. The other grating is impregnated with lead dioxide, PbO_2, which serves as the second electrode. The half-reactions of the lead storage cell are

$$PbSO_4 + 2\ e^- \rightarrow Pb + SO_4^{2-} \qquad (-0.356\ V)$$

$$PbO_2 + 4\ H^+ + SO_4^{2-} + 2\ e^- \rightarrow PbSO_4 + 2\ H_2O \qquad (+1.70\ V)$$

The total emf of the cell is

$$+0.356 + (1.70) = 2.056\ V$$

An automobile battery has six of these cells in series producing slightly over 12 V.

The outstanding feature of the lead storage cell is that it can be recharged. The two half-reactions are reversible. During the discharge of electricity from the cell, the two half-reactions proceed as follows:

(At anode) $Pb + SO_4^{2-} \rightarrow PbSO_4 + 2\ e^-$

(At cathode) $PbO_2 + 4\ H^+ + SO_4^{2-} + 2\ e^- \rightarrow PbSO_4 + 2\ H_2O$

(Total) $Pb + PbO_2 + 4\ H^+ + 2\ SO_4^{2-} \rightarrow 2\ PbSO_4 + 2\ H_2O$

Note that both half-reactions produce lead sulfate, $PbSO_4$, which deposits as crystals on both gratings in the cell. During the recharge process, a source of electrical energy with an emf greater than the voltaic emf of the cell is attached to the cell, and electrical energy is forced back into it. Each of the half-reactions proceeds from right to left under the influence of this external emf. The $PbSO_4$ crystals dissolve, and Pb, PbO_2, and H_2SO_4 are regenerated in the cell (see Figure 11.4).

Another interesting feature of the lead storage cell is that its state of charge can readily be determined. In the discharge process, H_2SO_4, the electrolyte in the cell, is used up. During the charging process, it is regenerated. Sulfuric acid is a very dense liquid, and the specific gravity of the solution of electrolyte decreases as the cell is in use. At full charge, the specific gravity of the electrolyte is 1.28. When the cell is discharged, the specific gravity may drop to as low as 1.15. A *hydrometer*, which measures the specific gravity of liquids, reveals how fully charged or discharged the cell may be.

The crystals of lead sulfate may cause a lead storage cell to "go dead" if they grow too large or cover the entire electrode. Factors promoting the growth and deposition of these crystals are excessive heat, water loss, or a rapid "drain" on the cell when too much electrical energy is drawn from it.

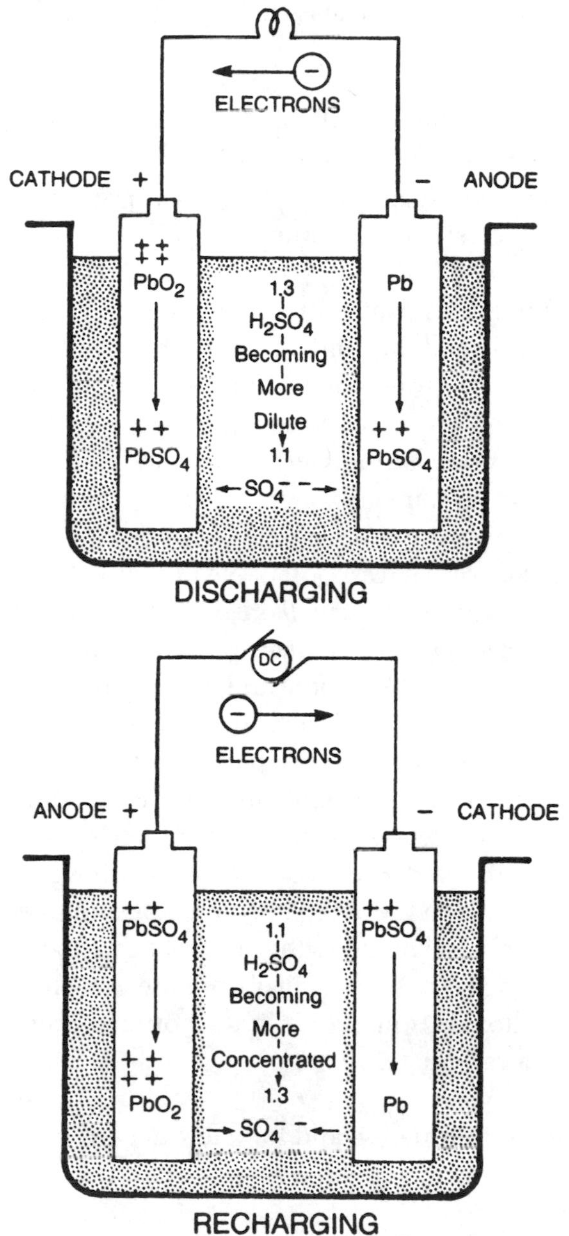

Figure 11.4 The Storage Battery

Scientists agree that so-called battery additives, which usually consist of mixtures of sodium and magnesium sulfates (Glauber's salt and Epsom salts), fail to prolong the life of a lead storage cell, since failure often is due to mechanical reasons such as hitting the curb too often and too hard.

When a lead storage battery is charged, the process of electrolysis breaks water down into hydrogen and oxygen gas. In older models of lead storage batteries, this water had to be replenished from time to time. Modern, maintenance-free lead storage batteries require no added water. They are made so that the negative plate has more capacity than the positive plate; thus, oxygen is released from the positive plate before hydrogen is formed on the negative plate. The oxygen travels to the spongy lead negative plate, where these reactions occur:

$$2\,Pb + O_2 \rightarrow 2\,PbO$$

$$PbO + H_2SO_4 \rightarrow PbSO_4 + H_2O$$

Thus, the water is re-formed and remains in the battery.

Nickel-Cadmium Cells

In the nickel-cadmium (also called nicad) cell, the anode is cadmium metal. The cathode contains nickel oxide, NiO_2. The half-reactions in a solution of potassium hydroxide are

(At anode) $Cd + 2\,OH^- \rightarrow Cd(OH)_2 + 2\,e^-$ $(+0.761\ V)$

(At cathode) $NiO_2 + 2\,H_2O + 2\,e^- \rightarrow Ni(OH)_2 + 2\,OH^-$ $(+0.49\ V)$

(Total) $Cd + NiO_2 + 2\,H_2O \rightarrow Cd(OH)_2 + Ni(OH)_2$ $(1.251\ V)$

The $Cd(OH)_2$ and $Ni(OH)_2$ are formed on the Cd and NiO_2 electrodes. Therefore, unlike dry cells, nickel-cadmium cells can be recharged and used repeatedly. In addition, unlike lead storage batteries, nicad cells can be made in a very small size, so that they are especially useful for small, portable electronic equipment such as hand calculators.

CORROSION

Not all the effects of voltaic action are useful. Tiny voltaic cells that form in metals are responsible for *corrosion,* the redox process that oxidizes metals, forming oxides and sulfides. The economic loss of corrosion totals billions of dollars each year. All commercial iron and steel, for example, contain impurities, principally carbon. These impurities are not uniformly distributed in the metal but are segregated at various points. A difference in electrical potential exists between the atoms of the metal and the atoms of the impurities. When the metal is in contact with moist air, a film of water forms on the surface of the metal. Carbon dioxide, CO_2, in the air dissolves in the water and forms a solution of an electrolyte, carbonic acid (H_2CO_3), through which ions can migrate. Because iron is more electropositive than its impurities, iron atoms act as the anode in the cell. They become oxidized and go into solution as ions. The liberated electrons pass through the metal to the atoms of the impurity, which act as a cathode. Hydrogen ions from the water migrate to the cathode, pick up electrons and form hydrogen atoms, and are oxidized by atmospheric oxygen to water. This reaction prevents polarization of the cell and permits the action to continue. The iron ions are further oxidized by atmospheric oxygen and react with the water to form the complex hydrated oxide $Fe_2O_3 \cdot (H_2O)_x$, commonly known as *rust.* Perfectly dry iron does not rust because voltaic action cannot take place in the absence of water. Much the same process is involved when a silver-plated fork tarnishes when used with eggs, except silver sulfide is formed instead.

EXPERIMENT 22: Thoroughly clean four nails with sandpaper. Then:

(a) Place the first in a glass of tap water so that the nail is completely covered with water, then cover the glass to prevent excessive evaporation.

(b) Place the second nail in a glass of water so that the nail is only partly covered with water. Again, cover the glass to prevent excessive evaporation.

(c) Boil some water for 5 minutes in a Pyrex dish, add the third nail, and continue to boil the water for a few more minutes. Then, remove the flame and quickly pour melted wax or Vaseline over the water to exclude air.

(d) Dissolve a teaspoon of lye in a cup of water. *Be very careful with this basic solution. Wear safety goggles and rubber gloves.* Put the solution in a soda bottle and cork the bottle. Shake the solution in the corked bottle thoroughly and then permit it to stand overnight. Boil a nail in water as you did in (c). Let the nail cool, then force it through the cork so that part of it is exposed to the air inside the bottle and part of it is exposed to the air outside the bottle. The lye solution will absorb the carbon dioxide from the air in the bottle, forming a carbonate.

Let all four nails stand for a few days, then observe the following:

Rusting in (a) where there is much water and little air (dissolved in the water).

Rusting in (b) where there is little water and much air.

No rusting in (c), where there is no air. (The boiling removed the air.)

No rusting inside the bottle in (d), where there is no CO_2, but rusting outside the bottle.

EXPERIMENT 23: From a hobby store or a chemical supply house obtain a few crystals of potassium ferricyanide. *Wear your safety goggles*. Heat 1 cup of water to boiling; add $1\frac{1}{2}$ teaspoons of clear gelatin, $\frac{1}{8}$ teaspoon of table salt, $\frac{1}{8}$ teaspoon of potassium ferricyanide crystals, and 8–10 drops of purple cabbage indicator (prepared in EXPERIMENT 5). Pour this solution over a thoroughly cleaned nail in a Pyrex dish. Let stand overnight. You will observe the nail being corroded. Where the iron goes into solution, deep blue iron(II) (ferrous) ferricyanide forms, indicating the anodes. Other parts of the nail act as cathodes, and hydroxide ions concentrate around them. This will cause the purple cabbage indicator to turn blue-green to green in those areas.

The battle against the corrosion of metals is a huge enterprise today. A number of methods are used, all designed to prevent or overcome localized voltaic action:

1. Progress has been made in fabricating more homogenous metals. When impurities are highly segregated, the cells can work in series and build up appreciable currents. Uniform distribution of impurities diminishes this effect and minimizes the corrosive action of the tiny voltaic cells.

2. Another method involves coating the metal with a film that prevents contact between the metal and moisture in the air, thus preventing voltaic action from occurring. Various types of anticorrosive paints are used, and they are effective as long as the film remains intact; however, if the metal is subjected to extreme temperature changes, it expands and contracts more than the coating, especially if the paint is dried out. The coating cracks and voltaic action begins. New paints are continually being developed to provide better surface protection.

3. Another anticorrosive process is galvanizing. In this process, iron is coated with a layer of zinc. Zinc is above iron in the activity series, but it oxidizes only superficially in the atmosphere. A thin film of zinc oxide, ZnO, forms as a surface layer. This film is so cohesive that it cannot be further penetrated by oxygen. Thus, the zinc protects the iron. The protection continues even when the zinc coating is broken. Moisture and carbon dioxide penetrate any crack that develops in the coating. They form a voltaic cell with the zinc and iron. Since zinc is more electropositive, it passes into solution as zinc ions. They react with the hydroxide ions of water to form zinc hydroxide, $Zn(OH)_2$; which, in turn, combines with the dissolved carbon dioxide to form a basic zinc carbonate, $Zn_2(OH)_2CO_3$. That compound, which is very soluble, forms a tight film similar to the ZnO film. It too is impervious to water and atmospheric gases. Thus, the crack is plugged, and the iron remains protected.

4. Another method is tin-plating. Tin, like zinc, oxidizes only superficially in the atmosphere. As long as the coat remains intact, the iron is adequately protected; but if a crack develops in the tin plate, moisture and carbon dioxide enter the crack and form a voltaic cell with tin and iron. Because iron is higher than iron in the activity series, it passes into solution by voltaic action. The tin then *accelerates* the corrosion of the iron. Tinplate is used in food containers rather than the more effective zinc because zinc may react with the food to produce poisonous compounds.

5. Iron and steel may also be protected from corrosion by a method that uses "sacrificial anodes." In this method, metals such as zinc or magnesium—called the *active metal*—are more electropositive than a *noble*

metal such as iron. When the active metal is wired to the noble metal, it exhibits a greater emf toward oxidizing agents. Consequently, it corrodes, and the noble metal is protected as long as any of the active metal remains. Thus, the active metal is "sacrificed" to protect the noble one. This method is particularly effective in protecting underground pipelines and underwater fittings, such as the propellers of ships.

The principle of selective corrosion of sacrificial anodes is similar to the situation we encountered earlier in this chapter in studying electrolysis. Recall that when more than one electrode reaction is possible, one preferentially occurs. A sacrificial anode is chosen based on the electrochemical consideration of Table 11.2 while keeping cost and other possible reactions in mind.

SUMMARY

Matter, being of electrical nature, is driven to undergo chemical change by either of two electrical phenomena: (1) forces of electrostatic attraction (ionic reactions, Chapter 9) or (2) the electromotive forces that drive redox reactions. Electrolysis is the utilization of electrical energy to produce chemical action. When an electric current passes through a solution of an electrolyte, the electrolyte is decomposed at the electrodes in an oxidation–reduction–type reaction. When more than one reaction is possible at an electrode, th=e one that requires the least amount of electrical energy takes place.

Faraday's laws reveal that the amount of a substance produced by the electrolysis of a substance depends on both the quantity of electricity used and the equivalent mass of the substance. The activity series of metals lists the metals in the order of their chemical reactivity in oxidation–reduction reactions. In solution, a metal higher on the list replaces the ions of metals lower on the list and liberates the lower metal in the free state. Oxidation–reduction reactions consist of two half-reactions: in one electrons are given off, and in the other, electrons combine with other species in solutions. An electromotive force accompanies each half-reaction. The voltage of the total reaction is the algebraic sum of the voltage of the two half-reactions.

A voltaic cell converts chemical energy into electrical energy. Several types of voltaic cells, such as dry cells, nickel–cadmium, and the lead storage cell, produce electrical energy in useful form.

The corrosion of metals is caused by the action of tiny voltaic cells that develop on the surface of the metal. The metal is oxidized by this action. Several methods of combating corrosion prevent or slow voltaic action on the surface of a metal.

PROBLEM SET NO. 13

1. In the electrolysis of a solution of sodium sulfate, Na_2SO_4, the following electrode reactions are possible:

   At anode:

   $SO_4^{2-} \rightarrow SO_4 + 2\,e^-$ (-1.9 V)

   $2\,H_2O \rightarrow O_2 + 4\,H^+ + 4\,e^-$ (-1.229 V)

   At cathode:

   $Na^+ + e^- \rightarrow Na$ (-2.711 V)

   $2\,H_2O + 2\,e^- \rightarrow H_2 + 2\,OH^-$ (-0.828 V)

(a) Which electrode reactions will occur in each case?

(b) What is the net reaction for the cell?

2. In the electrolysis of fused NaCl, the electrode reactions are

At anode: $2\ Cl^- \rightarrow Cl_2 + 2\ e^-$ $(-1.359\ V)$

At cathode: $Na^+ + e^- \rightarrow Na$ $(-2.711\ V)$

(a) What is the minimum voltage required of a battery to cause the electrolysis of this cell?

(b) Would a single dry cell produce sufficient voltage to cause this electrolysis?

(c) Would an automobile battery produce sufficient voltage to cause this electrolysis?

3. A current of 30 A is passed through a bath of fused calcium chloride for 1 hour. What weight of metallic calcium will be deposited on the cathode?

4. How long will it take a current of 20 A to deposit 40 g of metallic sodium from fused NaCl?

5. What volume of Cl_2 at standard conditions will be liberated by a current of 15 A passing through fused NaCl for 1 hour?

6. Which of the following reactions will take place? Write balanced equations for those that do take place.

(a) $Mg + NiCl_2$ (d) $Ag + HCl$

(b) $H_2 + AuCl_3$ (e) $Al + CuSO_4$

(c) $Cu + ZnCl_2$ (f) $Cu + AgNO_3$

7. A voltaic cell consists of aluminum metal in $Al(NO_3)_3$ solution joined to lead metal in $Pb(NO_3)_2$ solution.

(a) What two half-reactions are involved?

(b) What is the total voltage of the cell?

(c) What is the oxidizing half-reaction?

8. A voltaic cell consists of lead metal in $Pb(NO_3)_2$ solution joined to silver metal in $AgNO_3$ solution.

(a) What two half-reactions are involved?

(b) What is the total voltage of the cell?

(c) What is the reducing half-reaction?

9. In the Edison storage cell, the electrodes consist essentially of iron and nickel dioxide. The half-reactions are

$Fe(OH)_2 + 2\ e^- \rightarrow Fe + 2\ OH^-$ $(-0.877\ V)$

$NiO_2 + 2\ H_2O + 2\ e^-$
$\rightarrow Ni(OH)_2 + 2\ OH^-$ $(+0.49\ V)$

The electrolyte in this cell is a solution of potassium hydroxide, KOH.

(a) Which electrode is the anode?

(b) What is the total voltage of the cell?

10. For the balanced oxidation–reduction reaction

$Cr_2O_7^{2-} + 6\ Fe^{2+} + 14\ H^+$
$\rightarrow 2\ Cr^{3+} + 6\ Fe^{3+} + 7\ H_2O$

(a) Use Table 11.2 to write the half-reactions involved.

(b) What is the total emf (voltage) driving this reaction?

CHAPTER 12

THE ATMOSPHERE

KEY TERMS

atmosphere, relative humidity, dew point, fractional distillation, allotropic form, sublimation, combustion

Surrounding Earth is a sea of gas known as the *atmosphere* or air. The individual gases in the atmosphere are invisible, but we feel their presence when we swing our hand through air or breathe it deeply.

CHARACTERISTICS

The principal gases present in the atmosphere are nitrogen, oxygen, carbon dioxide, and water vapor. Except for water vapor, which varies considerably in amount, the atmosphere has a remarkably constant composition. Table 12.1 summarizes the composition of the atmosphere exclusive of water vapor.

Up to 50 miles (80 km) above the surface of the Earth, the composition of the air remains about the same, but the air is much thinner and contains more ozone, particularly above 45 miles (about 70 km).

EXPERIMENT 24: Cut a cork in half the long way. Drop wax from a birthday candle onto the flat cut surface of the cork and then stand a $\frac{1}{2}$-inch length of a small birthday candle in the wax. This is a "boat." Float it in a small pot of water, light the candle, and then cover the entire boat with a large glass so that the open end of the glass extends well down into the water.

You will observe that the water immediately rises inside the inverted glass, and that the candle goes out when about one-fifth the air in the glass is replaced with water. One-fifth of our atmosphere is oxygen, as was one-fifth of the air in the glass. When the oxygen was consumed as the candle burned, the space formerly occupied by the oxygen was then replaced by water, and the water level inside the inverted glass rose.

Substance	Percentage by Volume at Sea Level
nitrogen	78.08
oxygen	20.95
argon	0.93
carbon dioxide	0.03
neon	0.0018
helium	0.0005
methane	0.0002
krypton	0.00011
hydrogen	0.00005
nitrous oxide	0.00005
xenon	0.000009
ozone	trace
radon	trace

Table 12.1 Composition of Dry Air

In addition to water vapor, other variable components of the atmosphere are microorganisms, dust, pollen, ammonia, oxides of nitrogen and sulfur, hydrogen sulfide, ozone, hydrocarbons, and miscellaneous substances resulting from geologic, vegetative, and industrial processes.

The atmosphere is divided into three layers. That part of it in contact with the ground and extending upward about 6 miles is called the *troposphere.* About half the total weight of the atmosphere lies in the troposphere. All the water vapor in the atmosphere is found there. Because water vapor and weather are intimately related, all weather phenomena take place in the troposphere. Between 6 miles and 35 miles up lies the layer known as the *stratosphere.* Aircraft on long flights normally use the lower potion of the stratosphere because it is free from clouds, storms, lightning, thunder, and other forms of weather. Most meteorites burn up in the stratosphere. In the lower part of the stratosphere, about 12 to 15 miles up, ozone attains its maximum atmospheric concentration. This gas absorbs much harmful solar radiation. Above the stratosphere is an area in which the air mixture is extremely thin. This region is known as the *ionosphere.* The aurora borealis, or northern lights, stages its show in this region. In the Southern Hemisphere, the same phenomenon is called the aurora australis. In the ionosphere, matter is so highly energized by solar radiation that it ionizes. Several ionic layers exist, and they reflect radio waves back to the surface, thus making possible long-range radio transmission. The ionosphere may extend upward to 200 miles above Earth.

The physical properties of the constituents of air are listed in Table 12.2, except for those of the noble gases (see Table 4.1). Note that air has such a constant composition that it has some physical properties of its own. The molecular mass listed for air is the weight in grams of 22.4 L of air at standard conditions (0°C and 1 atm).

Water vapor may vary from a mere trace on a cold dry day to well over 7% on a hot, humid

Substance	**Formula**	**Molecular Mass, g/mol**	**Melting Point, °C**	**Boiling Point, °C**	**Solubility in Water (at 0°C), cm^3/100 ml**	**Density (at 0°C), g/L**
nitrogen	N_2	28.02	−209.86	−195.8	2.33	1.251
oxygen	O_2	32.00	−218.4	−183.0	4.89	1.429
carbon dioxide	CO_2	44.01	−56.6*	−78.5†	171.3	1.977
water vapor	H_2O	18.02	——	——	——	0.804
hydrogen	H_2	2.02	−259.14	−252.8	2.14	0.090
ozone	O_3	48.00	−192.5	−112	49	2.144
air	——	29.00	——	——	——	1.293

*At a pressure of 5.2 atm

†Sublimes (passes directly from solid to gas)

Table 12.2 Physical Properties of Components of Air

day. Normally, the amount of water vapor in air is described as the *relative humidity,* which is the ratio of the partial pressure of water vapor in air to the vapor pressure of water at the temperature of the air, expressed as a percent. For example, if the partial pressure of water vapor in air at 77°F (25 °C) is 16.5 Torr, we can use Table 6.2 to compute the relative humidity of the air. In that table we find that the saturation vapor pressure of water at 25 °C is 23.8 Torr. The relative humidity therefore is

$$\frac{\text{Partial pressure of water vapor}}{\text{Vapor pressure of water (at saturation)}} \times 100$$

$$= \text{relative humidity}$$

$$\frac{16.5 \text{ Torr}}{23.8 \text{ Torr}} \times 100 = 69.3\%$$

Notice in Table 6.2 that if the temperature dropped from 25 °C to 19 °C, the air would then be saturated with water vapor, and the relative humidity would be 100%. The temperature to which air must be lowered to saturate it with water vapor is the *dew point.* The dew point of air is easy to measure. If a wet piece of cloth is attached to the bulb of a thermometer, the cooling caused by evaporation of water from the cloth lowers the temperature reading to the value of the dew point. A second thermometer can then be used to measure the actual temperature of the air. With these two temperatures and Table 6.2, the relative humidity can be calculated. The vapor pressure associated with the dew point is the partial pressure of water vapor in the atmosphere.

The vapor pressure associated with the actual air temperature is the saturation vapor pressure.

The relative humidity is then computed from the following relationship:

$$\frac{\text{Vapor pressure at dew point}}{\text{Saturation vapor pressure at actual temperature}} \times 100$$

$$= \text{relative humidity}$$

For example, suppose the dew point reading is 15 °C, but the air temperature is 23 °C. Using values from Table 6.2, we can compute the relative humidity as

$$\frac{12.8 \text{ Torr}}{21.1 \text{ Torr}} \times 100 = 60.7\%$$

Relative humidity is the ratio of the partial pressure of water vapor in air to the vapor pressure of water at the temperature of the air, expressed as a percent. The temperature to which air must be lowered to saturate it with water vapor is the dew point.

The relative humidity of air is an important factor in human comfort. The expression "It's not the heat, it's the humidity," is based on fact. If the relative humidity is below 50%, air temperatures as high as 80 °F (27 °C) feel comfortable. Because the air is relatively dry, evaporation of moisture from the skin cools the body. But air feels uncomfortably hot at 80 °F when the relative humidity rises toward 90% or more. We have developed ways to keep indoor air comfortable with air conditioning. The steps in the air conditioning process as it was first invented are as follows:

1. *Dehumidifying* of the air by passing it over a desiccant (drying agent) such as silica gel or calcium chloride

2. *Chilling* the air to the temperature of the desired dew point.

3. *Saturating* of the air with water vapor at this low temperature by bubbling it through water.

4. *Warming* the air back to the desired room temperature.

The amount of moisture in the air can also be controlled by a different process, based on the properties of a saturated solution. Air of the desired room temperature is bubbled through a saturated solution of a salt that is in equilibrium with excess solute. As you know, a saturated solution has a definite vapor pressure at a given temperature. If the partial pressure of the moisture in the air is *greater* than the vapor pressure of the saturated solution, the solution will *absorb* moisture from the air; but if the partial pressure of the moisture in the air is *less* than the vapor pressure of the saturated solution, the solution will *add* moisture to the air. Thus, through the selection of the proper salt solution, exact humidity control can be maintained.

Let us now consider in more detail some of the components of the atmosphere.

COMPONENTS

Despite its relatively constant composition, air is a mixture. Each component of air retains its own unique physical and chemical properties, and air can be separated into its components by physical means, using the different boiling points of the gases. The separation process is *fractional distillation*. When liquid air is heated, the gas with the lowest boiling point boils away first, leaving other gases behind. As the temperature continues to rise, individual gases boil away at different times, and they can be collected and compressed into tanks.

Nitrogen

Occurrence. The atmosphere is the only important source of free (elemental) nitrogen; however, since the atmosphere is about 78% nitrogen, the supply is abundant. The principal source of combined nitrogen (in compounds) is the guano deposits found along the coast of Chile. These bird droppings, which have accumulated in large quantity over centuries, are as much as 50% sodium nitrate, $NaNO_3$.

Preparation. Commercially, nearly all nitrogen is obtained from the fractional distillation of liquid air. This nitrogen is contaminated by the noble gases present in trace quantities in air but is sufficiently pure for most uses. In the laboratory nitrogen is obtained by heating a mixture of sodium nitrite, $NaNO_2$, and ammonium chloride, NH_4Cl, in solution. The essential reaction is

$$NH_4^+ + NO_2^- \rightarrow N_2(g) + 2\ H_2O(l)$$

Physical Properties. These are summarized in Table 12.2.

Chemical Properties. The outstanding characteristics of nitrogen are its *high stability* and *relative inertness*. It combines with other elements only with difficulty. The process of inducing nitrogen to combine chemically with other substances is known as the nitrogen fixation and is accomplished in the following ways:

1. The very active metals combine directly with nitrogen at high temperatures to form nitrides:

$$3\ Mg(s) + N_2(g) \rightarrow Mg_3N_2(s)$$

2. In the cyanamide process, nitrogen and hot calcium carbide react to give

calcium cyanamide and carbon:

$$CaC_2(s) + N_2(g) \rightarrow CaCN_2(s) + C(s)$$

The hot cyanamide then reacts with steam under pressure to form ammonia:

$$CaCN_2(s) + 3\ H_2O(g) \rightarrow CaCO_3(s) + 2\ NH_3(g)$$

3. In the Haber process, nitrogen and hydrogen combine directly in the presence of a finely divided metallic catalyst at high temperature (about 500 °C) and extreme pressure (about 500 atm):

$$N_2(g) + 3\ H_2(g) \rightarrow 2\ NH_3(g)$$

4. The nitrogen cycle in nature provides the essential nitrogen compounds for plant life. Bacteria in the roots of leguminous plants like clover and peas oxidize nitrogen into proteins, which in turn decompose to form nitrates in the soil. Other bacteria cause plant and animal tissue to decay, producing ammonia and free nitrogen, which return to the atmosphere. The ammonia is oxidized to nitrates by still other bacteria. In addition, much nitrogen combines directly with oxygen in the atmosphere during lightning storms. The oxides fall to the ground dissolved in rain and are converted to nitrates by bacteria.

Uses. Large quantities of nitrogen are used in the fertilizer and explosives industries, as well as in the production of many drugs and dyes. Nitrogen is used as an inert atmosphere in metallurgical operations, in rooms used for the storage of flammable or explosive materials, and in electric lightbulbs to lengthen the life of the filament by preventing its oxidation. Argon is sometimes mixed with nitrogen in filling lightbulbs.

Principal Compounds. Ammonia, NH_3, is a colorless gas with a sharp, irritating odor. It is highly soluble in water: at 0 °C, 1176 cm^3 of it dissolves in 1 mL of water. It is usually made commercially by the Haber process. As a dilute aqueous solution, it is a familiar household cleaning agent, and its vapors are a well-known stimulant. Large amounts of ammonia are used in commercial refrigeration plants for cooling and in the production of ice, because it absorbs energy when permitted to expand suddenly (see Figure 12.1). Ammonia is the starting material in the manufacture of most other compounds of nitrogen.

Another important nitrogen compound is nitric acid, HNO_3. It is a colorless, volatile liquid with a piercing odor. It is used in the manufacture of sulfuric acid, nitrates, fertilizers, dyes, and explosives. It mixes completely (is miscible) with water. It boils at 86 °C and freezes at −47 °C. Nitric acid is a powerful oxidizing agent. In water solution, it is a strong acid. It combines with oxides and hydroxides to form nitrates. Similarly, it forms nitrates with most metals. Nitric acid oxidizes nonmetals such as sulfur or phosphorus to sulfates and phosphates, respectively. In the

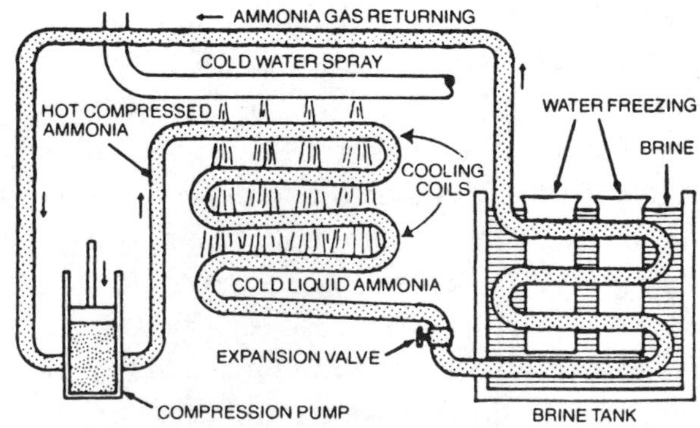

Figure 12.1 Use of Ammonia in Refrigeration

Ostwald process, nitric acid is made by oxidizing ammonia:

$$4\ NH_3(g) + 5\ O_2(g) \rightarrow 6\ H_2O(l) + 4\ NO(g)$$
(nitric oxide gas)

$$2\ NO(g) + O_2(g) \rightarrow 2\ NO_2(g)$$
(nitrogen dioxide gas)

$$3\ NO_2(g) + H_2O(l) \rightarrow 2\ HNO_3(aq) + NO(g)$$

Another commercial process, the arc process, involves direct combination of oxygen and nitrogen under the influence of an electric arc to form nitric oxide, NO:

$$N_2(g) + O_2(g) \rightarrow 2\ NO(g)$$

The nitric oxide is then converted to nitric acid, as in the last two steps of the Ostwald process.

A third important category of nitrogen compounds is the oxides. Nitric oxide, NO, is a colorless gas. Nitrogen dioxide, NO_2, is a reddish brown gas. Both are used in the manufacture of nitric and sulfuric acids. Nitrous oxide, N_2O, is prepared by heating ammonium nitrate:

$$NH_4NO_3(s) \rightarrow 2\ H_2O(l) + N_2O(g)$$

This gas, when inhaled, reduces consciousness and diminishes sensitivity to pain. Hence, it is used as an anesthetic in minor surgery and dentistry. It is claimed that the inhaling of small amounts of this gas can produce hysterical laughter. Hence, nitrous oxide is sometimes called "laughing gas."

Oxygen

Occurrence. Oxygen is the most abundant and widely distributed element on Earth. Oxygen makes up about 20% of the atmosphere, 50% of the solid crust of Earth, and 89% of the water. It is essential to nearly all forms of plant and animal life.

Preparation. Commercially, oxygen is obtained by the fractional distillation of liquid air. The liquid remaining after the other atmospheric gases have boiled away from liquid air is essentially pure oxygen. It is bottled in tanks under pressure after separation from the other gases. Oxygen is also obtained commercially as a by-product from the industrial electrolysis of water solutions. In the laboratory, oxygen may be prepared in the following ways:

1. By heating certain metallic oxides, especially oxides of metals below copper in the activity series:

$$2\ HgO(s) \rightarrow O_2(g) + 2\ Hg(l)$$

2. By heating certain oxygen-bearing salts such as potassium chlorate:

$$2\ KClO_3(s) \rightarrow 2\ KCl(s) + 3\ O_2(g)$$

3. By the reaction between water and sodium peroxide:

$$2\ H_2O(l) + 2\ Na_2O_2(s) \rightarrow 4\ NaOH(aq) + O_2(g).$$

Physical Properties. These are summarized in Table 12.2.

Chemical Properties. At room temperature, oxygen is only mildly reactive, but at elevated temperatures it combines with most elements and many compounds, especially those containing carbon and hydrogen, to form oxides of all the elements. For example:

(magnesium)

$$2\ Mg(s) + O_2(g) \rightarrow 2\ MgO(s)$$
(brilliant white flame)

(copper)
$2\ Cu(s) + O_2(g) \rightarrow 2\ CuO(s)$ (greenish flame)

(sulfur)
$S(s) + O_2(g) \rightarrow SO_2(g)$ (blue flame)

(ethyl alcohol)
$C_2H_5OH(l) + 3\ O_2(g) \rightarrow 2\ CO_2(g) + 3\ H_2O(g)$ (yellow flame)

(carbon tetrachloride)
$CCl_4(l) + O_2(g) \rightarrow$ (No reaction; CCl_4 is nonflammable.)

Uses. Oxygen is necessary to sustain nearly all forms of animal life. Cylinders of oxygen gas are carried aboard planes or on mountain climbing expeditions to permit breathing at high elevations. In medicine, oxygen assists patients recovering from lung diseases and pneumonia. In industry, oxygen makes possible, for example, the high temperatures of oxyhydrogen and oxyacetylene torches.

Principal Compounds. The metallic oxides, when combined with water, produce hydroxides:

$$CaO(s) + H_2O(l) \rightarrow Ca(OH)_2(aq)$$

The metallic oxides are therefore known as *basic anhydrides* (meaning "water remover"). The oxides of nonmetals, when combined with water, produce acids:

$$SO_2(g) + H_2O(l) \rightarrow H_2SO_3(aq)$$

These oxides are therefore known as *acid anhydrides*

Ozone

Occurrence. Ozone is an *allotropic form* of oxygen; that is, ozone is the same substance as oxygen but in different molecular form, possessing different properties. A molecule of ozone contains three atoms of oxygen instead of two found in a molecule of ordinary oxygen. Small amounts of ozone occur in the atmosphere and near lightning storms and sparking electrical equipment.

Preparation. Ozone is prepared by passing oxygen or air through an electrical discharging apparatus known as an ozonizer (see Figure 12.2). About 19% of the oxygen is converted by this process. The equation is

$$3\ O_2(g) \rightarrow 2\ O_3(g)$$

Physical Properties. These are summarized in Table 12.2.

Ozone is the same substance as oxygen but in different molecular form, possessing different properties.

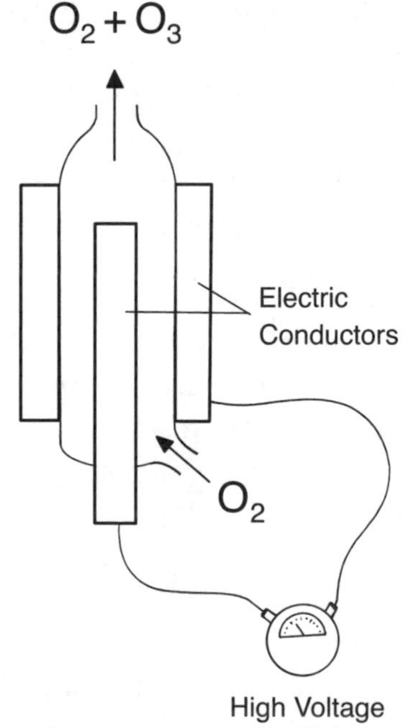

Figure 12.2 Ozonizer

Chemical Properties. Ozone, a pale blue gas, is a more powerful oxidizing agent than ordinary oxygen. If air with a trace of ozone is bubbled through a solution of iodide ion, the iodide ion is oxidized to free iodine:

$$O_3(g) + 2I^- + H_2O(l) \rightarrow O_2(g) + I_2(s) + 2\ OH^-$$

If starch is also dissolved in the water, an intense blue color forms as the free iodine is liberated. That chemical reaction serves as a test for the presence of ozone.

Uses. The uses of ozone depend on its oxidizing powers. It is a powerful bleaching agent, for it oxidizes colored substances. It is also used to kill bacteria in air and drinking water. Although ozone has a penetrating odor—you can smell it during thunderstorms—it is an excellent deodorizer.

Hydrogen

Occurrence. Because of its high reactivity, free hydrogen is rare in the atmosphere; but in its combined forms, hydrogen makes up about 1% by weight of Earth's crust. It forms about 11% by weight of water, and it is found in all petroleum products, acids and bases, and animal and plant life-forms. In total number of atoms, hydrogen is second only to oxygen in abundance on Earth but is by far the most abundant element in the universe.

Preparation. Commercially, hydrogen is prepared from water. Three different processes are commonly used:

1. The electrolysis of water solutions.

2. The action of steam on hot iron:

$$3\ Fe(s) + 4\ H_2O(g) \rightarrow Fe_3O_4(s) + 4\ H_2(g)$$

3. The water-gas method. The mixture of hydrogen and carbon monoxide is known as *water gas*. Both of these gases are combustible, so the mixture is used as a gaseous fuel. In this method, steam is passed over hot carbon in the form of coke or coal:

$$H_2O(g) + C(s) \rightarrow CO(g) + H_2(g)$$

In the laboratory, hydrogen is usually prepared by the action of active metals on an acid:

$$Zn(s) + 2\ HCl(aq) \rightarrow ZnCl_2(aq) + H_2(g)$$

It may also be obtained from the action of very active metals on water:

$$Ca(s) + 2\ H_2O(l) \rightarrow Ca(OH)_2(s) + H_2(g)$$

Physical Properties. These are summarized in Table 12.2.

Chemical Properties. Hydrogen is not very reactive at room temperature, but at higher temperatures it burns vigorously—and often explosively—in air or oxygen to form water.

$$2\ H_2(g) + O_2(g) \rightarrow 2\ H_2O(g)$$

Hydrogen is a moderately strong reducing agent, as can be seen from its position in Table 11.2. For example, it can reduce the oxides of metals less active than manganese to free metals:

$$H_2(g) + CuO(s) \rightarrow H_2O(l) + Cu(s)$$

Hydrogen combines directly with the very active metals to form ionic hydrides, in which the valence number of hydrogen is -1:

$$Ca(s) + H_2(g) \rightarrow CaH_2(s)$$

These solids react vigorously with water to give off hydrogen and form the metallic hydroxide.

Uses. In the presence of a catalyst and under pressure, hydrogen combines with vegetable oils to form solid fats such as margarine and shortening. The process is known as *hydrogenation*. This process is also used extensively in refining petroleum products to increase the yields of gasoline. Hydrogen is also used in the production of ammonia and many other chemical substances. It is used as a fuel to propel rocket engines and in fuel cells to generate electricity. Its use in inflating balloons was stopped after the German dirigible *Hindenburg* burned in 1937.

Principal Compounds.

1. *Acids* and *bases*. See Chapters 9 and 13–15.

2. *Organic compounds*. Chapter 22 deals with a few of the millions of known compounds of carbon. Most also contain hydrogen.

3. *Hydrogen peroxide*, H_2O_2. Pure hydrogen peroxide is an oily liquid that freezes at $-0.89°$ C and boils at 151.4°C. It is quite unstable, decomposing with violent force when brought in contact with dust or organic material:

$$2\,H_2O_2(l) \rightarrow 2\,H_2O(l) + O_2(g)$$

It is made by the electrolysis of a concentrated solution of potassium hydrogen sulfate (potassium bisulfate), $KHSO_4$, with a strong current of electricity to form potassium peroxydisulfate, $K_2S_2O_8$. When heated with steam, that compound forms hydrogen peroxide and regenerates the potassium bisulfate:

$$K_2S_2O_8(s) + 2\,H_2O(l) \rightarrow 2\,KHSO_4(s) + H_2O_2(l)$$

The peroxide is then concentrated by distillation at very low pressure.

Dilute solutions of hydrogen peroxide may also be made from the reaction of barium peroxide and ice-cold sulfuric acid:

$$BaO_2(s) + H_2SO_4(l) \rightarrow BaSO_4(s) + H_2O_2(l)$$

Uses. Hydrogen peroxide of 90% or more concentration is used in liquid-fuel rockets as an oxidizing agent and as a propellant. Aqueous solutions of 30% hydrogen peroxide are used as bleaches for fabrics, ivory, feathers, and other substances. A 3% solution is used as a disinfectant, a mouthwash, and a mild bleaching agent.

Water

Occurrence. Water is the most important chemical compound on Earth. It is also the most abundant and is essential to all life processes. About 70% of the human body is water. Animals and plants have correspondingly high percentages of water in their makeup.

Water circulates throughout the environment in the water cycle. Water evaporates from bodies of water (lakes and streams, for example), entering the atmosphere and leaving behind contaminants (salts, pollutants, and others). Eventually, this water condenses and falls back to Earth as rain or snow, continuing the cycle of nature's purification and reuse of water.

Although about 75% of Earth's surface is covered by water, and despite a saturated rock layer, the *water table*, that lies just below much of the land surface, lack of water makes vast land areas nearly uninhabitable, and serious water shortages threaten human agriculture and industry in many places. Two factors contribute to the shortage of water:

1. The extensive solvent action of water renders much of it unfit for use. The oceans are vast solutions whose solute concentration is so great that the water is poisonous to humans and their crops.

2. Too much of the fresh water that humans take for their own use is wasted.

Preparation. Commercially water is not "prepared." It is, instead, freed of undesirable substances carried or dissolved in it. Freshwater from lakes and streams is purified by *sedimentation* and *filtration* to remove suspended clay, sand, and organic material. Small amounts of chlorine are then added to kill bacteria.

In the laboratory, and for relatively small-scale consumption, water may be separated from its solutes by either distillation or deionization. In distillation, the water is heated until it is converted into water vapor. This gaseous water is then condensed back into liquid water—a laboratory imitation of nature's water cycle.

In the deionization process, water flows through beds of cationic and anionic polymer resins. The cationic resin exchanges its own hydrogen ions for positively charged ions in the water, such as Ca^{2+}, Mg^{2+}, and Fe^{2+}. The anionic resin exchanges its own hydroxyl ions for the negatively charged ions in the water such as Cl^- and SO_4^{2-}. Although deionization does not remove nonionic solutes, it produces water of relatively high purity. The two types of resins can be regenerated when they have exchanged their full capacity of ions. The hydrogen type is regenerated with a solution of hydrochloric acid; and the hydroxide type is regenerated with a solution of sodium hydroxide.

Physical Properties. Water is a clear, odorless liquid that freezes at 0 °C and boils at 100 °C. In many ways, it is the most unusual of liquids. That water is even a liquid is a mystery. All other compounds of similar molecular weight and structure are gases at room temperature (H_2S, NO, NO_2, N_2O, NH_3, CH_4, HF, HCl, etc.). Water has the highest specific heat, heat of fusion, and heat of vaporization of any liquid at room temperature. It also has the greatest solvent action. Its temperature of maximum density is 4 °C, so it expands at both lower and higher temperatures. It is the only common substance that is a liquid at room temperature that expands with cooling. As a result of the expansion of water at temperatures below 4 °C, colder water in lakes and rivers rises to the surface, and freezing takes place only on the surface. Ice, which is less dense than water, floats and actually insulates the water below it. As a result, the deep water of lakes and rivers remains liquid in cold weather. This protects many forms of aquatic life during the winter.

Chemical Properties. Water is an essential part of most chemical changes, either as a reactant or as a medium in which reactions take place. Water is an important weak electrolyte, and it reacts with many elements and compounds.

Uses. Water has countless residential, commercial, industrial, and agricultural applications. One little-known use is the flooding of agricultural fields to protect plants, such as the cranberry, against freezing temperatures. During the winter, when air temperature falls considerably below 0 °C, the surface of the water freezes; but beneath the ice, the liquid water and, consequently, the plants are held at temperatures slightly above the freezing point.

TRIVIA: If you want clear, not cloudy, ice cubes, start with *hot* water. The dissolved air in water forms bubbles as the ice freezes and gives rise to the cloudy condition; but air is much less soluble in hot water than in cold water, so starting with hot water reduces the number of air bubbles tremendously, resulting in clear ice cubes.

Carbon Dioxide

Occurrence. In addition to forming about 0.03% of the atmosphere, carbon dioxide, CO_2, is dissolved in all natural waters. It is also present in a combined form in carbonate minerals and rocks, the most abundant of which is limestone, $CaCO_3$.

Preparation. Commercially, carbon dioxide is prepared by heating limestone:

$$CaCO_3(s) \rightarrow CaO(s) + CO_2(g)$$

Carbon dioxide is also made by the combustion (rapid burning) of coke or natural gas, or by the fermentation of sugars to make alcohol. The gas is collected, compressed to a liquid, and stored in tanks. The other product, CaO, or lime, is also commercially important. Large amounts of carbon dioxide are added to the atmosphere from the combustion of carbon or carbonaceous material with excess oxygen, contributing to the greenhouse effect.

In the laboratory, carbon dioxide is generally prepared by the action of acids on metallic carbonates:

$$2\ HCl(aq) + MgCO_3(s) \rightarrow MgCl_2(aq) + H_2O(l) + CO_2(g)$$

In nature, the carbon cycle (analogous to the water cycle) tends to keep the amount of carbon dioxide in the atmosphere relatively constant. Plant life, in the presence of sunlight and through the catalytic action of chlorophyll, traps carbon from the carbon dioxide of the atmosphere to form cellulose, sugar, starch, and protein in plant cells. Pure oxygen is returned to the air in this process, which is called *photosynthesis*. Animal life acquires carbon from eating plant life or animals that have eaten plants. Animals inhale oxygen from the air and exhale carbon dioxide as a waste product.

Physical Properties. Like water, carbon dioxide has some unusual physical properties. At room temperature, it is a colorless, heavy gas. If it is cooled to $-78.5\,°C$, it condenses not into a liquid, but directly to a solid. Solid carbon dioxide is known as *dry ice*. Similarly, the solid passes directly from the solid to the gaseous state as it warms up. This phenomenon is known as *sublimation*. Carbon dioxide must be compressed to a pressure of at least of 5.2 atm before a liquid state forms. At that pressure, the freezing point of carbon dioxide is $-56.6\,°C$. Notice that the *normal boiling point of carbon dioxide is below its freezing point!* This unusual physical property accounts for sublimation. Other physical properties of carbon dioxide are summarized in Table 12.2.

Chemical Properties. Carbon dioxide is a very stable compound; it neither burns nor supports combustion. At high temperatures, it can be reduced to carbon monoxide, CO, by hot carbon or zinc:

$$C(s) + CO_2(g) \rightarrow 2\ CO(g)$$

It combines with the oxides and hydroxides of the very active metals to form carbonates:

$$CaO(s) + CO_2(g) \rightarrow CaCO_3(s)$$

$$Ca(OH)_2(s) + CO_2(g) \rightarrow CaCO_3(s) + H_2O(l)$$

It dissolves in water to form the weak electrolyte carbonic acid, H_2CO_3.

Uses. Large quantities of carbon dioxide are used in the manufacture of white lead, sodium carbonate, and sodium bicarbonate (baking soda). It is also used in the manufacture of carbonated beverages, being dissolved in these liquids under pressure. Removal of the cap from a bottle of soda permits the excess carbon dioxide to bubble free. One type of fire extinguisher contains liquid carbon dioxide under pressure. Solid carbon dioxide (known as *dry ice*) is used as a refrigerant.

Noble Gases

Occurrence. Helium, neon, argon, krypton, and xenon are found in the atmosphere. Helium is also found in natural gas deposits in the southwestern part of the United States. Radon is found associated with radium-bearing materials.

Preparation. Radon is radioactive and is a gaseous product of the radioactive decay of radium. All the other noble gases are obtained from the fractional distillation of liquid air. Helium is also obtained from natural gas deposits by liquefying all other constituents and collecting the gaseous helium. The percentage of helium in these deposits ranges from 1% to almost 2%.

Physical Properties. The physical properties of the noble gases are summarized in Table 4.1.

Uses. Helium is added to oxygen to replace the nitrogen in air used by deep-sea divers. Nitrogen dissolves in blood under the pressures required for this use, and when the pressure is reduced as the diver rises from depth the nitrogen comes out of solution and forms bubbles in the bloodstream. This causes the painful and sometimes fatal ailment known as the bends. The name of the condition is descriptive, as it causes an afflicted diver to double over from pain. The less soluble helium reduces the possibility of danger from that source. Helium is also used to inflate balloons and blimps.

Argon is used with nitrogen in filling electric lightbulbs. Argon together with helium, is frequently used as an inert atmosphere in scientific work. "Neon lights" are another common use. Table 12.3 lists supplies some data about the gases used in these lights. Radon was once used in hospitals to combat cancer because of its radioactive properties. Krypton is used as a gas in incandescent and fluorescent lamps. Xenon is used as a gas in electron and luminescent tubes, flashlamps, and lamps used to excite ruby lasers.

COMBUSTION

One of the most important chemical reactions that occurs in air is combustion. *Combustion* is the rapid combination of a substance with

Color	Gas mixture	Pressure, Torr	Color of Glass
white	helium	3–4	clear
yellow	helium	3–4	amber
light green	neon–argon–mercury	10–20	green
dark green	neon–argon–mercury	10–20	amber
light blue	neon–argon–mercury	10–20	clear
dark blue	neon–argon–mercury	10–20	purple
red	neon	10–18	clear
deep red	neon	10–18	red

Table 12.3 Neon Signs

oxygen accompanied by the release of energy in the forms of heat and light. If the rate of reaction is slow, and only heat energy is given off, the process is called *slow oxidation*. The rusting of iron is a slow oxidation. The burning of wood is combustion. The flame produced by combustion consists of burning gases vaporized from the combustible substance by the heat of the reaction. The flame may be colored by energized ions emitting light energy, as well as by bits of solid material heated to incandescence by the reaction energy. Note that the total amount of energy released by the oxidation of a substance is the same regardless of the rate of the oxidation reaction.

Both corrosion and combustion are reactions with oxygen. Corrosion is slow oxidation. Combustion is rapid oxidation accompanied by the heat and light.

Before a substance can burst into flame, it must be heated to a definite temperature. This minimum temperature is known as the *kindling temperature*. Each combustible substance has its own kindling temperature.

EXPERIMENT 25: *Be extremely careful! Wear your safety goggles. Have a fire extinguisher or baking soda shaker close by!* Select a 9-in. pie tin. On the inside, just at the edge of the sloping side, place at equally spaced intervals the following items: the head of a match, shavings from a cork, torn bits of paper, a small piece of cotton dipped in oil, a small piece of cotton wet with lighter fluid, a small piece of cotton wet with turpentine. Center the pan directly over the element or burner on the stove and turn on the burner or element. The objects in the pan will catch fire in the order of increasing kindling temperature.

Spontaneous combustion is the rapid initiation of the combustion process. It is most likely to occur when a combustible material that is a poor conductor of heat is stored in still air. Oxygen in the air begins slowly to oxidize the combustible material and thereby to generate heat. The heat is not conducted away but accumulates around the material. Eventually, the temperature rises to the kindling temperature, and active combustion begins. To prevent spontaneous combustion, oily or paint-stained rags should never be permitted to accumulate, especially in the corners of closets or cabinets. If they must be kept, safety dictates that they be stored in closed metal containers in a well-ventilated spot.

SUMMARY

The atmosphere is a shell of gases surrounding Earth. It contains nitrogen, oxygen, argon, carbon dioxide, hydrogen, and the noble gases in relatively constant amounts. Water vapor is an important variable component of air. Relative humidity is the ratio of the partial pressure of the water vapor in the air to the vapor pressure of water at the temperature of the air. The dew point is the temperature at which the air would be saturated with water vapor.

Nitrogen is obtained from the atmosphere by the fractional distillation of liquid air. The fixation of nitrogen is accomplished in the cyanamide process, in the Haber process, and in the nitrogen cycle in nature. The principal compounds of nitrogen are ammonia, nitric acid, and its oxides.

Oxygen is likewise obtained from the atmosphere by the fractional distillation of liquid air. It is very active chemically, especially at high temperature. Acid and basic anhydrides are among the important compounds of oxygen. Ozone is an allotropic form of oxygen and is a more powerful oxidizing agent.

Hydrogen is obtained from water or acids. It burns in oxygen and is a good reducing agent. Hydrogen peroxide is an important compound of hydrogen used as a source of oxygen for rocket fuels.

Water is the most important chemical compound. Its unique physical properties contribute to its usefulness. It is involved in most chemical reactions either as a reactant or as a medium in which the reaction takes place.

Carbon dioxide is obtained by heating limestone. It recycles in nature through the carbon cycle. Solid carbon dioxide sublimes because its boiling point is below its freezing point. The noble gases are usually obtained from air. Helium is found as part of some deposits of natural gas, and radon is a product of the radioactive decay of radium. Combustion is rapid oxidation accompanied by the evolution of heat and light energy. The kindling temperature of a substance is the lowest temperature at which it will burst into flame. Spontaneous combustion may occur if a combustible material that is a poor conductor of heat is stored in still air.

PROBLEM SET NO. 14

1. Compute the densities of the substances in Table 12.2 relative to air = 1.293
2. The air temperature is 20 °C and the dew point is 10 °C. Find the relative humidity.
3. The air temperature on a hot day is 29 °C (84.2 °F). The relative humidity is 89%. Find the dew point (both Celsius and Fahrenheit).
4. Is it easier for our body to rid itself of heat by sweating on a humid day or on a dry day if the temperature is the same in both cases?
5. What is the valence of sulfur in potassium peroxydisulfate, $K_2S_2O_8$?
6. How many free elements are you likely to inhale in your next breath?
7. At standard conditions, 22.4 L of helium weighs 4 g. Is the formula for a molecule of helium He, He_2, or He_3?
8. The following statement is made: "When I swing my hand rapidly through air, it cools off. If I ride a bicycle fast, I cool off. Therefore, anything that moves through air cools." Comment on the statement, referring to meteorites.
9. Why does a bottle of soda fizz violently after being shaken?
10. An Alka-Seltzer® tablet contains calcium phosphate, aspirin, citric acid, and sodium bicarbonate. What gas effervesces from the solution when a tablet dissolves? Why?

CHAPTER 13 THE HALOGENS

KEY TERMS

halogens, activity series of nonmetals, hydrogen halides, flux

The elements in group VIIA (17) of the periodic table form an important family of elements, the *halogens*. Recall from Chapter 3 that elements in each family are similar in their electron structure. Because chemical behavior depends on electron structure, we find similarity of chemical behavior among elements of the same family.

The term halogen (from the Greek *hals*, "salts") refers to the tendency of these elements to form salts. Tables 13.1 and 13.2 summarize the physical and chemical characteristics of the halogen family. (The element astatine [At] is so rare that it has been omitted from the tables.) These tables show not only similarities but also the gradual differences that appear as we proceed from the lighter to the heavier elements. Note that fluorine is the most active member of the halogen family and that it forms the most stable compounds.

It is a general principle of chemistry that *the more active the element, the more stable its compounds*. The replacement properties of the halogens hint—correctly—that an *activity series of nonmetals*, similar to the activity series of metals, exists. In order of decreasing activity, the activity of halogens is as follows: fluorine, chlorine, bromine, and iodine. As in the case of the activity series of metals, the more active

Characteristic	Fluorine	Chlorine	Bromine	Iodine
atomic number	9	17	35	53
electron configuration	$1s^22s^22p^5$	$[Ne]3s^23p^5$	$[Ar]3d^{10}4s^24p^5$	$[Kr]4d^{10}5s^25p^5$
atomic mass, g/mol	19.00	35.45	79.90	126.9
physical state	gas	gas	liquid	solid
color	pale yellow	greenish yellow	dark red	bluish black
density, g/cm^3	1.14 (liq)	1.51 (liq)	3.12 (liq)	4.93 (solid)
boiling point, °C	−188	−35	59	114
freezing point, °C	−220	−101	−7	114
solubility, g/100 ml water at 0°C	decomposes to $HF + O_3$	1.46	4.17	0.03

Table 13.1 Physical Properties of the Halogens

Characteristic	Fluorine	Chlorine	Bromine	Iodine
general activity	extremely active	very active	less active	least active
activity with hydrogen	violent, even in dark	slow in dark, violent in light	must be heated	slow and incomplete even when heated
formula of hydrogen halide	HF	HCl	HBr	HI
stability of hydrogen halide	extremely stable	very stable	less stable	least stable
oxidizing power	most powerful	very powerful	less powerful	least powerful
replacement of nonmetals	replaces Cl, Br, O, I, S	replaces Br, O, I, S	replaces O, I, S	replaces S only
reaction with water	decomposes it to HF + O_3	rapidly forms HCl + HOCl	slowly forms HBr + HOBr	no reaction

Table 13.2 Chemical Properties of the Halogens

nonmetallic elements are capable of replacing the less active nonmetals from salt solutions. For example

$$F_2(g) + 2\ NaBr(aq) \rightarrow 2\ NaF(aq) + Br_2(l)$$

Hydrogen halides are compounds made of hydrogen combined with a halogen, such as HF, HCl, or HBr. Halides, like free halogens, are very poisonous because of their great chemical activity.

OCCURRENCE AND PREPARATION

Because the halogens are so active, none is found free (uncombined) in nature; but in the chemically combined state, the halogens are both abundant and widely distributed. By far the most abundant of these elements is chlorine. Fluorine, bromine, and iodine follow in that order. Table 13.3 summarizes the occurrence of the halogens.

The more active the element, the more stable its compounds. Among the halogens, fluorine is the most active, and its compounds are the most stable.

Free halogens are prepared through the vigorous oxidation of compounds that contain halogen ions, either by powerful oxidation reactions or by the oxidizing effect of an electric current. Review Table 11.2 for the oxidation–reduction potentials of the halogen half-reactions:

$$2\ I^- \rightarrow I_2(s) + 2\ e^- \qquad -0.5355\ V$$

$$2\ Br^- \rightarrow Br_2(l) + 2\ e^- \qquad -1.087\ V$$

$$2\ Cl^- \rightarrow Cl_2 + 2\ e^- \qquad -1.359\ V$$

$$2\ F^- \rightarrow F_2 + 2\ e^- \qquad -2.87\ V$$

Element	Occurrence
fluorine	the mineral fluorite, CaF_2; the mineral cryolite, Na_3AlF_6
chlorine	as chloride ion, Cl^-, in seawater (2%); in rock salt beds of the mineral halite, NaCl; in salt beds of KCl, $MgCl_2$, and $CaCl_2$; in human gastric juices as HCl (0.05%)
bromine	as bromide ion, Br^-, in seawater (0.008%); as NaBr or $MgBr_2 \cdot KBr \cdot 6H_2O$ in salt beds
iodine	as iodide ion, I^-, in seawater (0.000004%); as sodium iodate, $NaIO_3$, in Chilean nitrate deposits; in seaweed such as kelp

Table 13.3 Occurrence of the Halogens

TRIVIA: The element iodine was discovered in 1811 by the French scientist Bernard Courtois, who was searching for a cheap source of potassium for his gunpowder factory. He was cleaning a seaweed tank with acid and noticed violet vapors being given off. These vapors condensed into metallic-looking crystals, which were determined to be I_2.

Because the fluoride half-reaction appears at the bottom of the list, we can tell that it is the most powerful oxidizing agent. In fact, no other chemical can oxidize fluoride ions to elemental fluorine gas. Therefore, fluorine can be prepared only by electrolysis. Substances such as the dichromate ion, $Cr_2O_7^{2-}$, and the permanganate ion, MnO_4^-, as well as fluorine gas can oxidize any of the other three halogens because they are lower in Table 11.2. With the aid of that table, you can write equations for a number of reactions that liberate chlorine, bromine, or iodine from their ionic solutions in the laboratory. Remember that the oxidizing half-reaction proceeds from right to left as written in Table 11.2.

Commercially, the halogens are prepared as follows:

1. *Fluorine.* The salt potassium hydrogen fluoride, KHF_2, is melted in a copper container fitted with graphite electrodes. Electrolysis of the fused salt liberates fluorine at the anode; it liberates both hydrogen and metallic potassium at the cathode. The latter two elements are valuable by-products. Copper is used as a container because although it is attacked by free fluorine, the resulting coating of copper fluoride, CuF_2, forms a protective layer on the metal that prevents further reaction. Lead, nickel, and magnesium behave in a similar way with fluorine and may be used in place of the copper in the electrolysis cell.

2. *Chlorine.* All commercial chlorine now comes from electrolysis. About 90% of commercial chlorine is produced by the electrolysis of brine, a solution of common table salt (see Figure 13.1). In the process, chlorine is collected at the anode and hydrogen gas at the cathode. The resulting solution contains sodium hydroxide, which is obtained as a by-product by the evaporation of the solution. The overall reaction for the process is

$$2\ NaCl(aq) + 2\ H_2O(l) \rightarrow Cl_2(g) + H_2(g) + 2\ NaOH(aq)$$

The electrolysis of molten sodium chloride also yields chlorine gas. The cathode

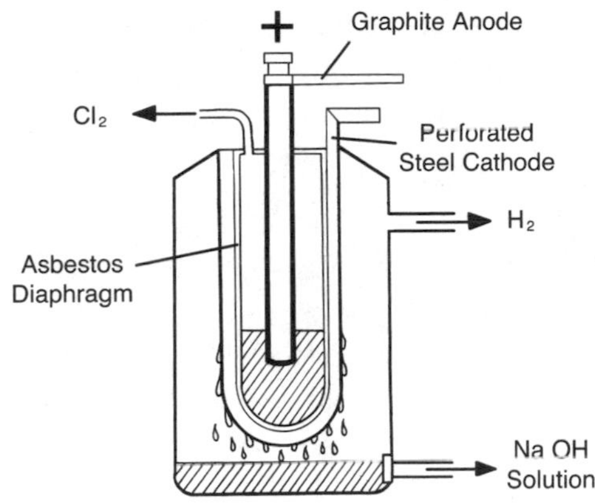

Figure 13.1 Preparation of Chlorine from Salt Solution in Nelson Cell

product in this case is pure sodium metal. The commercial demand for metallic sodium determines the extent to which this more costly process is used.

3. *Bromine.* Most bromine is now prepared from seawater. The water is first acidified with sulfuric acid, and the bromide ions present are then oxidized to elemental bromine by passing chlorine gas into the solution. The reaction is

$$2\ Br^-(aq) + Cl_2(g) \rightarrow 2\ Cl^-(aq) + Br_2(l)$$

4. *Iodine.* Most commercial iodine is obtained from mineral deposits in Chile and subterranean brines in Michigan and Japan. Many years ago, another source was seaweed, which tends to concentrate the naturally occurring iodine in seawater. The iodine ions are oxidized by a solution of sulfuric acid and sodium nitrite. The reaction is

$$2\ I^-(aq) + 4\ H^+(aq) + 2\ NO_2^-(aq) \rightarrow 2\ H_2O(l) + 2\ NO(g) + I_2(s)$$

In the laboratory, fluorine is prepared by the same method as is used commercially. The other halogens may be conveniently oxidized from acidified solutions of their ions by the oxidizing action of such ions as dichromate or permanganate, or by the action of such compounds as manganese dioxide, MnO_2. Typical reactions are as follows, where the symbol X^- represents any of the three halide ions, Cl^-, Br^-, or I^-:

$$6\ X^-(aq) + Cr_2O_7^{2-}(aq) + 14\ H^+(aq) \rightarrow 2\ Cr^{3+}(aq) + 3\ X_2 + 7\ H_2O(l)$$

$$10\ X^- + 2\ MnO_4^-(aq) + 16\ H^+(aq) \rightarrow 2\ Mn^{2+}(aq) + 5\ X_2 + 8\ H_2O(l)$$

$$2\ X^- + MnO_2(s) + 4\ H^+(aq) \rightarrow Mn^{2+}(aq) + X_2 + 2\ H_2O(l)$$

PROPERTIES

All halogens form diatomic molecules in which the two atoms are covalently bonded. All have sharp, disagreeable odors and attack the skin and mucous membranes of the nose and throat. Although iodine is a solid at room temperature, it readily sublimes because of its high vapor pressure. Iodine vapor is deep violet in color. The solubility of the halogens in non-aqueous solvents such as carbon tetrachloride, CCl_4, and carbon disulfide, CS_2, increases with increasing atomic mass. Iodine is about 650 times more soluble in CS_2 than in water.

EXPERIMENT 26: Place a few drops of tincture of iodine in a glass half filled with water. Add a few grains of cornstarch. Stir to dissolve the starch. The blue color that develops is a complex compound of iodine and starch. This is a sensitive test for the presence of iodine.

USES

1. *Fluorine.* Despite the toxic nature of both fluorine gas and the fluoride ion, the uses of fluorine are rapidly increasing. Freon, CCl_2F_2, was once used as a refrigerant, especially in automobile air conditioning units, (where it was called R-12); however, it was discovered that Freon and other chlorofluorocarbons (CFCs) played a role in the reduction of the atmospheric ozone layer and so it was banned and replaced by other compounds. In automobiles, products such as Enviro-Safe, Freeze 12, and R-22 are now used instead. Cryolite, which is sodium aluminum fluoride, Na_3AlF_6, is produced synthetically and is a vital *flux* (a material that aids in the melting of another material) in the electrolytic production of aluminum metal. Compounds of fluorine and carbon such as Halon are important because of their heat and fire resistance, but they, too, are destructive to ozone and were banned in 1994. Hydrochlorofluorocarbons (HCFCs) are similar to CFCs, but are less destructive to ozone. They, too, are to be phased out by 2020, when they are expected to be replaced by hydrofluorocarbons (HFCs). Many important plastics contain fluorine. Drinking water that contains about 1 part per million (ppm) of fluorine protects teeth against decay; however, if the concentration of fluorine in the water is greater than 3 ppm, teeth become mottled with brown spots. Fluorine compounds are used as insecticides and wood preservatives. Both lithium and sodium fluorides serve as a flux in the soldering of aluminum.

2. *Chlorine.* Large quantities of chlorine are used to bleach wood pulp for the paper industry, and cotton and linen fabrics in the textile industry. Many cities add small amounts of chlorine to drinking water to kill bacteria. Water in swimming pools is chlorinated either directly with chlorine or by adding the compounds calcium hypochlorite, $Ca(OCl)_2$, or sodium hypochlorite, $NaOCl$ (bleach).

3. *Bromine.* Bromine is used chiefly in the petroleum, drug, and photographic industries. The compound ethylene dibromide, $C_2H_4Br_2$, was once used as an additive in leaded antiknock gasoline and as a fumigant for grains and fruit. Many dyes and drugs contain bromine. Some bromine compounds have a nerve-soothing effect. Silver bromide is used to coat photographic film and plates.

4. *Iodine.* An alcoholic solution of iodine, known as tincture of iodine, once enjoyed great popularity as an antiseptic. Other compounds of iodine such as iodoform, CHI_3, are also used in the drug industry. Small amounts of iodine are essential in the diet to ensure proper functioning of the thyroid gland. A deficiency causes goiter. Iodized salt, containing a small amount of sodium iodide, NaI, and seafood are the chief food sources of iodine.

PRINCIPAL COMPOUNDS

Hydrogen Halides

The *hydrogen halides* (compounds of hydrogen and a halogen), HF, HCl, HBr, and HI, are all covalently bonded gases at room temperature. They are colorless and have penetrating, sharp odors. When perfectly dry, the hydrogen halides are nonconductors of electricity; however, they are extremely soluble in water, and in solution they dissociate into ions in the manner of all electrolytes. In water solution, they are known as the hydrohalic acids. Hydrofluoric acid is a weak electrolyte. The others are strong electrolytes. All the hydrogen halides fume in moist air because they dissolve in the moisture of the air and condense as droplets of acid solution. Table 13.4 summarizes the physical properties of the dry hydrogen halides.

The dry halogen halides are relatively inert, but in the presence of even a trace of water they take on their acid characteristics and become vigorously active. Hydrofluoric acid attacks glass and silicate materials like porcelain. Therefore, it must be stored in plastic bottles. Hydrochloric acid undergoes all the typical acid reactions with metals and bases. Hydrobromic and hydriodic acids are typical strong acids and good reducing agents.

When a fluoride or chloride salt is treated with concentrated sulfuric acid, the more volatile hydrogen halide forms and passes from the reaction chamber. Typical reactions are

$$CaF_2(s) + H_2SO_4(l) \rightarrow 2\ HF(g) + CaSO_4(s)$$

$$2\ NaCl(s) + H_2SO_4(l) \rightarrow 2\ HCl(g) + Na_2SO_4(s)$$

Acids

When bromides or iodides react with sulfuric acid, they are oxidized by the sulfuric acid completely to the free halogens. The reactions are

$$2\ NaBr(s) + H_2SO_4(l) \rightarrow Br_2(l) + SO_2(g) + 2\ NaOH(s)$$

$$2\ NaI(s) + H_2SO_4(l) \rightarrow I_2(s) + SO_2(g) + 2\ NaOH(s)$$

For that reason, hydrobromic and hydriodic acids must be prepared by an alternative method. The most common involves first reacting the halogen with phosphorus to form a phosphorus trihalide, then reacting it with water to produce the acid. For bromine, the reactions are

$$2\ P(s) + 3\ Br_2(l) \rightarrow 2\ PBr_3(s)$$

$$PBr_3(s) + 3\ H_2O(l) \rightarrow 3\ HBr(aq) + H_3PO_3(aq)$$

Property	Hydrogen Fluoride	Hydrogen Chloride	Hydrogen Bromide	Hydrogen Iodide
molecular mass, g/mol	20.01	36.46	80.91	127.91
decomposes at (°C)	unknown	1500	800	180
density, g/ml, 0°C	0.00090	0.00100	0.00350	0.00566
critical temperatures, °C	188	51.4	90	150
critical pressure, atm	64	82.1	84.5	81.9
boiling point, °C	19.5	−84.9	−67.0	−35.4
freezing point, °C	−83.1	−114.8	−88.5	−50.8

Table 13.4 Physical Properties of Hydrogen Halides

The principal use of hydrofluoric acid is for etching glass. The area to be etched is first covered with paraffin, which does not react with HF. Then, the design is scratched through the paraffin. The glass is next dipped into a hydrofluoric acid solution, then washed. Finally, the paraffin is removed. The glass in electric lightbulbs is "frosted" by treatment with this acid.

Hydrochloric acid, known commercially as muriatic acid, is widely used in the manufacture of textiles, glucose, soap, glue, dyes, and many other products. It removes scale (mineral deposits) from the surface of metals, and dilute solutions remove mortar from brick or stone walls.

Hydrobromic acid is used in the preparation of bromides and other chemicals. Hydriodic acid is used both in the preparation of many organic chemicals and as a reducing agent.

When chlorine gas dissolves in water, a chemical reaction takes place, forming a mixture of acids called chlorine water:

$$Cl_2(g) + H_2O(l) \rightarrow HCl(aq) + HOCl(aq)$$

The compound HOCl is called hypochlorous acid. It is a strong oxidizing agent. It is responsible for both the bleaching and the disinfecting action of chlorine in water. Incidentally, chlorine bleach cannot be used on silk or wool, because the acids HCl and HOCl in it destroy the fibers.

Hypochlorous acid is a weak electrolyte, so its salts such as sodium hypochlorite, NaOCl, and calcium hypochlorite, $Ca(OCl)_2$, hydrolyze in water solution to form molecules of the acid. Consequently, solutions of these salts are effective bleaching agents. Household bleaches are about 5% solutions of sodium hypochlorite. The oxidizing action of hypochlorous acid in these solutions gives them their germicidal and disinfecting properties. Calcium hypochlorite, sold under the trade name "HTH" (High Test Hypochlorite), is used to disinfect drinking water on ships and water in swimming pools. A related compound, bleaching powder, (Ca^{2+}, OCl^-, Cl^-), is an effective bleaching agent. It is prepared by passing chlorine gas over moist slaked lime (calcium hydroxide):

$$Ca(OH)_2(s) + Cl_2(g) \rightarrow Ca(OCl)Cl(s) + H_2O(l)$$

This compound is also known as chloride of lime.

In the laboratory, hypochlorous acid solutions and solutions of hypochlorites are stored in brown bottles, preferably in the dark, to prevent decomposition of the acid by sunlight. In direct sunlight, the acid decomposes:

$$2\ HOCl(aq) \rightarrow 2\ HCl(aq) + O_2(g)$$

SUMMARY

The halogens are the elements in group VIIA (17) in the periodic table. They are all active chemically, but they decrease in activity with increasing atomic number. The more active an element, the more stable its compounds.

The activity series of nonmetals in order of decreasing activity include: F, Cl, Br, and I. The halogens are too active to occur free in nature, but their compounds are abundant and widely distributed. The halogens are prepared by oxidizing them—either electrically or by oxidizing agents—from their compounds. Free halogens and their compounds formed with hydrogen, the halides, are very poisonous because they are so highly reactive.

The hydrogen halides are covalent compounds when perfectly dry, and relatively inert. They dissolve even in traces of water to become ionic hydrohalic acids, which are vigorously active. The bleaching action of chlorine and its compounds is a result of the formation of hypochlorous acid in water.

PROBLEM SET NO. 15

1. Would you expect the halogen astatine, atomic number 85, to be vigorously active chemically or relatively inert?

2. Write a balanced equation for the electrolysis of KHF_2.

3. It is often stated that hydrofluoric acid is much "stronger" than hydrochloric acid because it attacks glass. How would you explain that statement?

4. What are the anode and cathode products in the electrolysis of a solution of sodium bromide, NaBr?

5. Why do some synthetic fibers turn yellow when treated with chlorine bleach?

THE SULFUR FAMILY

KEY TERMS

sulfur family, metalloid

The elements below oxygen in group VIA (16) of the periodic table are often called the *sulfur family*. All have 6 electrons in the outermost shell. Of the four family members—sulfur, selenium, tellurium, and polonium—sulfur is the most abundant and the most important.

Recall from the previous chapter that the halogens are nonmetals, but their activity decreases moving down through group VII of the periodic table. In a sense, the halogens become more like metals as their atomic weight increases. In the sulfur family, where the number of electrons in the outermost shell is fewer than in the halogens, this tendency toward increasingly metallic characteristics is even more pronounced.

From a chemical point of view, nonmetals gain electrons when they react, and substances that gain electrons behave like nonmetals. Substances that lose electrons in their reactions behave like metals. Although sulfur is a definite nonmetal, it exhibits positive valences in many of its compounds, just as metals do. Selenium and tellurium, in both appearance and properties, are so much like metals that they are often called *metalloids*. Polonium, a very rare radioactive element, is definitely metallic in both properties and behavior.

TRIVIA: Farts smell because of small amounts of hydrogen sulfide gas and mercaptans, both of which contain sulfur.

OCCURRENCE AND PREPARATION

Since sulfur is one of the least active nonmetals at ordinary temperatures, vast amounts of it occur free (uncombined) in nature, particularly in underground beds in Texas and Louisiana. Impure sulfur is also found in the volcanic regions of Mexico, Japan, and Sicily. Chemically combined, sulfur is found

- as sulfide ores: FeS_2, Cu_2S, PbS, As_2S_3, Sb_2S_3, and Bi_2S_3;
- as sulfate ores: $CaSO_4$, $SrSO_4$, and $BaSO_4$;
- in organic plant and animal life, particularly substances possessing strong odors and tastes like garlic, onions, horseradish, mustard, and eggs;
- in petroleum.

Both selenium and tellurium are usually found as impurities in deposits of chemically combined sulfur ores. For example, small amounts of selenide or telluride compounds such as PbSe, Cu_2Se, or PbTe occur in sulfide ores of copper and lead. Tellurium is most commonly associated with bismuth as Bi_2Te_3. Selenium is more abundant than tellurium. Polonium is very rare and may be found in

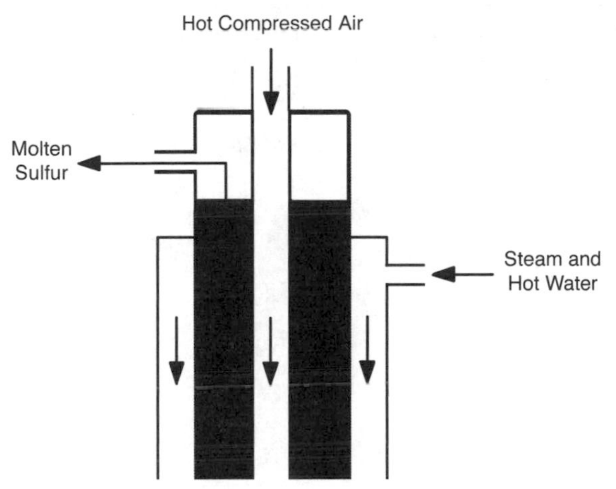

Figure 14.1 The Frasch Process

association with other radioactive elements in pitchblende.

Free sulfur is obtained from underground deposits by the Frasch process. Figure 14.1 illustrates the concentric pipes that are inserted down into the sulfur beds in this process. The steam and hot water in the outer pipe melt the sulfur below, and the hot compressed air forced down through the inside pipe exerts the pressure that forces the molten sulfur up through the middle pipe. Because the outer and inner pipes are hot, the molten sulfur stays liquid until it flows into huge vats where it cools and solidifies. This sulfur is so pure that it needs no further refining for most uses.

Selenium, with some tellurium, is one of the main components of the anode slimes that form during the electrolysis of copper from copper sulfate solution. Selenides, which form some of the impurities in the copper metal being refined, deposit as free selenium on the anode.

Tellurium is usually found with bismuth ores. After the bismuth has been removed, the telluride residue is dissolved in HCl solution and then treated with a sodium sulfite, Na_2SO_3, solution that oxidizes any telluride present to free tellurium. The tellurium precipitates from solution.

> In both the halogen and the sulfur families, the elements become more like metals as their atomic weight increases.

PROPERTIES

Table 14.1 summarizes the physical properties of the sulfur family of elements.

All these elements are solid at room temperature, and all are very brittle. Sulfur forms molecules of the formula S_8, whereas selenium and tellurium form endless-chain molecules. Like oxygen, these elements all have various allotropic forms. Sulfur has three: rhombic sulfur, which is stable at room temperature; monoclinic sulfur, which is stable between 95.6°C and the melting point of sulfur; and amorphous sulfur, a plastic rubbery form obtained by pouring molten sulfur heated almost to the boiling point into water. All forms slowly revert to the rhombic form when allowed to stand at room temperature. Selenium has two principal allotropic forms: red selenium, which is amorphous and soluble in carbon disulfide; and gray selenium, which is metallic in appearance and insoluble in carbon disulfide. Tellurium also has two principal allotropic forms: crystalline tellurium, which is silver-white and completely metallic in appearance; and amorphous tellurium, which is a brownish black powder.

When heated in air or oxygen, all the elements in the sulfur family burn with a blue flame to produce the corresponding dioxide: SO_2,

Property	Sulfur	Selenium	Tellurium
atomic number	16	34	52
atomic mass, g/mol	32.03	78.96	127.6
color	yellow	red to lead-gray	black to silver-white
density, g/cm^3	2.07	4.3–4.8	6.25
melting point, °C	112.8	217	452
boiling point, °C	444.7	685	1390
electrical conductivity	very poor	poor in dark; good in light	good

Table 14.1 Physical Properties of the Sulfur Family

SeO_2, or TeO_2. All other chemical properties are similar among these elements, although the activity decreases slightly as the atomic mass increases. Thus, the chemical properties of sulfur serve to illustrate the properties of all the members of this family:

1. When heated, sulfur combines with metals to form sulfides:

   $$Zn(s) + S(s) \rightarrow ZnS(s) \quad \text{(zinc sulfide)}$$

2. Sulfides react with hydrochloric acid to form hydrogen sulfide:

   $$ZnS(s) + 2\ HCl(aq) \rightarrow ZnCl_2(aq) + H_2S(aq)$$

   Hydrogen sulfide, hydrogen selenide, and hydrogen telluride are poisonous gases, and in solution they are slightly acidic and very weak electrolytes.

3. Sulfur dioxide dissolves in water to form sulfurous acid:

   $$SO_2(g) + H_2O(l) \rightarrow H_2SO_3(aq)$$

   Sulfurous, selenous, and tellurous acids are weak electrolytes. They are easily oxidized to sulfuric acid, H_2SO_4; selenic acid, H_2SeO_4; and telluric acid, H_2TeO_4, respectively.

USES

Sulfur is the starting point in many important processes. It is used to make sulfur dioxide, sulfuric acid, matches, black gunpowder, insecticides, and many organic compounds, including the sulfa drugs. It is used in great quantities by the rubber industry in vulcanization. In the vulcanizing process, the addition of sulfur to rubber with heat improves the material's stability at extreme temperatures and its resistance to wear.

Selenium is used in making red glass for traffic lights and enamels. The electrical conductivity of selenium is low in the dark but dramatically increases when exposed to light. This peculiar electrical conduction property makes it valuable in various types of light meters and photoelectric cells. The Xerox photocopying process depends on the photoconductivity of selenium. Tellurium is sometimes mixed with lead to increase its tensile strength.

PRINCIPAL COMPOUNDS

Hydrogen Sulfide

Hydrogen sulfide, H_2S, is a colorless gas at room temperature with the characteristic foul odor of rotten eggs. It is extremely poisonous. One volume of the gas in 200 volumes of air can be fatal when breathed for even a short period of time. It is prepared by treating sulfides with hydrochloric acid. Hydrogen sulfide is used mainly in analytical chemistry to separate mixtures of metallic ions.

Sulfur Dioxide

Sulfur dioxide, SO_2, is commercially very important. It may be prepared by the following processes:

1. Burning sulfur in air:

$$S(s) + O_2(g) \rightarrow SO_2(g)$$

2. Heating sulfide ores in air. This process is known as *roasting*:

$$2\ ZnS(s) + 3\ O_2(g) \rightarrow 2\ ZnO(s) + 2\ SO_2(g)$$

Roasting produces valuable sulfur dioxide, which is collected in the roasting furnaces. Roasting also converts metallic ore to an oxide, which facilitates the extraction of the free metal.

3. The action of strong acids on sulfites. This is the usual method of preparing SO_2 in the laboratory:

$$Na_2SO_3(s) + 2\ HCl(aq) \rightarrow 2\ NaCl(aq) + H_2O(l) + SO_2(g)$$

Sulfur dioxide is a colorless gas at room temperature with a strong, pungent odor. When dry, it is relatively inert. It neither burns nor supports combustion. In the presence of catalysts, sulfur dioxide combines with oxygen to form another gas, sulfur trioxide:

$$2\ SO_2(g) + O_2(g) \rightarrow 2\ SO_3(g)$$

Both of these oxides of sulfur are acid anhydrides, combining with water as they dissolve in it to form acids:

$$SO_2(g) + H_2O(l) \rightarrow H_2SO_3(aq)\ \text{(sulfurous acid)}$$
$$SO_3(g) + H_2O(l) \rightarrow H_2SO_4(aq)\ \text{(sulfuric acid)}$$

The most important use of sulfur dioxide is as a starting point in the manufacture of sulfuric acid, but it also has many other uses. When liquefied under pressure, sulfur dioxide is a valuable solvent used in the refining of lubricating oils. Its water solution, sulfurous acid, is used as a bleaching agent for straw, wool, silk, sponges, and other materials that would be injured by chlorine. This solution is also used to preserve some types of food. Salts of sulfurous acid (sulfites) are used in the manufacture of paper from wood pulp and in the production of the important photographic fixing agent "hypo," which is sodium thiosulfate:

$$Na_2SO_3(aq) + S(s) \rightarrow Na_2S_2O_3(aq)$$

EXPERIMENT 27: Obtain a small package of "hypo" crystals from a photographic supply dealer. A solution will also work. *Be sure to wear safety goggles and gloves*. Prepare the blue starch–iodine complex exactly as in Experiment 26. To the blue solution, add a few crystals (or drops) of hypo. Stir. The blue color will disappear because the iodine has been reduced to iodide ion by the reducing agent, sodium thiosulfate. The iodide ion does not combine with starch. This reduction has extensive application in analytical chemistry.

Sulfuric Acid

Sulfuric acid, H_2SO_4, is generally prepared by one of two major processes.

1. *The lead-chamber process.* Steam, oxygen from air, sulfur dioxide, and oxides of nitrogen (which serve as catalysts) are introduced into lead-lined chambers where the following reactions take place:

$$2\ NO(g) + O_2(g) \rightarrow 2\ NO_2(g)$$
$$NO_2(g) + SO_2(g) + H_2O(l) \rightarrow H_2SO_4(aq) + NO(g)$$

   A coating of lead sulfate forms on the lead linings of the chambers and protects the metal from the acid. The acid produced by this process is dilute and impure. The lead-chamber process now supplies less than 3% of the sulfuric acid made.

2. *The contact process.* About 97% of sulfuric acid is made by this process. Filtered and washed sulfur dioxide and air are passed over a catalyst of finely divided platinum or vanadium pentoxide, V_2O_5, at about 400°C. Sulfur trioxide forms:

$$2\ SO_2(g) + O_2(g) \rightarrow 2\ SO_3(g)$$

   This gas is then absorbed in concentrated sulfuric acid, forming fuming sulfuric acid, or oleum:

$$SO_3(g) + H_2SO_4(l) \rightarrow H_2S_2O_7(l)$$

   Water is then added:

$$H_2S_2O_7(l) + H_2O(l) \rightarrow 2\ H_2SO_4(aq)$$

The product marketed from the contact process is about 98% pure sulfuric acid. Chemically, sulfuric acid is both an acid and an oxidizing agent. In dilute solutions, it is a strong acid, slightly less active than hydrochloric and nitric acids. It reacts with metals above hydrogen in the activity series of metals to liberate hydrogen, and it neutralizes bases. Concentrated sulfuric acid oxidizes both metals and nonmetals, forming sulfates and liberating SO_2. Concentrated sulfuric acid is also an important dehydrating agent or desiccant.

Sulfuric acid, known commercially as *oil of vitriol,* is possibly the most important chemical substance in industry. It is used in the manufacture of hydrochloric and nitric acids:

$$2\ NaCl(s) + H_2SO_4(aq) \rightarrow Na_2SO_4(aq) + 2\ HCl(aq)$$
$$2\ NaNO_3(aq) + H_2SO_4(aq) \rightarrow Na_2SO_4(aq) + 2\ HNO_3(aq)$$

It is used in the manufacture of a host of other chemical substances and in the production of phosphate fertilizers, explosives, dyes, drugs, synthetic flavors, and paint pigments. It is used as an oxidizing agent in petroleum refining, as an electrolyte in storage batteries, and as a bath in electroplating

Concentrated sulfuric acid is a thick, syrupy liquid that boils at about 290°C and has a specific gravity of 1.84 g/cm^3. It is miscible with water in all proportions. When sulfuric acid is mixed with water, tremendous amounts of heat are liberated. If a small amount of water is dropped into concentrated sulfuric acid, the liberated heat boils the water and vigorously spatters acid from the container. *This is extremely dangerous!* Therefore, when concentrated sulfuric acid (or any concentrated acid) is to be diluted, *acid is added to water—never water to acid.* In addition, the acid is mixed with water slowly and in small amounts, with constant stirring to dissipate heat. Remember, when working with acids, always wear safety goggles, an apron, and gloves, and always add *acid to water!*

SUMMARY

The elements below oxygen in group VIA (16) of the periodic table form the sulfur family. These elements are less active than the halogens and become more metallic with increasing atomic weight.

Sulfur is the most abundant and most important member of this family. The other members in order of decreasing abundance are selenium, tellurium, and polonium. All these elements have several allotropic forms.

Sulfur occurs free in nature as well as chemically combined and is extracted from underground beds by the Frasch process. Hydrogen sulfide is an extremely poisonous gas. Sulfur dioxide is prepared commercially by burning sulfur or by roasting sulfide ores. Its principal use is in the manufacture of sulfuric acid. Sulfuric acid is the most important manufactured chemical substance. Relatively impure sulfuric acid is made in the lead chamber process. Pure sulfuric acid is made in the contact process. For safety reasons, water is never added to acid. Sulfuric acid solutions are prepared by adding acid to water—slowly and with constant stirring.

PROBLEM SET NO. 16

1. What are the formulas of the following compounds:

   (a) zinc telluride (c) selenic acid

   (b) hydrogen selenide (d) sodium thiosulfate

2. Write balanced equations for the preparation of hydrogen telluride from zinc, tellurium, and hydrochloric acid.

3. What is meant by "allotropic forms" of elements?

4. Which of the following acids are strong and which are weak? Why?

   (a) Sulfurous, H_2SO_3 (d) Selenic, H_2SeO_4

   (b) Sulfuric, H_2SO_4 (e) Tellurous, H_2TeO_3

   (c) Selenous, H_2SeO_3 (f) Telluric, H_2TeO_4

5. Which substance is oxidized and which is reduced in the following reactions?

   (a) $S(g) + O_2(g) \rightarrow SO_2(g)$

   (b) $2\ H_2SO_3(aq) + O_2(g) \rightarrow 2\ H_2SO_4(aq)$

   (c) $Na_2SO_3(aq) + S(s) \rightarrow Na_2S_2O_3(aq)$

CHAPTER 15

THE PHOSPHORUS FAMILY

KEY TERMS

phosphorus family, phosphates

The elements below nitrogen in group VA (15) of the periodic table are known as the *phosphorus family* of elements. Because these elements have 5 electrons in their outermost shell, their most common valence numbers are +3 and −5. Phosphorus is the only definite nonmetal in the group. Arsenic is a metalloid. Antimony and bismuth are definite metals.

OCCURRENCE AND PREPARATION

Phosphorus is too active to be found free in nature. Nearly all is combined with oxygen in the phosphate ion, PO_4^{3-}. Such compounds are called *phosphates*. Its principal ores, phosphorite and apatite, contain calcium phosphate, $Ca_3(PO_4)_2$. Phosphorus is also an essential constituent of bones, teeth, muscle, brain, and nerve tissue. The phosphorus in the body needs to be renewed continuously, so phosphorus-rich foods such as eggs, beans, fish, milk, and whole wheat are important in one's diet. Plants also require phosphorus, which they absorb from the soil. When the soil is low in phosphate, farmers and gardeners may apply phosphate fertilizers.

Small amounts of arsenic are found free in nature, but normally arsenic is found as the sulfide As_2S_3 in the mineral realgar. Antimony occurs as the sulfide in the mineral stibnite, Sb_2S_3, the largest known deposits of which are in the Hunan province of China. Bismuth is the least abundant of the group and is found as the sulfide Bi_2S_3, the telluride Bi_2Te_3, or the oxide Bi_2O_3, and as an impurity in lead, tin, and copper ores.

Phosphorus is obtained by heating a mixture of calcium phosphate, sand, and coke to a high temperature, usually in an electric furnace. (See Figure 15.1.) The overall reaction is:

$$Ca_3(PO_4)_2(s) + 2\,SiO_2(s) + 5\,C(s) \rightarrow 3\,CaSiO_3(s) + 5\,CO(s) + 2\,P(s)$$

The volatile phosphorus distills off and is condensed under water to protect it from oxidation.

Arsenic and antimony are obtained from sulfide ores by first roasting the ores to the oxides

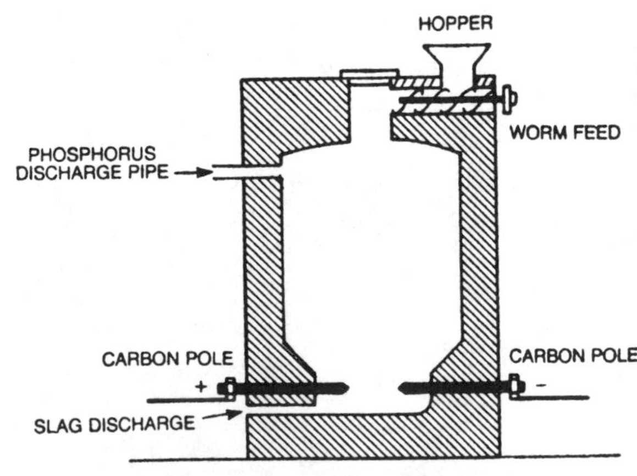

Figure 15.1 Preparation of Phosphorus

and then reducing the oxides by heating with carbon. The reactions for arsenic are

$$2\ As_2S_3(s) + 9\ O_2(g) \rightarrow 2\ As_2O_3(s) + 6\ SO_2(g)$$
$$As_2O_3(s) + 3\ C(s) \rightarrow 3\ CO(g) + 2\ As(s)$$

The reactions for antimony are similar. Antimony may also be obtained by the direct reduction of its sulfide by iron:

$$Sb_2S_3(s) + 3\ Fe(s) \rightarrow 3\ FeS(s) + 2\ Sb(s)$$

Bismuth is most commonly obtained as an anode by-product in the electrolytic refining of lead. It may also be reduced from its oxide ore by heating with carbon in a sloping furnace:

$$Bi_2O_3(s) + 3\ C(s) \rightarrow 3\ CO(g) + 2\ Bi(s)$$

The molten bismuth then flows from the furnace into molds.

PROPERTIES

Table 15.1 summarizes the physical properties of the members of the phosphorus family. Phosphorus has two principal allotropic forms, white and red phosphorus. White phosphorus is a soft, waxy, translucent solid that becomes brittle at 5.5°C. It is crystalline and insoluble in water but very soluble in carbon disulfide. It is extremely poisonous. It must be stored under water because if it is exposed to atmospheric oxygen, it catches fire spontaneously. In the presence of light, it becomes yellow because a film of red phosphorus forms on its surface. White phosphorus has a molecular formula of P_4. Red phosphorus is formed by heating white phosphorus in the absence of air to approximately 250°C. Red phosphorus is generally considered amorphous, but it is really a mixture of several other crystalline allotropic forms of phosphorus. It is insoluble in both water and carbon disulfide, and it is not poisonous. It is safe to handle because it does not burn spontaneously at room temperature.

Arsenic has several allotropic forms. The most common is a gray crystalline form with a metallic appearance. It is very poisonous. The body cannot eliminate it, so if small quantities are taken internally over time, it accumulates until a fatal dose is reached. Both antimony and bismuth are brittle, crystalline, metallic

Property	Phosphorus	Arsenic	Antimony	Bismuth
atomic number	15	33	51	83
atomic mass, g/mol	30.97	74.92	121.8	209.0
electron configuration	$[Ne]3s^2 3p^3$	$[Ar]3d^{10}4s^2 4p^3$	$[Kr]4d^{10}5s^2 5p^3$	$[Xe]4f^{14}5d^{10}6s^2 6p^3$
density, g/cm^3	1.8	5.7	6.7	9.8
melting point, °C	44.1	817*	631	271
boiling point, °C	280	sublimes	1700	1560
color	white	gray	silver-white	gray-white

*at 28 atm and 613°C

Table 15.1 Physical Properties of the Phosphorus Family

solids. Both expand on solidification just as ice does.

Chemically, all these elements combine with oxygen to form oxides. All form the trioxide when burned with a limited amount of oxygen, and all except bismuth form the pentoxide when burned in excess oxygen. For antimony the equations are

$$4\ Sb(s) + 3\ O_2(g) \rightarrow 2\ Sb_2O_3(s)$$
$$4\ Sb(s) + 5\ O_2(g) \rightarrow 2\ Sb_2O_5(s)$$

The oxides of phosphorus, arsenic, and antimony combine with water to form acids. The acidity of these compounds decreases rapidly with the increased atomic mass of the element. Bismuth trioxide combines with water to form a base, bismuth hydroxide. All these compounds are weak electrolytes. Typical reactions are

$$P_2O_3(s) + 3\ H_2O(l) \rightarrow 2\ H_3PO_3(aq) \quad \text{(phosphorous acid)}$$
$$P_2O_5(s) + 3\ H_2O(l) \rightarrow 2\ H_3PO_4(aq) \quad \text{(phosphoric acid)}$$
$$Bi_2O_3(s) + 3\ H_2O(l) \rightarrow 2\ Bi(OH)_3(aq) \quad \text{(bismuth hydroxide)}$$

Arsenous, arsenic, antimonous, and antimonic acids are formed in the same manner as phosphorous and phosphoric acids. (Note the difference in spelling between *phosphorous* acid and the element *phosphorus*.)

All these elements combine with hydrogen to form gases similar to ammonia. They are extremely poisonous and possess increasingly repulsive, garliclike odors. These compounds are arsine (AsH_3), phosphine (PH_3), stibine (SbH_3), and bismuthine (BiH_3).

TRIVIA: Fluorescent materials called optical brighteners are added to laundry detergents to make clothes appear cleaner and brighter than they actually are. These compounds are dyes that adhere to clothes and absorb the ultraviolet rays of sunlight. The compounds then emit that energy as blue light that masks the yellow color of old fabric.

USES

Most phosphorus is used in making matches. Red phosphorus is used directly, and white phosphorus is used in making tetraphosphorus trisulfide, P_4S_3, which has replaced white phosphorus in the striking tips of matches. White phosphorus is also used in some types of rat poisons and in the manufacture of military incendiary grenades and bombs. Arsenic was once used in many insecticides such as paris green (copper arsenite–acetate, $(CuO)_3As_2O_3 \cdot Cu(C_2H_3O_2)_2$) and acid lead arsenate, $PbHAsO_4$. These were especially effective against leaf-eating insects, but concern with the indiscriminate use of these toxic substances prompted their removal from the market. Arsenic is used in some drugs, improves the quality of brass, and when mixed with lead, hardens it for use in shot. Antimony is used in manufacturing flameproofing compounds, paints, ceramic enamels, glass, and pottery; it is also used in alloys to greatly increase the hardness and mechanical strength of the metal. Mixtures of antimony or bismuth with metals expand on solidifying and form sharp, detailed castings. Bismuth is used chiefly in making low-melting metal mixtures for use in fuses, sprinkler systems, and fire alarms. Some bismuth compounds are used in drugs.

Phosphorus is too active to be found free in nature. Nearly all is combined with oxygen in the phosphate ion, PO_4^{3-}. Such compounds are called phosphates.

PRINCIPAL COMPOUNDS

Phosphates

The phosphates of sodium and calcium are by far the most important compounds of this family. Tricalcium phosphate, $Ca_3(PO_4)_2$, present in phosphate minerals and rock, is not suitable as a fertilizer because it is insoluble; however, this compound is converted to the much more soluble monocalcium phosphate by treatment with sulfuric acid:

$$Ca_3(PO_4)_2(s) + 2\ H_2SO_4(aq) \rightarrow Ca(H_2PO_4)_2(aq) + 2\ CaSO_4(s)$$

The mixture of monocalcium phosphate and calcium sulfate so formed is marketed as "superphosphate" fertilizer.

Sodium forms a series of orthophosphates when sodium hydroxide neutralizes a solution of phosphoric acid:

$$NaOH(aq) + H_3PO_4(aq) \rightarrow H_2O(l) + NaH_2PO_4(aq)$$
(monosodium phosphate)

$$2\ NaOH(aq) + H_3PO_4(aq) \rightarrow 2\ H_2O(l) + Na_2HPO_4(aq)$$
(disodium phosphate)

$$3\ NaOH(aq) + H_3PO_4(aq) \rightarrow 3\ H_2O(l) + Na_3PO_4(aq)$$
(trisodium phosphate)

Monosodium phosphate is a source of hydrogen ions in solution, and this acidic characteristic makes it a valuable ingredient in baking powders. Disodium phosphate is used in boilers of ships to soften the water by precipitating calcium and magnesium. Trisodium phosphate hydrolyzes extensively to produce an alkaline solution that reacts with grease and thus is used as a cleaning agent. Many of the first synthetic detergents contained high levels of phosphates. These phosphates acted as extremely good fertilizers and created uncontrolled algal blooms in lakes and streams. The detergents were reformulated to greatly reduce the level of phosphate, thus correcting the problem. Phosphate is also used to soften water because it precipitates the calcium ion from solution.

Sulfides

Whereas tetraphosphorus trisulfide is used mostly in matches, the sulfides of the other members of this family are important paint pigments. Arsenic trisulfide, As_2S_3, is yellow; antimony trisulfide, Sb_2S_3, is red; and bismuth trisulfide, Bi_2S_3, is black.

SUMMARY

The phosphorus family is composed of phosphorus, arsenic, antimony, and bismuth. Phosphorus is the most important member of this family. It is found as compounds such as phosphates in minerals, bones, teeth, muscles, and in brain and nerve tissue.

The members of the phosphorus family can be prepared by heating ores containing these elements and then reducing them to the elemental form. Bismuth can also be isolated in the electrolysis of lead ores containing bismuth compounds.

Phosphorus exists in two principal forms, white and red. The white form is extremely dangerous because it spontaneously ignites when exposed to the oxygen in the air. Arsenic is highly toxic.

Chemically, the members of this family form various oxides that can dissolve in water to form acids and react with hydrogen to form poisonous gases. Phosphates are used as fertilizers.

PROBLEM SET NO. 17

1. Fish have been described as "brain food." Why?

2. Write the balanced equations for the preparation of antimony from stibnite by roasting and reduction with coke.

3. Write balanced equations for the preparation of arsenic acid, H_3AsO_4, from arsenic, oxygen, and water.

4. What are the formulas of the following compounds?

   (a) Arsenous acid

   (b) Antimonous acid

   (c) Antimonic acid

5. Baking powders contain sodium hydrogen carbonate (also called sodium bicarbonate), $NaHCO_3$, and monosodium phosphate, NaH_2PO_4. Explain the role of these compounds in the baking process.

CHAPTER 16

CARBON, SILICON, AND BORON

KEY TERMS

silicates, hydrocarbons, destructive distillation, glass

Carbon, silicon, and boron are the remaining nonmetals among the elements. Carbon and silicon are the first two elements in group IV of the periodic table. They have 4 electrons in their outermost shell. Boron is the first element in group III of the periodic table; it has 3 electrons in its outermost shell. It is the only nonmetal that has fewer than 4 electrons in its outermost shell.

OCCURRENCE AND PREPARATION

Although silicon is less familiar than many other elements, it is second only to oxygen in abundance on Earth. It makes up about 25% of the weight of the crust of Earth, with about 70% of the entire landmass consisting of silicon-bearing rocks. Silicon never occurs free but always combined with oxygen. There are two principal kinds of silicon-containing substances:

1. The compound *silicon dioxide*, SiO_2. This compound is found in many forms such as ordinary sand, quartz, sandstone, flint, agate, and the semiprecious stone amethyst.

2. *Silicate* rocks or minerals (containing silicon and oxygen in varying proportions). In a variety of silicate rocks, metallic ions are combined with many different silicate ions, each containing different proportions of silicon and oxygen. These rocks vary from extremely hard substances like garnet to very soft substances like asbestos. Other silicate materials are clay, mica, talc, zircon, beryl, feldspar, ultramarine, and zeolite, all of which are important economically.

Although carbon is less abundant than many other elements, it is one of the most readily available. It occurs free as the mineral graphite and as diamonds. Carbon is an essential constituent of all plant and animal life. Coal forms as plants decay. The formation of coal proceeds in the following steps:

dry plant material	→	dry peat	→	lignite	→	bituminous (soft coal)	→	anthracite (hard coal)
44% C		50% C		57% C		78% C		81% C

Petroleum and natural gas contain compounds of carbon and hydrogen called *hydrocarbons*. These compounds and their derivatives are so numerous that a special field of chemistry, organic chemistry, is devoted to their study. Carbon is also found as carbon dioxide in the atmosphere and in several forms of carbonate rock, especially limestone, $CaCO_3$.

Boron is relatively rare. It is never found free but, like silicon, is always found combined with oxygen in borates. The most important source is borax, sodium tetraborate,

$Na_2B_4O_7 \cdot 10\ H_2O$. This compound is mined in the desert regions of California. Another compound, boric acid, H_3BO_3, is mined in the volcanic regions of Italy.

Both silicon and boron are relatively difficult to obtain in the free state. Both may be prepared either in the laboratory or commercially by the action of powerful reducing agents on their oxides at high temperature:

$$SiO_2(s) + 2\ C(s) \rightarrow 2\ CO(g) + Si(s)$$
$$SiO_2(s) + 2\ Mg(s) \rightarrow 2\ MgO(s) + Si(s)$$
$$B_2O_3(s) + 3\ Mg(s) \rightarrow MgO(s) + 2\ B(s)$$

The various common forms of carbon are obtained as follows. Graphite in massive crystals is mined or is obtained by heating coke and pitch in furnaces to very high temperatures. Volatiles are driven off, and large graphite crystals grow in the furnace. Charcoal, which consists of tiny graphite crystals, is prepared by the destructive distillation of wood. *Destructive distillation* means heating in the absence of air. The process decomposes the complex carbon compounds in the wood, forming volatile substances and leaving a residue of free carbon. The volatiles in charcoal include wood alcohol, acetic acid, acetone, turpentine (from pine), and many other valuable by-products. Bone black, also consisting of tiny graphite crystals, is prepared by the destructive distillation of bones and other packinghouse wastes. Coke, which contains 90–95% graphitic carbon, is made by the destructive distillation of soft coal in huge coke ovens. Coal tar and ammonia are valuable by-products. Carbon black, or lampblack, is made by burning a hydrocarbon fuel like natural gas in a limited supply of air. The sooty flame strikes a cold surface, depositing graphitic carbon, which is then scraped off. Gem-quality diamonds are mined mostly in South Africa. Industrial diamonds used in cutting tools come from South America or are created synthetically.

PROPERTIES

The physical properties of these three elements are summarized in Table 16.1.

Property	Boron	Carbon	Silicon
atomic number	5	6	14
atomic mass, g/mol	10.81	12.01	28.09
electron configuration	$1s^22s^22p^1$	$1s^22s^22p^2$	$[Ne]\ 3s^23p^2$
density, g/cm^3	2.3	2.25–3.5	2.3
melting point, °C	2300	3600	1415
boiling point, °C	2550	sublimes	2355

Table 16.1 Physical Properties of Boron, Carbon, and Silicon

Silicon is second only to oxygen in abundance. It makes up about 25% of the weight of Earth's crust, and about 70% of the entire landmass consists of silicon-bearing rocks.

Carbon has two allotropic forms, diamond and graphite. Diamond consists of transparent, octahedral (eight-sided) crystals that, when pure, are colorless. Impurities may change the color to range from pale blue to jet black. Diamond is the hardest known substance and a poor conductor of heat and electricity. Diamond's high refractive index is responsible for its brilliance, or "fire." Graphite is black. Graphite crystals are layered sheets of carbon atoms. The layered structure makes graphite very soft, for the sheets slip easily over one another. Graphite is a good conductor of electricity. Both graphite and diamond are somewhat soluble in molten iron. Such solutions, when solidified to room temperature, are known as steel. Trace amounts of other substances may be added to produce different types of steel.

Free silicon occurs only as a crystalline solid, either as massive crystals or as a powder consisting of tiny crystals too small to be seen with an optical microscope. Boron has two allotropic forms, a crystalline variety and an amorphous form. Both silicon and boron are hard and brittle.

At ordinary temperatures, carbon is chemically inert. It combines with oxygen at moderately high temperatures, either directly with elemental oxygen or with oxygen in oxides. Thus, carbon is an important reducing agent, particularly for metallic oxides. At extremely high temperatures—attained normally only in an electric furnace—carbon reacts with other elements to form carbides. In all its compounds, carbon forms covalent bonds with other elements.

Chemically, boron is very much like carbon. It combines with oxygen only at moderately high temperatures and with other elements at extremely high temperatures. Its compounds are covalent. Silicon is somewhat more active than carbon and boron. It combines with the more powerful oxidizing agents and with strong bases at ordinary temperatures in the reaction

$$Si(s) + 2\ OH^- + H_2O(l) \rightarrow 2\ H_2(g) + SiO_3^{2-}$$
(silicate ion)

USES

Free carbon in each of its forms has a variety of important uses. Graphite is used as a dry lubricant. Molded mixtures of clay and graphite form the "lead" used in pencils. Harder pencil leads contain more clay. Graphite is used as the electrode in dry cells and in electric arc furnaces. It is an important ingredient in stove polish and in some paints. Because charcoal becomes porous when it loses volatiles, it can absorb gases onto its surface. Thus, it has important uses in gas masks and in water purification. It is also an important decolorizer in the refining of sugar. Charcoal is a common fuel, and is the form of carbon most often used in reducing some metallic ores and in the manufacture of gunpowder. Bone black is also an important decolorizer. Coke is essential in the reduction of iron from its ore. It is also an important fuel. Carbon black is used in the manufacture of carbon paper, printer's ink, shoe polish, and paint. It is an important additive to rubber used in automobile tires. Diamonds are extensively used as abrasive material, in cutting and drilling tools, and in jewelry making.

Silicon has two important uses. It is the backbone of semiconductors—the "chips" that form the logic base for calculators and computers. Silicon computer chips are being made increasingly smaller, while their power is increasing and increasing. Also, silicon is the basis for photoelectric or solar cells that transform light into electricity. Silicon solar cells are used on rooftops of homes, in spacecraft, in railroad signaling equipment, in light buoys, and to supply electricity to remote villages.

Free boron is used in flares to provide a green color; it is also used as an igniter in rockets. Because boron filaments are ex-tremely hard and have a high melting point, they are used in exotic building materials requiring great strength and heat resistance. Small amounts of boron are added to silicon semiconductors to enhance conductivity.

PRINCIPAL COMPOUNDS

Oxides

Carbon has two important oxides. Carbon dioxide was described in Chapter 12. Carbon monoxide, CO, is an important fuel gas. It is formed by heating metallic oxides with carbon:

$$ZnO(s) + C(s) \rightarrow Zn(s) + CO(g)$$

The gas is then collected and used as a fuel. Carbon monoxide may also be prepared by passing steam over white-hot coke:

$$H_2O(g) + C(s) \rightarrow H_2(g) + CO(g)$$

The mixture of hydrogen and carbon monoxide is known as *water gas*, and it is an important fuel mixture. In general, carbon monoxide is formed whenever carbon is burned in a limited supply of oxygen:

$$2\ C(s) + O_2(g) \rightarrow 2\ CO(g)$$

Carbon monoxide is a colorless, odorless, and tasteless gas. It is very poisonous. Breathing one volume of it mixed with 800 volumes of air for half an hour can be fatal. It burns with a blue flame to form carbon dioxide:

$$2\ CO(g) + O_2(g) \rightarrow 2\ CO_2(g)$$

Carbon monoxide is an important ingredient in many fuel-gas mixtures. Table 16.2 describes some of the important fuel gases.

Silicon dioxide, SiO_2, is frequently called *silica*. It is one of the most abundant natural compounds. When molten silica is cooled it does not crystallize but instead slowly solidifies to *glass*, a rigid, supercooled liquid with its atoms randomly arranged. Other oxides may be added to silica to produce variation in the properties of glasses. Table 16.3 summarizes the composition of some common kinds of glass.

Silica and glass are inert to all acids except hydrofluoric, which etches glass. Basic

		Percentage Composition						
Fuel	Manufacture	CO	H_2	CH_4	Other Hydrocarbons	CO_2	N_2	Heating Value, cal/L
water gas	heating steam and coke	40	50	1.2	none	4.4	3.8	2,700
producer gas	heating steam, air, and coke	20	21	4	none	6.8	3.8	1,600
coal gas	destructive distillation of coal	4.3	44.8	41	6.0	1.1	2.3	6,300
natural gas	gas wells	none	none	85	14	none	1.0	10,500

Table 16.2 Fuel Gases

Type of Glass	SiO_2	B_2O_3	Na_2O	K_2O	MgO	CaO	ZnO	PbO	Al_2O_3	Fe_2O_3
window	70.6		17		0.1	10.6			0.8	0.1
bottle	74		17		3.5	5			1.5	0.4
crown optical	74.6		9	11		5				
borosilicate optical	68.1	3.5	5	16			7			
light flint optical	54.3	1.5	3	8				33		
heavy flint optical	38			5				57		
pyrex chemical resistant	80.5	12	3.8	0.4					2.2	
vycor low expansion	96	4								

Table 16.3 Composition of Glasses

substances attack silica, especially at elevated temperatures:

$$SiO_2(s) + 2\ NaOH(aq) \rightarrow H_2O(l) + Na_2SiO_3\ (aq)\ \text{(sodium silicate)}$$

Sodium silicate is a glass that dissolves in water, so it is often called *water glass*. It is used in soaps and cleansers, in fireproofing, as an adhesive, and in preserving eggs. Other silicates form very important minerals, such as mica; silicates containing aluminum are heated with limestone to make Portland cement. Asbestos was once an important silicate that was used extensively; however, it has since been banned because of its health hazards. When soluble silicates are dissolved in dilute acid solutions, a stiff jelly forms. Dehydration (removal of water) from the jelly produces silica gel, a porous variety of silica. It is an important dehumidifying agent in commercial air-conditioning units and is used by the chemical industry as a catalyst. It absorbs gases as charcoal does, and thus is used to trap industrial vapors that can then be reused.

Boric oxide, B_2O_3, prepared by dehydrating boric acid with heat, is an important additive in chemical-resistant and thermal-resistant glasses. Heating an acidified solution of borax yields boric acid, H_3BO_3:

$$Na_2B_4O_7(aq) + 2\ HCl(aq) + 5\ H_2O(l) \rightarrow 2\ NaCl(aq) + 4\ H_3BO_3(s)$$

This compound is a weak electrolyte, only slightly soluble in water. It is used as an eyewash and as an antiseptic. Borates, such as borax, hydrolyze to form mildly alkaline solutions. Thus, borax is a good cleansing agent. Because calcium and magnesium borates are insoluble, borax is a good water-softening agent, precipitating the calcium and magnesium ion in hard water. It is also a mild antiseptic, a good flux, and a glass additive.

Halides

Carbon tetrachloride, CCl_4, is the most important halide of carbon. It is prepared indirectly because carbon and chlorine do not combine directly. Carbon and sulfur combine readily at high temperatures to form carbon disulfide:

$$C(s) + 2\,S(s) \rightarrow CS_2(l)$$

This liquid is then treated with chlorine gas:

$$CS_2(l) + 3\,Cl_2(g) \rightarrow S_2Cl_2(l) + CCl_4(l)$$

Carbon tetrachloride is then distilled from the sulfur monochloride, which is a valuable by-product used in the rubber industry. Carbon tetrachloride is a colorless, nonflammable liquid that was once used extensively in dry-cleaning clothes and in fire extinguishers; however, because of its toxic effects on the liver, it has been banned from such uses, but it is still an important industrial solvent where it is used in a closely controlled fashion.

Silicon tetrafluoride, SiF_4, and silicon tetrachloride, $SiCl_4$, fume in moist air. They are gases and—when mixed with ammonia—are used in making smoke screens and for skywriting. Boron trifluoride, BF_3, is a gas and is a catalyst in the low-temperature manufacture of synthetic rubber.

Carbides

Calcium carbide, CaC_2, is prepared by heating lime with coke in an electric furnace.

$$CaO(s) + 3\,C(s) \rightarrow CO(g) + CaC_2(s)$$

It reacts with water at room temperature to produce acetylene, C_2H_2:

$$CaC_2(s) + 2\,H_2O(l) \rightarrow Ca(OH)_2(aq) + C_2H_2(g)$$

This gas is an important starting material in the manufacture of many organic chemicals. It is also the fuel used in oxyacetylene torches. Buoys marking sea-lanes are charged with calcium carbide. It reacts with seawater to produce acetylene, which is periodically ignited to provide flashing lights.

Iron carbide, Fe_3C, is present in all steel and is responsible for its hardness and other properties of steel. Silicon carbide, SiC, and tungsten carbide, WC, are among the hardest known substances. They are used as abrasives and in cutting, grinding, and polishing implements.

Hydrogen Cyanide

Hydrogen cyanide, HCN, occurs in several plant products, particularly peach kernels and laurel leaves. It is a poisonous gas. Hydrogen cyanide is a weak electrolyte, so cyanide salts hydrolyze to produce it. Sodium and potassium cyanides are used in solution as electroplating baths; and molten sodium cyanide is used in heat-treating steels.

SUMMARY

Carbon, silicon, and boron are nonmetals. Carbon and silicon are relatively common elements. Silicon is found as silicates and silicon dioxide, and carbon is found in the elemental state as coal, graphite, and diamond. Petroleum and natural gas are primarily composed of carbon-containing compounds. Boron is somewhat rare and is found primarily in nature in borax.

Carbon is used as a lubricant in the form of graphite, and coal is used in the making of steel. Silicon is used extensively in the making of semiconductors and solar cells. Boron is also used in the production of certain semiconductors.

All three of these elements form oxides, such as carbon dioxide, silicon dioxide, and boric oxide. They also react with hydrogen to form halides.

PROBLEM SET NO. 18

1. What compound form if carbon is burned in
   (a) a limited supply of oxygen?
   (b) an excess of oxygen?

2. Account for the fact that producer gas has a lower heating value than water gas.

3. How many different elements are present in ordinary window glass?

4. Why would you caution a child not to eat the kernel inside a peach stone?

THE ALKALI METALS

KEY TERMS

metals, alkali metals, alloys

Of the 118 elements listed, in the periodic table, 88 are *metals*. Metals are elements that tend to lose electrons in reactions, have a luster, and are malleable (can be hammered into a thin sheet), and ductile (can be drawn into a thin wire). Table 17.1 contrasts the general physical properties of metals and nonmetals. Metals are divided into several major families, including the alkali metals, alkaline earth metals, iron and steel mixtures, non-iron metals, and the noble or rare metals. We will look at each of these groups separately in Chapters 17–21.

The *alkali metals* are the elements of group IA (1) of the periodic table (omitting hydrogen). They are called alkali metals because their hydroxides are strongly alkaline, and because many of their salts easily hydrolyze to form alkaline solutions. Sodium and potassium are the two most important members of this family. Lithium, rubidium, cesium, and the extremely rare francium are the others. All have 1 electron in their outermost shell. They are extremely active chemically, giving up the electron readily to form stable ionic compounds.

METALLURGY

The history of civilization shows how humans increasingly learned to use the remarkable properties of metals. The Stone Age gave way to the Bronze Age, then the Iron and Steel Ages as people discovered methods of extracting pure

Metals	Nonmetals
1. All except Hg are solids.	1. At least half are gases.
2. Malleable and ductile solids	2. Brittle solids
3. Good conductors of heat and electricity	3. Poor conductors of heat and electricity
4. Shiny luster; good reflectors	4. No luster; poor reflectors
5. Have only a few electrons in their outermost shell	5. Have many electrons in outermost shell
6. Lose electrons easily to form cations	6. Gain electrons to form anions
7. Good reducing agents	7. Good oxidizing agents
8. Hydroxides are basic.	8. Hydroxides are acidic.

Table 17.1 Physical Characteristics of Metals and Nonmetals

metals from ores. Only a few metals—such as gold, silver, platinum, and copper—occur free in nature. They are called *noble metals*. All others are normally found chemically combined in mineral deposits. If a mineral deposit contains sufficient metal to make it economically profitable to extract the metal from it, it is called an *ore*. The rock material associated with ore, usually silicate rock, is called *gangue*.

The extraction of a metal from its ore usually requires several steps. First, the ore is crushed and concentrated, often by some flotation process. The crushed ore is placed in a bath containing a wetting agent (a liquid that selectively wets the ore but not the gangue) and a foaming agent (a solute that forms a stiff foam with the wetting agent when agitated). The foam rises, carrying ore with it and leaving the gangue at the bottom of the bath. By separating the ore-bearing foam from the rest of the bath, the ore is concentrated.

Next, the metal is reduced from the ore by some process. The extremely active metals must be reduced by electrolysis, where they deposit as pure metal on the cathode of an electrolysis cell. Slightly less active metals may be liberated from their ores by the action of powerful reducing agents such as metallic aluminum. Metals like zinc, iron, and lead may be reduced from their oxides by carbon. If the ores are sulfides or carbonates, they are roasted to the oxide before reduction with carbon. The least active metals may be precipitated from solution by adding a more active metal to the solution.

Alloys are solid solutions of two or more metals and have metallic characteristics. They are economically more important than pure metals. They are generally produced by melting metals together and permitting the mass to solidify. Amalgams are alloys of mercury. Cermets are

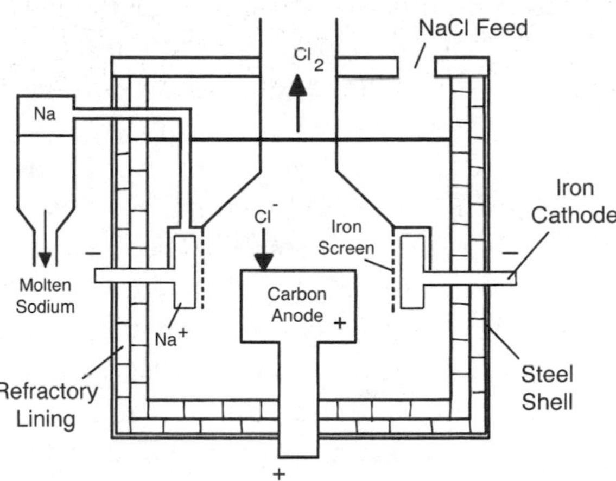

Figure 17.1 The Downs Cell Used to Manufacture Sodium

"alloys" of silicate materials with metals. They combine the strength and toughness of metals with the heat and oxidation resistance of the silicate. They are used in high-temperature applications, such as in gas turbines, jet aircraft parts, and in rockets, where the toughness of metals and temperature resistance are required in a single substance.

The alkali metals are usually prepared by the electrolysis of their salts. The metals deposit on the cathode of the cell. In the Downs process, molten salt, NaCl, is the electrolyte, as shown in Figure 17.1. The overall equation is

$$2\ NaCl(s) \rightarrow 2\ Na(s) + Cl_2(g)$$

Potassium and the other members of this family are obtained from their molten chlorides in a similar way.

OCCURRENCE

The alkali metals are too active to occur uncombined in nature. Sodium and potassium rank sixth and seventh in abundance of elements. Sodium ion makes up about 1.14% of seawater. Large deposits of sodium salts are also found in desert regions, particularly as $NaNO_3$, Na_2SO_4, Na_2CO_3, and $Na_2B_4O_7$. Vast deposits of rock salt, NaCl, also occur. Many complex silicate rocks also contain sodium. Potassium makes up about 0.04% of seawater. It is found in vast beds of KCl and K_2CO_3. It is even more common than sodium in complex silicate rocks. The other alkali metals are rare and are normally found only in complex silicate rock or as impurities in sodium and potassium salt beds.

PROPERTIES

The first three members of this family—lithium, sodium, and potassium—are the only three metals with densities less than that of water. These and other physical properties of the alkali metals are summarized in Table 17.2. All alkali metals are soft, silvery metals. Softness increases with increasing atomic mass. They are all excellent conductors of heat and electricity. When exposed to light, these metals, particularly cesium, emit electrons in an evacuated chamber. That property makes these metals useful in photoelectric cells. When sufficiently heated the alkali metals emit light radiation of characteristic colors:

Li—*red* **Na**—*yellow* **K**—*violet* **Rb**—*red* **Cs**—*blue*

The alkali metals are intensely reactive. The pure metals tarnish immediately in air, forming a hydroxide film. Their activity increases (they become chemically more metallic) with increasing atomic mass. The alkali metals react so vigorously with water to form hydroxides that large chunks of them dropped into water explode. Hydrogen gas is liberated from the water in that reaction. So much heat is generated that potassium burns in contact with water. Cesium will burst into flame even in moist air! These metals must be stored under kerosene, inside a paraffin coating, or in sealed, evacuated flasks.

The alkali metals react explosively with dilute acids to liberate hydrogen. They burn vigorously when heated in oxygen or air. In fact, they are so active that they, especially sodium, are often amalgamated with mercury to reduce the intensity of their reactions. In the amalgam they retain their chemical properties, but to a lesser degree.

Property	Lithium	Sodium	Potassium	Rubidium	Cesium
atomic number	3	11	19	37	55
atomic mass, g/mol	6.94	22.99	39.10	85.47	132.9
electron configuration	$1s^2 2s^1$	$[Ne]3s^1$	$[Ar]4s^1$	$[Kr]5s^1$	$[Xe]6s^1$
melting point, °C	180.5	97.8	63.3	38.9	28.4
boiling point, °C	1,342	883	760	686	670
density, g/cm^3	0.53	0.97	0.86	1.53	1.9

Table 17.2 Physical Properties of Alkali Metals

USES

Years ago most of the metallic sodium made was used to manufacture tetraethyl lead, an antiknock additive in gasoline that was banned in the United States in the 1980s because it released toxic lead into the environment. Metallic sodium is used in the manufacture of a variety of organic chemicals. Its use in the yellow sodium-vapor lights along highways is well known. It is also used in the reduction of titanium metal. Alloyed with potassium, sodium forms an important heat-transfer medium in nuclear power plants, such as the ones in the U.S. Navy's nuclear-powered submarines. All alkali metals are excellent electron sources in photoelectric cells and in cyclotrons, where they are used to investigate of the structure of atoms.

PRINCIPAL COMPOUNDS

Halides

Sodium chloride, NaCl, is one of the most important chemical compounds. It is either mined directly from salt beds or extracted from them by dissolving the salt with water and forcing the brine to the surface with pipes arranged somewhat like those in the Frasch process of extracting sulfur. It is also obtained from seawater by evaporation and recrystallization. Small amounts of salt are essential in the diet, and it occurs in all the body's tissues and fluids. Sodium chloride is used in the preservation of meats and fish and in the preparation of other foods. Ordinary table salt is highly purified sodium chloride. When perfectly pure, it does not absorb moisture, but when slight traces of calcium or magnesium chlorides are present, these deliquescent compounds—which tend to absorb moisture from the air and dissolve in it—can cause the table salt to cake in humid weather. Sodium chloride is also used in the manufacture of many other chemicals, in soap, caustic soda, baking soda, and in the glazing of ceramic ware. Potassium chloride, KCl, is mined from beds in the earth. It is a fertilizer and is used in the preparation of other potassium compounds. The bromides and iodides of sodium and potassium are used in medicines and in the photographic industry.

> Many sodium compounds are both familiar and useful. Sodium chloride, NaCl, is ordinary table salt. Sodium hydrogen carbonate, $NaHCO_3$, is also called sodium bicarbonate. It is the common baking soda used for raising bakery products and for treating acid indigestion.

Hydroxides

Sodium hydroxide, NaOH, also known as caustic soda, is prepared by the electrolysis of brine (sodium chloride solution). Hydrogen is the cathode by-product, and chlorine is the anode by-product. The sodium hydroxide, which forms in the electrolyte solution, is taken from the solution by evaporation after the electrolysis is complete. The overall reaction is

$$2\ NaCl(aq) + 2\ H_2O(l) \rightarrow H_2(g) + Cl_2(g) + 2\ NaOH(aq)$$

Concentrated solutions of this strong base injure the skin and dissolve animal fibers such as wool or silk. Vegetable fibers such as cotton or linen are not affected. Cotton treated with sodium hydroxide takes on a sheen and is known as *mercerized* cotton. Sodium hydroxide is used in the manufacture of soap and many sodium compounds. As lye, it is a common household cleanser. It is also used in refining petroleum.

Potassium hydroxide, KOH, likewise known as caustic potash, is prepared electrolytically from a solution of potassium chloride in the same way as sodium hydroxide. Potassium hydroxide is used in the preparation of many other potassium compounds and in the manufacture of fine soaps. It is the electrolyte in the Edison storage cell.

Carbonates

Sodium carbonate, Na_2CO_3, and sodium hydrogen carbonate (also called bicarbonate), $NaHCO_3$, are the important carbonate compounds of this family. They are prepared by the Solvay process. The raw materials in this process are limestone, ammonia, water, and sodium chloride. The limestone is heated to produce carbon dioxide:

$$CaCO_3(s) \rightarrow CaO(s) + CO_2(g)$$

The carbon dioxide and ammonia are then bubbled through water, causing the following series of reactions:

$$CO_2(g) + H_2O(l) \rightarrow H_2CO_3(aq)$$

$$NH_3(g) + H_2O(l) \rightarrow NH_4OH(aq)$$

$$NH_4OH(aq) + H_2CO_3(aq) \rightarrow H_2O(l) + NH_4HCO_3(aq)$$

The water is saturated with salt, which reacts with the ammonium hydrogen carbonate (or bicarbonate), NH_4HCO_3:

$$NH_4HCO_3(aq) + NaCl(aq) \rightarrow NH_4Cl(aq) + NaHCO_3(s)$$

The sodium hydrogen carbonate precipitates from the solution and is filtered from it and dried. The summary equation for the process is

$$CO_2(g) + NH_3(g) + H_2O(l) + NaCl(aq) \rightarrow NH_4Cl(aq) + NaHCO_3(s)$$

The dried sodium bicarbonate is then heated strongly to produce sodium carbonate:

$$2\ NaHCO_3(s) \rightarrow CO_2(g) + H_2O(l) + Na_2CO_3(s)$$

The Solvay process is remarkable for its lack of waste of materials. The lime formed in the first reaction is added to water to form slaked lime, $Ca(OH)_2$:

$$CaO(s) + H_2O(l) \rightarrow Ca(OH)_2(s)$$

When this product is added to the filtrate that results from the reaction of salt and ammonium hydrogen carbonate, the following reaction occurs:

$$Ca(OH)_2(s) + 2\ NH_4Cl(aq) \rightarrow CaCl_2(aq) + 2\ NH_4OH(aq)$$

When the ammonium hydroxide is heated, ammonia is recovered from the decomposition:

$$NH_4OH(aq) \rightarrow H_2O(l) + NH_3(g)$$

CO_2 and NH_3 are recovered and reused in the process. Calcium chloride is the only by-product formed.

Sodium hydrogen carbonate, often called *baking soda*, is used in making baking powders and antacids. Dry sodium carbonate, also known as *soda ash*, is widely used as a source of alkalinity in boiler water and to assist in the softening of water. Soaps, soap powders, cleansing agents, photographic developers, and many other products are produced from sodium carbonate. Hydrated crystals of sodium carbonate, $Na_2CO_3 \cdot 10\ H_2O$, are known as washing soda and are employed as a common household cleanser.

Nitrates

Both sodium and potassium nitrates are used as meat preservatives, in the manufacture of explosives and fireworks, and as nitrate fertilizers. Sodium nitrate, $NaNO_3$, called Chile saltpeter, is mined in the desert regions of Chile. It is used to prepare potassium nitrate, KNO_3, which is known as saltpeter. To prepare it, saturated solutions of sodium nitrate and potassium chloride are mixed. Because KNO_3 is the least soluble of the four possible salts at low temperatures, it precipitates. The reaction is

$$(Na^+, NO_3^-) + (K^+, Cl^-) \rightarrow NaCl(aq) + KNO_3\,(s)$$

Peroxides

When the alkali metals are burned in oxygen or air, the metallic peroxide is formed:

$$2\ Na(s) + O_2(g) \rightarrow Na_2O_2(s)$$

These peroxides are all powerful oxidizing agents. They are used as bleaching agents and in the preparation of hydrogen peroxide.

Other Compounds

Many other important compounds of sodium and potassium are used industrially. The chlorates, like $KClO_3$, are used in the manufacture of fireworks, flares, and matches. Most soap is sodium stearate, $NaC_{18}H_{35}O_2$, although potassium stearate is often used in fine soaps. Sodium zeolite, $NaAlSi_2O_6$, is a water-softening agent.

Ammonium Compounds

The ammonium ion, NH_4^+, behaves chemically just like sodium or potassium ions, combining readily with negative ions in solution to form compounds that are recovered when the solutions are evaporated. The ammonium ion is formed when ammonia gas dissolves in water to form the weak electrolyte ammonium hydroxide:

$$NH_3(g) + H_2O(l) = NH_4OH(aq)$$

Neutralizing this base with the proper acid forms the corresponding ammonium salts. For example:

$$NH_4OH(aq) + HCl(aq) \rightarrow H_2O(l) + NH_4Cl(aq)$$
(ammonium chloride)

$$2\ NH_4OH(aq) + H_2SO_4(aq) \rightarrow 2\ H_2O(l) + (NH_4)_2SO_4(aq)$$
(ammonium sulfate)

Many ammonium compounds are industrially important. Ammonium nitrate, NH_4NO_3, is a valuable fertilizer. It is also used in the manufacture of explosives. Ammonium hydrogen carbonate, NH_4HCO_3, is a source of CO_2 and is used in some baking powders. Ammonium chloride, NH_4Cl, also called *sal ammoniac*, dissociates into ammonia and hydrogen chloride when hot. This property makes it a valuable soldering flux, for it cleans the surface of the metals being soldered. It is also an important fertilizer, and it is used in the manufacture of dry cells.

SUMMARY

Most of the elements are metals. Most metals have only a few electrons in their outermost shell, so they form compounds through the loss of electrons. They are good reducing agents and form basic hydroxides. They tend to be tough, malleable, and ductile rather than brittle, and they are good conductors of heat and electricity. The alkali metals are the elements of group IA (1) in the periodic table.

Metals can be extracted from ores in several ways, requiring several steps, including grinding, physical separation of metal from rock, and chemical purification. Alloys are solid solutions of two or more metals.

The alkali metals are soft, light metals. Having only 1 electron in their outermost shell, they are extremely active. They are always found chemically combined. Sodium and potassium are among the most common elements, occurring in seawater and in a variety of salt beds. Each of these metals is prepared by the electrolysis of a molten compound of the element. The metals have limited use because of their great activity, but their compounds are stable and used in many industries.

Chemically, the ammonium ion behaves like the ions of the alkali metals. Ammonium compounds are important in industry, especially as a source of nitrogen in fertilizers.

PROBLEM SET NO. 19

1. Write the balanced for the preparation of potassium from fused potassium hydroxide.
2. Explain why unlike the halogens, the alkali metals of higher atomic mass are more active than those of lower atomic mass.
3. What mass of hydrogen gas is evolved when 4.6 g of sodium metal is added to water?
4. Which raw materials are lost in the Solvay process, and what becomes of them?
5. Explain why heating a solution of ammonium hydroxide liberates the gas ammonia.

THE ALKALINE EARTH METALS AND ALUMINUM

KEY TERMS

alkaline earth metals, amphoteric, hard water, ion exchange, ceramics

The *alkaline earth metals,* so named because their oxides form mildly alkaline solutions, are the elements in group IIA (2) of the periodic table. They have 2 electrons in their outermost shell, which they readily lose to form compounds. These elements are only slightly less active than the alkali metals. In fact, the alkaline earth metals of high atomic mass are more active than the alkali metals of low atomic mass. Calcium and magnesium are the two most important metals in this family. Beryllium, barium, strontium, and radium are the others.

Aluminum is not an alkaline earth metal, but its activity is similar, and its physical properties resemble those of magnesium. On the periodic table, aluminum is in group III, just below boron. It has 3 electrons in its outermost shell that are lost as it forms compounds.

OCCURRENCE

None of the alkaline earth metals occur free in nature. All are found in compounds. Beryllium is rare. Its principal ore is beryl, a complex aluminosilicate, $Be_3Al_2Si_6O_{18}$. When this compound contains traces of chromium impurities, it is the green gemstone emerald. When beryl is contaminated with traces of iron, it is aquamarine. Magnesium is the eighth most abundant of all elements, constituting about 2% of Earth's crust. It makes up about 1.14% of seawater. Its chloride makes up part of the mineral carnallite, $MgCl_2 \cdot KCl \cdot 6\ H_2O$; and its sulfate, Epsom salt, $MgSO_4 \cdot 7\ H_2O$, is found in salt beds. Magnesium is found as carbonate deposits in magnesite, $MgCO_3$, and as dolomite, $CaMg(CO_3)_2$. Many complex silicates such as asbestos, $CaMg_3(SiO_3)_4$; talc, $H_2Mg_3(SiO_3)_4$; and meerschaum, $Mg_3Si_2O_5(OH)_4$, also contain magnesium.

Calcium is just behind aluminum and iron in the relative abundance of metals, and it makes up about 3% of Earth's crust. It is less abundant in seawater than is magnesium (0.05%), but its land deposits are extensive. Its carbonate, $CaCO_3$, is found in many varieties such as limestone, marble, chalk, seashells, and the crystalline mineral calcite. Calcium occurs with magnesium in dolomite. Fluorite, CaF_2; anhydrite, $CaSO_4$; and gypsum, $CaSO_4 \cdot 2\ H_2O$, are important calcium minerals. Apatite, $Ca_5(PO_4)_3(OH,F,Cl)$, and many complex silicate rocks are also sources of calcium.

The other alkaline earth metals are rare. Strontium is found as the carbonate mineral strontianite, $SrCO_3$, and the sulfate mineral celestite, $SrSO_4$. Barium is found in similar compounds: witherite, $BaCO_3$, and barite, $BaSO_4$. Radium occurs in minute quantities in uranium ores, in the oxide pitchblende, and in the vanadium compound carnotite.

Aluminum is the most abundant metal on earth, constituting more than 7% of Earth's crust. It occurs in a variety of silicate rocks, especially mica, $KH_2Al_3(SiO_4)_3$, and feldspar, $KAlSi_3O_8$. The action of moisture and carbon dioxide in the atmosphere disintegrates these rocks to form clay and ultimately aluminum oxide. That process is known as *weathering* and is responsible for the formation of the principal aluminum ore, bauxite, $Al_2O_3 \cdot 2\ H_2O$. Aluminum also occurs as cryolite, Na_3AlF_6.

METALLURGY

Of the pure metals in this family, magnesium is by far the most abundantly produced, although metallic beryllium, calcium, and barium are commercially important. All the alkaline earth metals may be prepared by the electrolysis of molten chlorides or by heating the oxides in a vacuum with a powerful reducing agent such as powdered aluminum or silicon (in the form of an alloy with iron known as ferrosilicon). Typical reactions are

$$CaCl_2(l) \rightarrow Cl_2(g) + Ca(s) \quad \text{(electrolysis)}$$

$$2\ MgO(s) + Si(s) \rightarrow SiO_2(s) + 2\ Mg(s) \quad \text{(heating in vacuum)}$$

> Aluminum is the most abundant metal on earth, making up more than 7% of Earth's crust.

Aluminum is prepared electrolytically by the Hall process (see Figure 18.1). Purified aluminum oxide, Al_2O_3, obtained from bauxite, is melted in an electric furnace with the aid of the fluxes cryolite and fluorite. (Fluxes are used

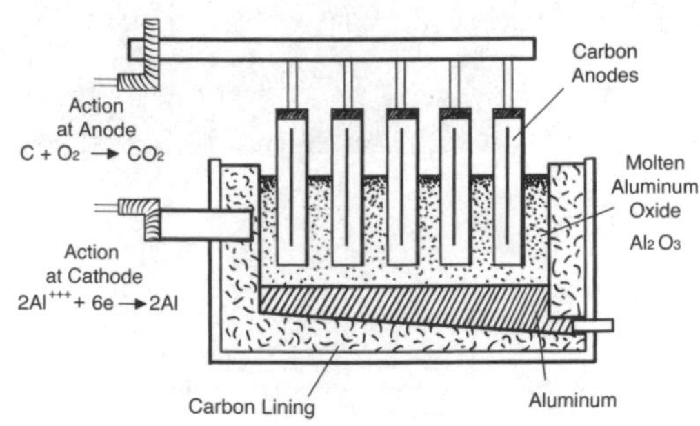

Figure 18.1 Hall Process for the Preparation of Aluminum

to reduce the melting point of a substance.) An electric current passes through the molten oxide and causes aluminum to be deposited on the carbon cathode lining the furnace. The resistance of the molten oxide to the flow of current generates heat, which maintains a high temperature inside the furnace. Oxygen is liberated at the carbon anodes immersed in the melt and attacks them, forming both carbon monoxide and carbon dioxide. Therefore, the anodes must be replaced regularly. The fundamental reaction of the Hall process is

$$2\ Al_2O_3(l) \rightarrow 3\ O_2(g) + 4\ Al(l)$$

A process that consumes less energy than the Hall process has been developed to make aluminum. Aluminum oxide is treated with chlorine:

$$2\ Al_2O_3(s) + 6\ Cl_2(g) \rightarrow 4\ AlCl_3(s) + 3\ O_2(g)$$

The aluminum chloride is electrolyzed:

$$2\ AlCl_3(s) \rightarrow 2\ Al(s) + 3\ Cl_2(g)$$

The chlorine liberated by electrolysis is recycled to produce more $AlCl_3$.

JOKE: What do you do with a dead chemist?— Barium.

Property	Be	Mg	Ca	Sr	Ba	Ra	Al
atomic number	4	12	20	38	56	88	13
atomic mass, g/mol	9.01	24.31	40.08	87.62	137.3	226.0	26.98
electron configuration	$1s^2 2s^2$	$[Ne]3s^2$	$[Ar]4s^2$	$[Kr]5s^2$	$[Xe]6s^2$	$[Rn]7s^2$	$[Ne]3s^2 3p^1$
melting pt., °C	1,278	649	839	769	725	700	660
boiling pt., °C	2970*	1090	1484	1384	1640	1140	2467
density, g/cm^3	1.85	1.74	1.55	2.6	3.5	5	2.7

*at 5 Torr

Table 18.1 Physical Properties of Alkaline Earth Metals and Aluminum

PROPERTIES

The physical properties of the alkaline earth metals and aluminum are summarized in Table 18.1.

All are silvery white metals and good conductors of heat and electricity. They are denser and melt at higher temperatures than the alkali metals. Like the alkali metals, barium is a good electron emitter. Calcium, strontium, and barium ions and vapors—when sufficiently hot—emit colored light as follows:

Ca—brick red Sr—brilliant crimson Ba—green

Of the metals in this group, only beryllium, magnesium, and aluminum do not appreciably tarnish on exposure to moist air. All the others form heavy hydroxide coats. A thin film of oxide forms on beryllium, magnesium, and aluminum, which then protects these metals from further attack. All these metals burn with a brilliant white light in air or oxygen. All are good reducing agents, and magnesium and aluminum are extensively used for that purpose. One such reducing reaction involving aluminum is known as the *thermite reaction*. In this reaction, other metals are reduced from their oxides. For example, igniting a mixture of powdered aluminum and ferric oxide reduces iron:

$$Fe_2O_3(s) + 2\ Al(s) \rightarrow Al_2O_3(s) + 2\ Fe(s)$$

The reaction generates enough heat to melt the iron and permit it to flow into cracks (see Figure 18.2). Thus, the thermite reaction can weld broken iron shafts.

The heavier alkaline earth metals combine readily with water at room temperature, liberating hydrogen and forming metallic hydroxides. Beryllium, magnesium, and aluminum

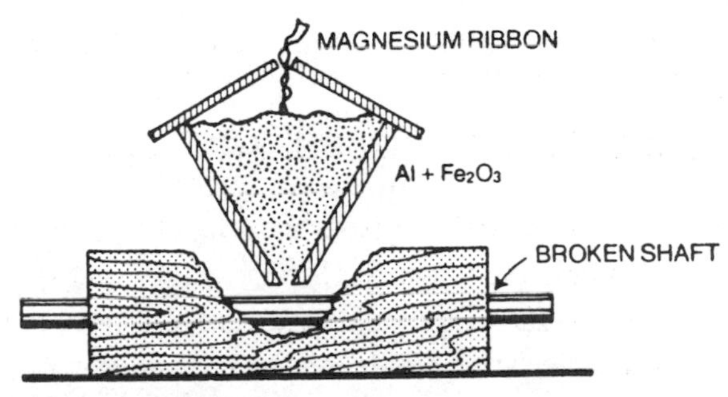

Figure 18.2 Thermite Welding

undergo the same reaction at elevated temperatures. All these metals react vigorously with acid to liberate hydrogen, but treating aluminum with concentrated nitric acid forms an impervious oxide film and stops the reaction.

USES

Beryllium is used primarily in alloys, especially to harden copper. One such copper alloy, known as a beryllium bronze, contains about 2% Be. It is used to form "nonsparking" tools. Another alloy is used in the formation of springs that will stretch a long way without breaking.

Magnesium is an important structural metal. It is used either in the pure form or alloyed with aluminum in the construction of lightweight aircraft. Magnesium turnings (small bits) were once burned in flashbulbs to produce brilliant light. Powdered magnesium is sometimes used in place of aluminum in the thermite reaction. The other alkaline earth metals have few important uses.

Aluminum has rapidly taken its place as one of the important structural metals. Its light weight, resistance to corrosion, and tensile strength make it an excellent outside cover for buildings. That it reflects light adds to its value as a covering material. Its lightness and strength make it valuable in the manufacture of truck bodies, wheels, railroad equipment and rolling stock, pistons, and many other objects. Because aluminum is so malleable (can be easily rolled into sheets), great quantities of it are used as aluminum foil. Because it is so ductile (can be easily drawn into wire), much of the wiring in cross-country electrical lines is made of aluminum. Aluminum powder, which consists of tiny flakes of the metal suspended in a suitable oil medium, finds extensive use as aluminum paint. We are all certainly aware of aluminum cans. Many of us have collected these cans and sold them to recycling places. These used cans are then melted and re-formed into new cans.

PRINCIPAL COMPOUNDS

Oxides

Oxides of aluminum and the alkaline earth metals have very high melting points. Consequently, they are used in the manufacture of firebrick and other refractory (high-temperature) materials. Calcium oxide, CaO, is known as lime or quicklime; and magnesium oxide, MgO, is known as magnesia. Both are prepared by the thermal decomposition of their carbonates. The process is known as *calcining*. In the preparation of lime, Figure 18.3, the process is known as the burning of limestone. Typical equations are

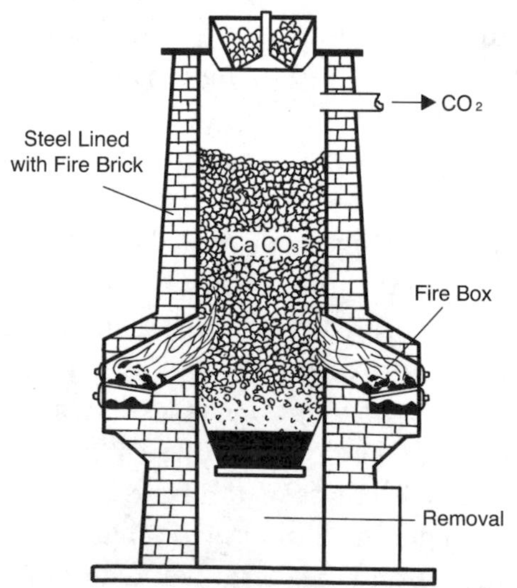

Figure 18.3 Preparation of Lime

$CaCO_3(s) \rightarrow CO_2(g) + CaO(s)$ (lime)
$MgCO_3(s) \rightarrow CO_2(g) + MgO(s)$ (magnesia)

Both oxides react with water to form slightly soluble hydroxides. When barium oxide is heated, it combines with oxygen to form barium peroxide. The equation is

$$2\ BaO(s) + O_2(g) \rightarrow 2\ BaO_2(s)$$

This compound is sometimes used as a source of hydrogen peroxide:

$BaO_2(s) + H_2SO_4(aq)$
$\rightarrow BaSO_4(s)\downarrow + H_2O_2(aq)$

In the preparation of lime, the carbon dioxide produced from calcining is a principal source of CO_2 for portable CO_2 fire extinguishers (see Figure 18.4).

Aluminum oxide, Al_2O_3, also known as alumina, occurs as crystals of corundum, ruby, and sapphire. It is extremely hard and has widespread use in making abrasive papers and grinding and cutting wheels. It is used in the manufacture of synthetic sapphires and rubies, which are used as jewels in timepieces and other delicate instruments.

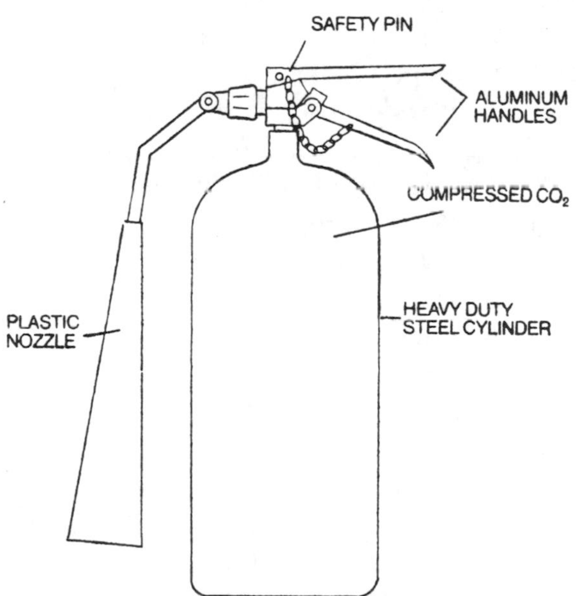

Figure 18.4 Carbon Dioxide Fire Extinguisher

Hydroxides

The hydroxides of the alkaline earth metals are strong bases, although they are only sparingly soluble in water. Barium hydroxide, $Ba(OH)_2$, is the most soluble and beryllium hydroxide, $Be(OH)_2$, the least. Beryllium and aluminum hydroxides are soluble in both acids and bases. Thus, they behave as bases toward acids and as acids toward bases. Such hydroxides are said to be *amphoteric*. The amphoteric reactions of aluminum hydroxide are

As base: $Al(OH)_3(aq) + 3\ HCl(aq)$
$\rightarrow 3\ H_2O(l) + AlCl_3(aq)$
As acid: $H_3AlO_3(aq) + 3\ NaOH(aq)$
$\rightarrow 3\ H_2O(l) + Na_3AlO_3(aq)$
(sodium aluminate)

Aluminum hydroxide, $Al(OH)_3$, is a sticky, porous material that is insoluble in water. It is an important intermediary in some dyeing processes because it clings to fabrics and absorbs coloring material that would otherwise not color the fabrics. When aluminum hydroxide has been precipitated on cloth it is called a *mordant*, and after it has adsorbed (collected on the surface) the dye, it is known as a *lake*. Because aluminum hydroxide collects great quantities of suspended material from solution, it is used extensively in the clarification of municipal drinking water. Because aluminum salts hydrolyze to produce aluminum hydroxide, they may be used as an indirect source of that compound.

Magnesium hydroxide, $Mg(OH)_2$, is known as milk of magnesia, a commonly used antacid and laxative. Calcium hydroxide, $Ca(OH)_2$, is prepared by adding water to lime. It is called slaked lime. It is used in processes requiring a mild alkali, such as in the removal of hair from hides, in the preparation of mortar and plaster, in water softening, and in the manufacture of many substances such as bleaching powder, paper, glass, and ammonia. Mortar is a paste of slaked lime, sand, and water. It sets and dries by absorbing carbon dioxide from the atmosphere, converting the slaked lime to calcium carbonate:

$$Ca(OH)_2(s) + CO_2(g) \rightarrow CaCO_3(s) + H_2O(l)$$

The setting of mortar requires time because the reaction proceeds from the outside to the interior of the mortar.

Sulfates

Magnesium sulfate, $MgSO_4 \cdot 7\ H_2O$, when purified, is the commonly used inexpensive laxative Epsom salt. Calcium sulfate, $CaSO_4 \cdot 2\ H_2O$, is the mineral gypsum. It is used as a fertilizer and in the manufacture of a variety of fireproof building materials. It is also used in the manufacture of plaster of paris. When gypsum is heated to about 120°C, it loses 75% of its water of crystallization:

$$2\ CaSO_4 \cdot 2\ H_2O(s) \rightarrow 3\ H_2O(l) + (CaSO_4)_2 \cdot H_2O(s)$$

(plaster of paris)

That reaction is reversed when the proper amount of water is added to plaster of paris, and the mass quickly solidifies as crystallized gypsum. A slight expansion accompanies the setting, which enables sharp reproduction of details in a mold. Plaster of paris is used in wall plaster, for plaster casts (although quick-setting alternatives have gained popularity in the medical community), and in casting and molding operations.

Aluminum sulfate, $Al_2(SO_4)_3 \cdot 18\ H_2O$, is widely used as a source of aluminum hydroxide (by hydrolysis), in the purification of water, in dyeing, and in paper making. It is also used in the preparation of alums. Alum is a double sulfate, consisting of the sulfates of a monovalent and a trivalent metal in a single crystalline compound. The general formula of alum may be written

$$M^+M^{3+}(SO_4)_2 \cdot 12\ H_2O$$

Many types of alum are known, consisting of monovalent sodium, potassium, or ammonium and trivalent aluminum, iron, or chromium. The aluminum alums are a source of aluminum hydroxide in essentially the same applications as those of aluminum sulfate.

Carbonates

Calcium carbonate, $CaCO_3$, is an abundant and important mineral. As limestone and marble it is widely used in construction. Marble is the principal stone used by sculptors. Limestone, as a source of both lime and carbon dioxide, is an important raw material in many manufacturing processes. Such varied products as mortar, baking soda, and toothpaste use limestone as a raw material. The glass industry uses great quantities of it, and the reduction of iron in a blast furnace requires limestone as a flux. It is also an important fertilizer. Limestone is a source of much scenic beauty, from the chalk cliffs of Dover to the magnificent underground limestone caverns, tourists visit throughout the world. The stalactites that hang from the ceilings of limestone caverns and the stalagmites that build up from the floors are limestone that precipitated from saturated solutions.

Cement and Concrete

When heated in a huge rotary kiln, a mixture of limestone with clay or shale, $H_4Al_2Si_2O_9$, forms a complex of lime, alumina, and silica. The clinkered mass of calcium aluminosilicates is powdered, and a small amount of gypsum is added to make Portland cement. When the cement is mixed with sand, gravel, and water, it sets to form concrete. The setting process involves the formation of crystals of calcium aluminate, calcium hydroxide, and silica. The crystallization process releases heat, so large batches of concrete must be cooled to ensure proper setting.

Hard Water

Water that contains calcium and magnesium ions in solution is known as *hard water* because the ions interfere with the cleansing action of soap. The ions with soap, sodium stearate, preventing the formation of a lather. The insoluble waxy salt formed from the soap and the ions is sometimes called "bathtub ring."

EXPERIMENT 28: *Be sure to wear your safety goggles!* Shave about $\frac{1}{8}$ inch of soap from the end of a small bar of white soap. Chop the shavings finely and add them to $\frac{1}{4}$ cup of alcohol. Stir thoroughly to dissolve as much of the soap as possible. Let the excess soap settle. Measure the following into small stoppered jars or bottles: $\frac{1}{4}$ cup of distilled water, $\frac{1}{4}$ cup of tap water, $\frac{1}{4}$ cup of tap water to which 5 crystals of Epsom salt have been added. To each of these liquids in turn, add the clear soap solution drop by drop, shaking the jar well after each addition of soap until a lather forms that persists for several minutes. The more soap required, the harder the water being tested. From your results you will be able to infer which water is hardest and which water is softest.

> Water that contains calcium and magnesium ions is known as "hard" water because the ions interfere with the cleansing action of soap. Removal of the ions "softens" the water.

The softening of hard water involves the removal of calcium and magnesium ions from solution. If the hard water contains sufficient bicarbonate ion, it may be softened by

- boiling:

$$Ca^{+2} + 2\,HCO_3^- \rightarrow CaCO_3(s)\downarrow + H_2O(l) + CO_2(g)$$

- or treatment with slaked lime:

$$Ca^{+2} + 2\,HCO_3^- + Ca(OH)_2(aq) \rightarrow 2\,CaCO_3(s)\downarrow + 2\,H_2O(l)$$

Water containing at least two moles of bicarbonate ion for every mole of calcium or magnesium ion to be removed is called *temporary* hard water. If the water contains less bicarbonate ion, it is known as *permanent* hard water. Adding soda ash, sodium phosphate, or borax may soften such water. These compounds precipitate calcium and magnesium ions as carbonates, phosphates, or borates respectively.

When large quantities of water are to be softened for industrial purposes, the ion-exchange properties of $NaAlSi_2O_6$, sodium zeolite, are often used. *Ion exchange* is a reversible process in which ions are released from an insoluble permanent material in exchange for other ions in a surrounding solution. Sodium zeolite can exchange its sodium ions for calcium and magnesium ions when in contact with hard water, thereby softening it. The sodium ions can be returned to the zeolite by treating it with a concentrated solution of sodium chloride (brine), which regenerates the zeolite for further use. Synthetic organic resins with ion-exchange properties are also widely used for this purpose. An acid resin exchanges hydrogen ions for all metallic ions, and a basic resin exchanges hydroxide ions for all negative ions. The result is the production of deionized water of high purity.

Hard water cannot be used in boilers because calcium and magnesium salts precipitate from boiling water on the walls of boiler tubes, forming boiler scale. This scale greatly reduces the heat transfer properties of the tubes and may ultimately cause them to blow out. Boiler scale usually consists of calcium sulfate and magnesium carbonates. You may have noticed this scale on the inside of coffee pots and teakettles in your own home.

Ceramics

Ceramics are nonmetallic materials made from clay and hardened by firing at high temperature. They contain silicate crystals suspended in a glassy cement and are commonly available as bricks, pottery, chinaware, and porcelain. All ceramics are made from clay, an impure aluminum silicate, or kaolin, a purer form of clay. Bricks are made by heating a paste of clay and sand until it is thoroughly dehydrated. Pottery, consisting of only clay and water, is first shaped on a potter's wheel and then fired in a kiln. It may later be glazed by coating it with a silicate glass. Chinaware is made from a paste of calcium phosphate, feldspar, and kaolin that is fired and later glazed. Porcelain is made from quartz, white kaolin, marble, and feldspar. The plastic paste is first molded and then fired.

SUMMARY

The metals of group IIA (2) of the periodic table are called the alkaline earth metals. Beryllium, magnesium, calcium, strontium, barium, and radium are the members of this family. Aluminum is in group III but resembles the alkaline earth metals chemically and physically. All of these silvery white metals are good conductors of heat and electricity, and all are quite active chemically.

Aluminum is the most abundant metal on Earth, and calcium and magnesium are also very abundant. The other alkaline earth metals are relatively rare. None of these metals is found free in nature. They are usually prepared by electrolysis of a molten compound. The Hall process for the electrolytic preparation of aluminum is of great industrial importance.

Beryllium is used as an alloy with copper to harden the copper. Magnesium and aluminum are important light structural metals. Their oxides are used in making refractory materials. Aluminum oxide is an important abrasive. Aluminum hydroxide is amphoteric and has adsorptive properties that are important in the dyeing industry. Calcium hydroxide is used in many industries both as a source of calcium and as a mild alkali.

The sulfates of these metals tend to be hydrated. Partially dehydrated calcium sulfate is plaster of paris. Aluminum sulfate is the source of many types of alum. Calcium and aluminum form part of the complex silicate structure of Portland cement. Calcium and magnesium ions are responsible for the hardness of water. Aluminum silicate is the basic ingredient in ceramics.

PROBLEM SET NO. 20

1. Carnallite, the double chloride of potassium and magnesium, is melted and then used as a bath for the electrolytic preparation of magnesium.

   (a) Will hydrogen, from the water of crystallization, be a by-product? Explain.

   (b) Using Table 11.2, compute the voltage needed to decompose KCl and $MgCl_2$ electrolytically.

   (c) Suggest a safeguard to ensure the deposition of magnesium free of potassium contaminants.

2. Show with ionization equations how solutions of aluminum sulfate hydrolyze.

3. (a) Write the formula of crystalline sodium aluminum alum.

   (b) Can an alkaline earth metal be present in an alum? Why?

4. Silicate ores are relatively undesirable as a source of a metal, not only because they are difficult to decompose chemically but also because of their low metal content. For example, what is the percentage by weight of pure beryllium that could be extracted from pure beryl ore, $Be_3Al_2Si_6O_{18}$?

5. The activity series of metals (Chapter 11) shows only magnesium and calcium from the alkaline earth family. Where would you expect strontium, barium, and radium to be placed in the table, and in what order?

CHAPTER 19 IRON AND THE STEEL ALLOY METALS

KEY TERMS

pig iron, cast iron, wrought iron, transition element, steel

Iron forms the skeleton that supports civilization. Pure iron is rarely encountered, but in many combined forms, it is the fundamental ingredient of modern tools and structures.

IRON

Occurrence

Iron is second only to aluminum in the list of abundant metals, constituting about 5% of Earth's crust. The ores containing the greatest concentration of iron are hematite, Fe_2O_3; magnetite, Fe_3O_4; and pyrite, FeS_2. Hematite once occurred in large deposits on the Mesabi Range in Minnesota, but its high-grade ores were depleted by the 1950s. Subsequently, methods were developed for recovering iron from the region's lower-grade taconite, a mixture of hematite and magnetite. Magnetite is magnetic iron oxide, also known as lodestone. Pyrite has a yellow color similar to gold and if often referred to as fool's gold. It is used as a source of sulfur dioxide (by roasting) in the manufacture of sulfuric acid.

Siderite, $FeCO_3$, is also worked for its iron content, and low-grade iron ores containing less than 40% iron are abundantly distributed over Earth. These ores will soon be our chief source of iron, since we are rapidly depleting those containing a higher percentage of iron. Meteorites contain free metallic iron, and it is believed that Earth's liquid core consists mostly of iron.

JOKE: What can you make from the elements potassium, nickel, and iron to protect yourself?—A KniFe.

Metallurgy

Iron is reduced from its oxide ore in a blast furnace (see Figure 19.1). The furnace is charged, or loaded, with ore (Fe_2O_3 and silicate impurities), coke, limestone, and hot air.

The solid ingredients are charged at the top of the furnace, with limestone laid in first, then

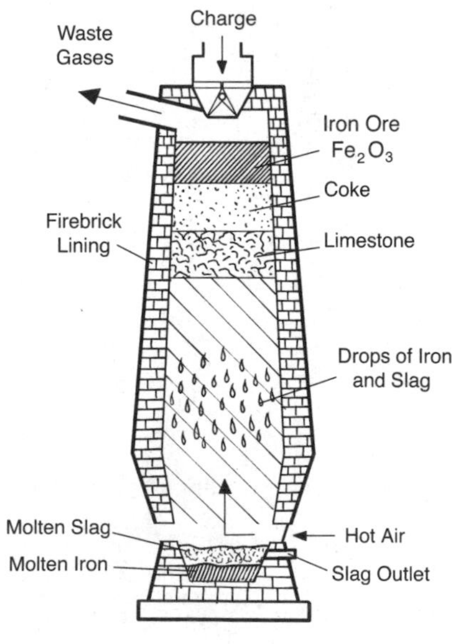

Figure 19.1 The Blast Furnace

a layer of coke, and then the ore in the top layer. Hot air is forced upward through the furnace from the bottom. The air oxidizes the coke to carbon monoxide:

$$2\ C(s) + O_2(g) \rightarrow 2\ CO(g)$$

The carbon monoxide is the reducing agent in the process, and the heat generated in its formation provides the high temperature required for the reduction of the iron:

$$Fe_2O_3(s) + 3\ CO(g) \rightarrow 3\ CO_2(g) + 2\ Fe(s)$$

Limestone acts as a flux and provides the lime that combines with silica impurities to form slag:

$$CaCO_3(s) \rightarrow CO_2(g) + CaO(g)$$
$$CaO(s) + SiO_2(s) \rightarrow CaSiO_3(s) \quad \text{(slag)}$$

Both iron and slag melt at the furnace's high temperatures and drop to the bottom. The two liquids are immiscible, forming separate layers, with the more dense iron at the bottom. Tap holes act as outlets for the liquid layers.

Solidified iron obtained from a blast furnace is known as *pig iron*. It is slightly over 90% pure and contains between 2% and 6% percent carbon, appreciable amounts of silicon and manganese, and small amounts of phosphorus and sulfur. The carbon is present as an iron carbide, Fe_3C, known as cementite, the compound responsible for the hardness and brittleness of pig iron.

If the pig iron is permitted to cool suddenly, all the carbon is retained as the carbide, and the product is known as *white cast iron*. If the pig iron is cooled slowly, some of the carbon separates from the carbide and forms tiny flakes of graphite. This product is known as *gray cast iron*. Because of the partial decomposition of the carbide, gray cast iron is softer and tougher than white cast iron, and it has a higher melting point. Because gray cast iron expands on solidifying, it makes better castings than white cast iron, which shrinks on solidifying. For that reason, gray cast iron is usually used in the making of stoves, radiators, and many other cast objects that are not subject to shock.

If pig iron is melted again in a furnace with additional iron oxide, Fe_2O_3, and stirred or puddled, the oxygen from the oxide combines with the carbon and removes it as carbon monoxide. It also combines with phosphorus and sulfur, forming oxides, and with silicon, forming a slag. When the puddling process is complete, the product is rolled or hammered at a temperature just below the solidification point to squeeze out the slag; however, some slag is retained, producing a fibrous structure in the metal. This product is known as *wrought iron*. It is soft, tough, and somewhat resistant to corrosion. It is used for making bolts, chains, anchors, iron pipe, fences, and grillwork.

Passing dry hydrogen gas over hot ferric oxide, Fe_2O_3 yields pure iron:

$$Fe_2O_3(s) + 3\ H_2(g) \rightarrow 3\ H_2O(g) + 2\ Fe(s)$$

Properties

Pure iron, atomic number 26 and atomic mass 55.85 g/mol, is a transition element. A *transition element* is one in which the d orbital is not full. Iron has two incomplete shells of electrons. Its shell electron configuration is $[Ar]3d^5 4s^1$. Transition elements may use not only the electrons in the outermost shell in forming compounds but also one or more electrons from the second outermost, or d, shell. Iron melts at 1535°C and boils at 3000°C. Its density is 7.86 g/cm^3. It is a soft, silvery white metal, very ductile and malleable, and a good conductor of heat and electricity. It is strongly magnetic.

Iron forms two series of compounds: iron(II), also called ferrous, compounds in which the valence of iron is 2; and iron(III), also called ferric, compounds in which the valence of iron is 3. When pure iron reacts with acids or with steam at high temperature to form hydrogen, the iron(II) (ferrous) compounds form. They may then be oxidized by atmospheric oxygen or other powerful oxidizing agents to the iron(III) (ferric) compounds. The rusting of iron is an electrochemical process (see Chapter 11). Dipping iron into concentrated nitric acid, a powerful oxidizing agent as well as an acid, renders iron passive because of the formation of a film of oxides. *Passive* iron does not rust unless scratching or striking with a sharp object breaks the film. Alkalies form a thin film of iron(II) (ferrous) hydroxide on iron that then prevents further attack on the metal by the alkalies. For that reason, iron forms a suitable container for melting sodium and potassium hydroxides in industry.

Uses

Iron is often used as a structural material in objects ranging from toys to skyscrapers. Telephones, electric motors, generators, and many other devices utilize its magnetic properties.

Compounds

Iron(III) (ferric) oxide, Fe_2O_3, the principal iron ore, has many direct uses. Because it is hard like alumina, it is a valuable abrasive, used especially in the polishing of glass. Its red color makes is an important pigment in the paint industry, and the cosmetic industry uses it to color foundations and eye shadows. Iron(III) (ferric) hydroxide, $Fe(OH)_3$, is reddish brown but otherwise has nearly the same properties as aluminum hydroxide. It is used as a mordant and in the clarification of drinking water. Magnetic oxide, Fe_3O_4, is an equimolar combination of FeO and Fe_2O_3. It forms when iron burns in air or oxygen and is an important source of iron.

STEEL

Steel is an alloy of iron that contains less that 2% carbon. If more carbon is present, the alloy is cast iron. To make steel, carbon is removed from pig iron. Three methods in use today are the oxygen furnace, open-hearth furnace, and electric furnace processes.

More than 60% of steel is made by the basic oxygen furnace process, which is the most rapid method. In the basic oxygen furnace, Figure 19.2, the egg-shaped converter is charged about two-thirds molten pig iron and one-third steel scrap. Fluxes such as lime are added to reduce the melting point and to combine with the impurities into a slag. A water-cooled oxygen lance is lowered into the furnace, and oxygen blown onto the top of the metal bath at supersonic speed creates intense heat of up to 3500°F. The oxygen combines with carbon and other impurities, converting the molten mass into high-quality steel, which can be used in manufacturing automobiles and other products. In less than 45 minutes, the cycle is complete, the converter is tilted, and the refined steel is poured into molds.

In the open-hearth furnace process, Figure 19.3, limestone and equivalent amounts of pig iron

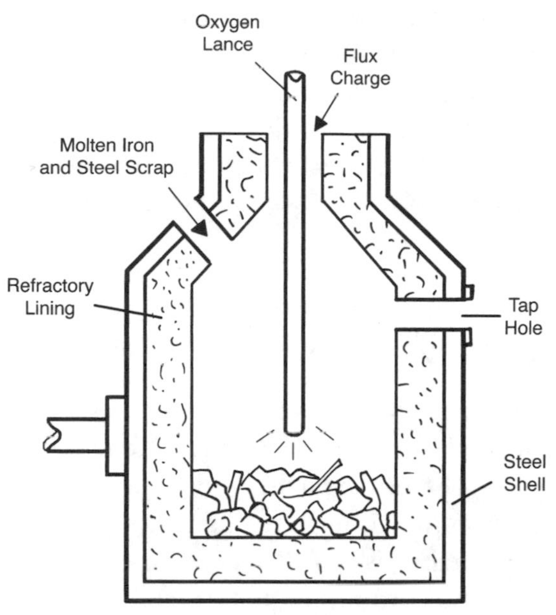

Figure 19.2 The Basic Oxygen Furnace

and scrap iron are melted in the huge furnace basin lined with basic magnesia brick. Gaseous fuel is burned with hot air or oxygen over the melt, and the heat is reflected downward onto the molten metal. An oxygen lance also helps combust the gaseous fuel to increase the temperature and hasten the melting. Sulfur and phosphorus are removed by direct combination with the basic lining of the furnace. Since the process requires 8 to 10 hours, there is ample time for careful chemical analysis of the metal being produced. The carbon and manganese contents are controlled by adding sufficient coke and spiegeleisen, a pig iron containing manganese that is used to raise the manganese content of steel. When the batch has sufficiently reacted and become homogeneous, it is poured into giant ladles and moved away for solidification.

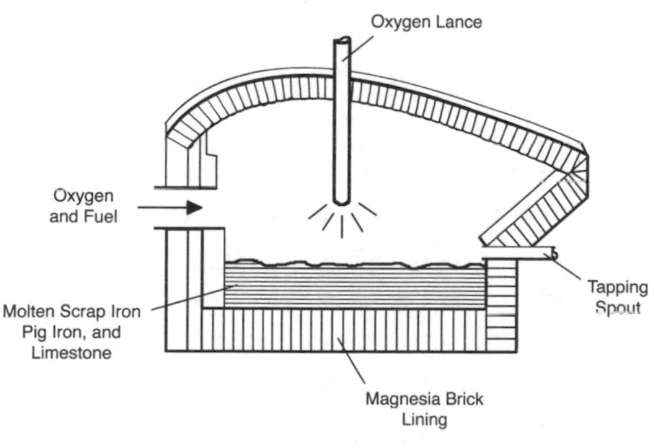

Figure 19.3 The Open-Hearth Furnace

> Steel is an alloy of iron that is less that 2% carbon. To make steel, carbon is removed from pig iron. More than 60% of steel is made by the basic oxygen furnace process.

In the electric-furnace process, Figure 19.4, almost perfect control over the composition of the steel can be attained. The charge is mostly scrap iron. Resistance of molten slag to the flow of electricity generates the heat. Three giant carbon electrodes supply the current. The basic slag removes the phosphorus and sulfur, and iron(II) (ferrous) oxide removes the carbon.

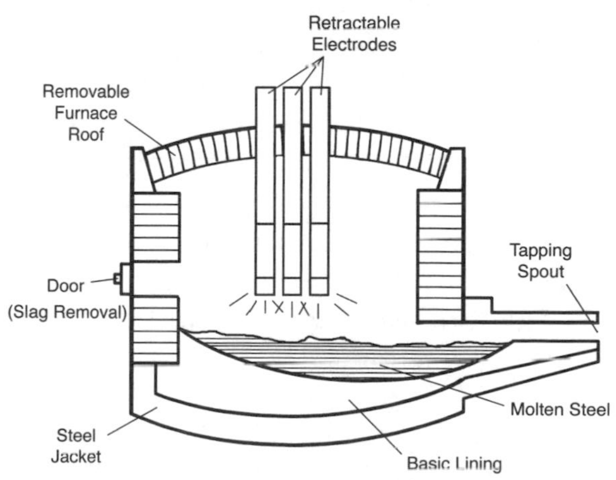

Figure 19.4 The Electric Furnace

Basic oxygen furnace and open-hearth furnace steels are high quality, and are the fundamental structural steels. Electric-furnace steel is the highest quality steel. It is used in applications in precision tools, crankshafts, and gears.

The optimum carbon content of hardenable (nonbrittle and possessing a high strength) steels is between 0.30% and 1.00%. The closer the carbon content is to 1.00%, the harder the steel and the easier it is to heat-treat. The heat treatment procedures annealing, quench hardening, and tempering can be used with all steels except those with very low carbon content (below 0.06% carbon).

In the annealing process, the steel is heated to a bright red color at 1225–1350°F and then cooled very slowly to produce a soft steel. If, however, the hot steel is plunged into cold water or oil—a process called *quench hardening*—the rapid chilling produces hard steel. Hardened steel is brittle because of internal strains, but tempering relieves those strains. Tempering involves reheating and holding the steel at a temperature below 1200°F to produce softer, but tougher steel.

Low-carbon steel (below 0.30% carbon) may be case-hardened by packing it in a mixture of barium carbonate, coke, and charcoal and heating to about 1700°F for several hours. In this process, called *pack carburizing*, carbon is absorbed from the carburizing mixture into the steel's outer layers, forming the carbide Fe_3C. This forms steel with a very hard shell and a tough interior:

$$BaCO_3(s) \rightarrow BaO(s) + CO_2(g)$$
$$CO_2(g) + C(s) \rightarrow 2\,CO(g)$$
$$3\,Fe(s) + 2\,CO(g) \rightarrow Fe_3C(s) + CO_2(g)$$

Case-hardened steels are also produced by gas carburizing in a furnace with a carbon-bearing atmosphere, or by liquid carburizing in a molten salt bath. Nitriding is a similar process that results in a hard shell containing nitride particles.

ALLOY STEELS

By alloying other metals with steel, a variety of alloy steels are made for many different applications. The principal alloying metals are the transition metals in the same period of the periodic table as iron. They include titanium, vanadium, chromium, manganese, cobalt, and nickel. The metals below chromium in the periodic table, molybdenum and tungsten, are also important steel alloy metals.

Table 19.1 shows the properties and uses of alloy steels. All the elements in Table 19.1 are transition metals except silicon.

Titanium and Zirconium

Titanium is an abundant metal, constituting about 0.6% of Earth's crust; but because it is thinly and evenly distributed, exploitation is difficult. It occurs as rutile, TiO_2, and ilmenite, $FeTiO_3$. Zirconium is rare, occurring as zirconia, ZrO_2, and zircon, $ZrSiO_4$.

Titanium metal is light, very strong, and corrosion resistant. Its use as a structural metal has increased in recent years, and it is used in everything from bicycles to replacement hip joints. At high temperatures, it is extremely reactive with both oxygen and nitrogen. For that reason, it is reduced from its dioxide in an electric furnace in which air has been replaced with a noble gas. In steel, as ferrotitanium,

Alloy Metal	Properties Imparted	Chief Uses
titanium	toughness and strength	gears, tools
vanadium	shock resistance; resistance to corrosion and metal fatigue	steel castings, axles, gears
chromium	corrosion resistance; hardness and toughness	stainless steel; tools; cooking utensils
manganese	toughness; wear resistance	safes; steam shovel teeth; crushers; railway frogs and switches
cobalt	magnetism	permanent magnets
nickel	hardness; toughness; tensile strength	automotive parts; armor plate
molybdenum	tensile strength; heat resistance	high-speed tools
tungsten	hardness; heat resistance	high-speed cutting tools
silicon	flexibility; electrical permeability	springs; transformer cores
zirconium	sparking	lighter flints
tantalum	hardness	tools

Table 19.1 Properties and Uses of Alloy Steels

titanium provides the wearing qualities necessary for curved railroad track.

Zirconium is reduced from its ores with magnesium. Zirconium is used in making the containers that hold the uranium fuel rods in nuclear power reactors because it does not absorb many neutrons.

Titanium dioxide, TiO_2, is a white paint pigment. It is used in glazing porcelain and whitening paper. A coating of $TiO_2 \cdot Ag \cdot TiO_2$ on the inside of incandescent lightbulbs reflects heat back to the filament to generate more light. This results in a 30–60% energy savings over standard incandescent lightbulbs. Titanium tetrachloride, $TiCl_4$, is a fuming liquid used in the making of smoke screens. Zirconium dioxide, ZrO_2, is used as a refractory material.

Vanadium, Niobium, and Tantalum

Among these rare metals, vanadium is the most important. It is found in carnotite, a uranium vanadate. In the extraction of uranium, vanadium is separated as the pentoxide, V_2O_5. Pure vanadium can then be reduced from the pentoxide by the thermite reaction (Chapter 18). Niobium and tantalum can be extracted the same way. Tantalum is extremely corrosion resistant. One of the most powerfully corrosive liquids—a mixture of one part concentrated nitric acid with three parts concentrated hydrochloric acid known as *aqua regia*—will not react with it. Surgeons use tantalum alloys as wire, pins, and plates to splice and support broken bones. Vanadium pentoxide, V_2O_5, is an important catalyst in the Haber process for making ammonia and in the contact process for making sulfuric acid (Chapter 14).

Chromium, Molybdenum, and Tungsten

Chromite, $Fe(CrO_2)_2$, is the chief ore of chromium, the most abundant of these three metals. Molybdenum is found as the sulfide molybdenite, MoS_2, a substance easily mistaken for graphite. Molybdenum is the least common of these metals. Tungsten occurs in scheelite, $CaWO_4$, and in $(FeMn)WO_4$, wolframite. Chromium may be prepared by reducing chromic oxide, Cr_2O_3, with aluminum in the thermite reaction, or by depositing it electrolytically from solution. Both molybdenum and tungsten are produced by reduction from their oxides, MoO_3 and WO_3, by hydrogen.

Chromium is a bluish white, brittle metal that is both hard and corrosion resistant. Chromium is used in the chromium-plating process to produce accessories on automobiles, bicycles, and furniture. It is not only one of the most important steel alloy metals, but it is also found in other important alloys. Nichrome, an alloy of chromium, nickel, iron, and manganese, is used in heating elements in toasters, irons, and ovens. Stellite, an alloy of cobalt, chromium, tungsten, and carbon, is used in surgical instruments, cutlery, and high-speed tools. Chromium compounds are brilliantly colored, and many of them are used as paint pigments. Table 19.2 lists some of them.

The chief use of molybdenum is in steel. Tungsten has the highest melting point of all metals, 3410°C. Tungsten wire drawn from hot compacted powdered tungsten is used as the filament in lightbulbs. Both tungsten and molybdenum are used as filaments in electron tubes and electric lightbulbs.

Sodium dichromate, $Na_2Cr_2O_7$, is used in tanning leather. Potassium dichromate, $K_2Cr_2O_7$, is a powerful oxidizing agent in acid solution. Ammonium molybdate, $(NH_4)_6Mo_7O_{24} \cdot 4\ H_2O$, is an important reagent in water analysis. Sodium tungstate, $Na_2WO_4 \cdot 2\ H_2O$ is used in fireproofing fabrics and cellulose.

Manganese

Although the United States has only limited deposits of manganese, it is abundant on Earth, especially in the millions of tons of manganese nodules that lie on the ocean floor. Their composition is about 28% manganese, 1.4% nickel, 1.2% copper, and

Compound	Formula	Color
chromic oxide	Cr_2O_3	green
zinc chromate (basic)	$ZnCrO_4 \cdot Zn(OH)_2$	yellow
lead chromate	$PbCrO_4$	yellow
basic lead chromate	$PbCrO_4 \cdot PbO$	red
mixed lead chromate	$PbCrO_4 + PbCrO_4 \cdot PbO$	orange

Table 19.2 Chromium Pigments

0.25% cobalt. Some companies are exploring the potential for seabed mining of those nodules.

On dry land, manganese occurs mainly as the dioxide ore pyrolusite, MnO_2. Manganese is prepared by reducing the dioxide with carbon in an electric furnace or with aluminum in the thermite reaction. It is a silvery metal with a reddish tint. It is hard and brittle. It is more active than the other transition metals, easily decomposing steam to liberate hydrogen. It is an essential constituent in all steel, because it removes oxygen and makes steel tough.

Manganese dioxide, MnO_2, is a powerful oxidizing agent and is used to decolorize glass and to depolarize dry cells (Chapter 11). The deep purple compound potassium permanganate, $KMnO_4$, is a powerful oxidizing agent in acid solution. Manganese is used as a disinfectant. A mixture of manganous sulfate, $MnSO_4$, and borax is used as a drier in paints.

Cobalt and Nickel

Both cobalt and nickel are relatively abundant, especially in Canada. Cobalt occurs chiefly as the arsenide ore smaltite, $CoAs_2$, which is associated with the nickel ore pentlandite, (Ni,Fe,Cu)S. Both cobalt and nickel ores are roasted to the oxide and then reduced in a blast furnace. Cobalt is refined from the blast furnace product electrolytically. Nickel is refined by the Mond process. In this process, the gas carbon monoxide is passed under pressure over impure nickel at 100 °C, forming the gaseous compound nickel carbonyl:

$$Ni(s) + 4\ CO(g) \rightarrow Ni(CO)_4(g)$$

In this way, the nickel is separated from its impurities. The gaseous carbonyl is then directed to a new chamber where—when the temperature is raised to 200 °C—it decomposes and precipitates pure metallic nickel:

$$Ni(CO)_4(g) \rightarrow 4\ CO(g) + Ni(s)$$

The carbon monoxide is then used again to refine more impure nickel.

Cobalt is a silvery metal with a bluish tint. Like iron, it is ferromagnetic. Alloyed in steel, it is used in making permanent magnets. Carboloy, an extremely hard alloy used in high-speed cutting operations, consists of tungsten carbide bonded with metallic cobalt. The alloy alnico, containing aluminum, nickel, and cobalt, is one of the most powerfully magnetic of all known substances.

Nickel is also ferromagnetic. It is a hard, white metal that resists tarnish and is one of the most important steel alloy metals. As nickel plate, it protects steel, copper, and brass from corrosion. When finely divided, it is an excellent catalyst, especially in the hydrogenation of fats and oils. The nickel coin contains 25% nickel metal alloys and 75% copper. Many other nickel alloys are important, especially Monel metal, which is used in kitchen and bathroom fixtures. It is made of nickel, copper, and iron.

Cobalt salts, such as cobaltous chloride, $CoCl_2$, readily form hydrates. When anhydrous, the cobalt ion is blue, and when hydrated it is pink. Porous paper or cloth impregnated with cobalt chloride becomes blue in dry weather and pink in rainy weather. Such papers were used in the past to measure the relative humidity. Nickel salts are brilliant light green in color. Nickel dioxide, NiO_2, is the cathode in the Edison storage battery (Chapter 11).

SUMMARY

Iron is the most important metallic element and is second to aluminum in abundance on Earth. Chemically, iron is typical of the transition elements that show variable valence because they have two incomplete shells of electrons. Iron forms two series of compounds exhibiting a valence of 2 and 3. It alloys primarily with other transition elements in alloy steels, substances that show a tremendous variety of properties and uses. Hematite, Fe_2O_3, its principal ore, is reduced to pig iron in a blast furnace. Solidified pig iron is cast iron.

Steel is made from pig iron by reducing the carbon content to less that 2%. It may be made by the basic oxygen furnace process, the open-hearth furnace process, or the electric furnace process. More than 60% of steel is made by the basic oxygen furnace process, which is the most rapid method. The finished steel may then be heat-treated to improve its properties.

Titanium is an important structural metal, and its dioxide is an important white pigment. Chromium is present in stainless steel and is extensively used as a plating on other metals. Many of its compounds are important pigments because of their brilliant color. Tungsten is used in making high-speed tool steels and filaments in lightbulbs. Manganese is essential in all steel as a deoxidizer. Cobalt improves the magnetic qualities of steel. Nickel, refined by the Mond process, is present in corrosion-resistant alloys and heating elements in appliances.

PROBLEM SET NO. 21

1. Which ore of iron contains the greatest percentage of iron?

2. How does iron rust?

3. Match the following steel objects with the type of heat treatment most likely used in their preparation:

   (a) file (1) tempering

   (b) razor blade (2) case hardening

   (c) gears (teeth) (3) quench hardening.

4. Write balanced equations for the thermite reduction of

   (a) chromium

   (b) vanadium

CHAPTER 20

NONFERROUS ALLOY METALS

KEY TERMS

smelting, photovoltaic cell

Copper, tin, zinc, and lead have been important metals since the beginning of civilization. For example, copper, tin and zinc were used to make the alloys we call bronze and brass. Copper is second only to iron among the metals in importance to civilization. Our civilization depends on the electrical conductivity properties of copper almost as much as on the structural strength of the steel that is created from iron. These four metals are the main ingredients in a great variety of alloys that are used in countless ways in commerce and industry.

COPPER

Copper is a fairly abundant metal. Some of it occurs free, but most copper is extracted from its ores, chalcocite, Cu_2O; chalcopyrite, $CuFeS_2$; and cuprite, Cu_2O. It also occurs as basic carbonates in green malachite, $Cu_2(OH)_2CO_3$, and $Cu_3(OH)_2(CO_3)_2$, blue azurite.

In the extraction of copper from sulfide ores, the ores are first concentrated by flotation and then roasted to the oxide before the reducing process begins. The roasted ore is then reduced with coke and air, forming blister copper, which is about 99% pure. The blister copper is then refined electrolytically in a solution of copper sulfate (see Figure 20.1),

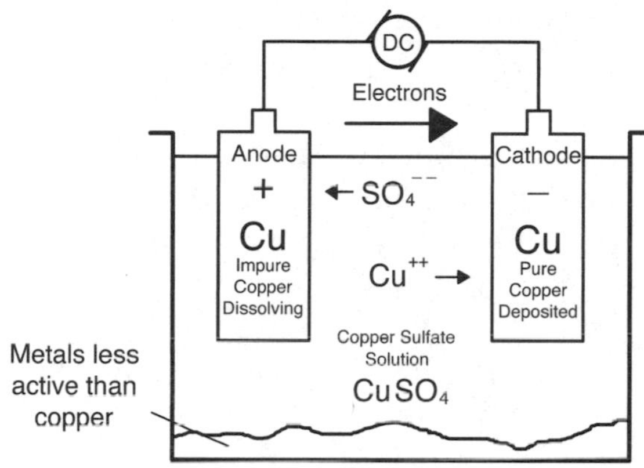

Figure 20.1 Electrolytic Refining of Copper

using the impure copper as the anode. Thin sheets of pure copper, called starter sheets, form the cathode. The copper in the impure anode dissolves during electrolysis and deposits on the pure cathode. The metals less active than copper in the anode fall to the bottom of the cell as a sludge called slimes and are recovered as by-products. The metals more active than copper pass into solution with the copper but do not deposit on the cathode because copper ions are more easily reduced. Thus, the electrolytic refining of copper, using copper electrodes, takes full advantage of the electrochemical properties of metals and their ions.

Copper is a soft, extremely ductile, and malleable metal with a characteristic reddish brown color. It is easily formed into wire, tubes, and sheets. Copper is an excellent conductor of electricity and is the metal most often used for that purpose. Because copper is below hydrogen in the activity series of metals, it is not very active chemically, but oxidizing acids react vigorously with it. When exposed for some time to moist air, copper becomes coated with the green copper carbonate malachite. The coating protects the metal from further corrosion. Copper is extensively used for electric wires and cables, including busbars, which are used in power plants to carry extremely high voltages. It is also used in drinking water pipes, and in cooking utensils. Its beauty makes it much used for ornamental purposes.

Copper alloys were the metals most used before the advent of steel. Alloys of copper and zinc are known as *brass*. Cartridges, musical instruments, hardware, and various types of pipes are made from brasses. The alloy *bronze* contains copper, tin, and zinc. It is used for statues and medals, and primitive civilizations used it for tools. Aluminum bronze is used for forgings, bolts, and gears. Beryllium bronze is used for nonsparking tools. Bell metal is made of copper and tin. Sterling silver contains 7.5% copper. Yellow gold and dental gold both contain copper. These are but a few of many important copper alloys.

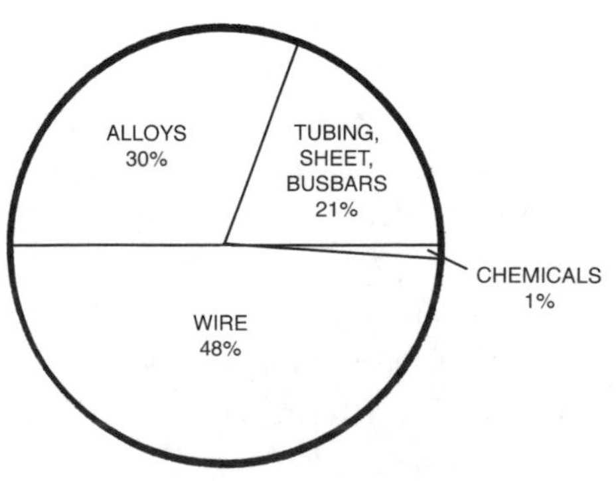

Figure 20.2 Uses of Copper

Copper is a transition metal and exhibits valences of 1 and 2 in its compounds. The latter valence is the more common. Because of its role in the electrolytic refining of copper, copper sulfate, $CuSO_4 \cdot 5\ H_2O$, is the most important copper compound. Also known as blue vitriol, it was once used to kill algae and fungi in reservoirs and swimming pools; however, its toxicity has curtailed its use.

ZINC, CADMIUM, AND MERCURY

Zinc is extracted from the ores sphalerite, ZnS; smithsonite, $ZnCO_3$; and zincite, ZnO. Cadmium is quite rare but is always present as an impurity in zinc ores. Some mercury is found free in nature, but it is normally found as the sulfide ore cinnabar, HgS.

Zinc ores are concentrated, roasted to the oxide, and then reduced by heating with carbon. The impure metal obtained by this process contains cadmium. The zinc is refined by distillation, with the more volatile cadmium coming off first. Mercury is so volatile that it distills off as the sulfide ore is being roasted. Its vapors are then collected and condensed:

$$HgS(s) + O_2(g) \rightarrow SO_2(g) + Hg(l)$$

Zinc and cadmium are both silvery white metals. Zinc melts at 420°C and boils at 907°C. Cadmium melts at 320°C and boils at 767°C. Both are above hydrogen in the activity series of metals, with zinc being more active. When zinc and cadium are exposed to air, surface oxide films form and protect them from further corrosion. Zinc hydroxide is amphoteric, but cadmium hydroxide is always basic. Mercury is a liquid at room temperature, freezing at

−39 °C. It boils at 357 °C. Its vapor is poisonous. Like all metals, it is an excellent conductor of electricity. Mercury is well below hydrogen in the activity series of metals and thus is not very active chemically, being attacked only by the oxidizing acids HNO_3 and concentrated H_2SO_4.

Great quantities of zinc are used in galvanized iron. The iron or steel is coated with zinc by dipping into molten zinc or by eletrodeposition. Water pails, guttering, and wire fencing are among the many objects made of galvanized iron. Zinc forms the case and negative pole of dry cells and is used in many types of electrical connectors. It is also an important constituent in many alloys. In brass and bronze, it is alloyed with copper. Zinc die-casting is an alloy with aluminum used in radio and automotive parts. Since 1983 copper pennies have been made from 97.5% zinc and 2.5% copper.

Because cadmium is even less active than zinc, it is an excellent plating metal. It is present in several low-melting alloys used in fire sprinkler systems and in wear-resistant alloys used as bearings.

Galvanized iron and steel are coated with zinc. The coating is applied either by dipping into molten zinc or by eletrodeposition. Water pails, guttering, and wire fencing are among the many everyday objects made of galvanized iron.

Mercury is used to fill thermometers and barometers used in scientific work. That it is a liquid at room temperature makes mercury ideal for certain types of electrical switches. Mercury vapor lamps are commonly used in lighting, and the vapor is commonly used in neon signs. Alloys of mercury are called *amalgams*. Mercury is commonly used as a cathode in industrial electrolytic cells, where is absorbs the metal being deposited as an amalgam and is later separated by distillation.

Zinc oxide, ZnO, is an important white paint pigment. It is used as a filler in making tire rubber, and its antiseptic properties are utilized in salves. A paste of zinc oxide and concentrated hydrochloric acid sets as a hard cement and is used in dentistry. Zinc chloride, $ZnCl_2$, is used as a wood preservative and as a flux in soldering and brazing. Cadmium sulfide, CdS, is a yellow paint pigment. Mercurous (Hg I) chloride, Hg_2Cl_2, known as *calomel*, was used

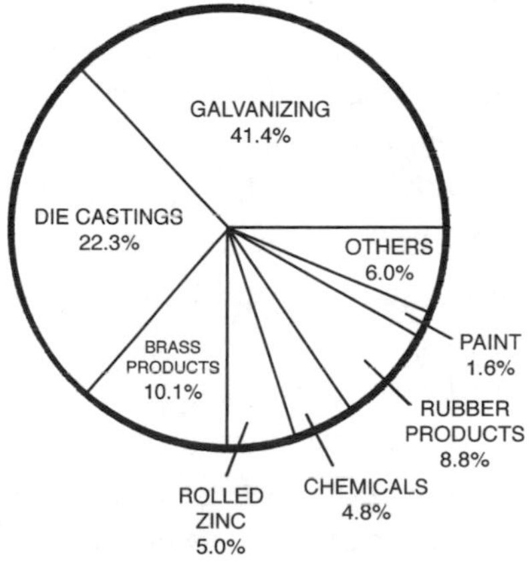

Figure 20.3 Uses of Zinc

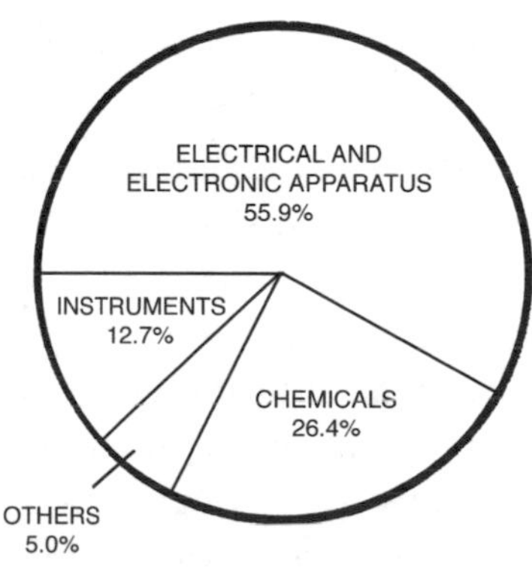

Figure 20.4 Uses of Mercury

in past centuries to stimulate secretion of bile from the liver. Mercuric (Hg II) chloride, $HgCl_2$, is known a *corrosive sublimate* and is a deadly poison. It has been used as a germicide. Mercury fulminate, $Hg(ONC)_2$, explodes when struck. It is used in blasting caps.

GALLIUM, INDIUM, AND THALLIUM

These three very rare metals have only recently been exploited. They are found as impurities in the sulfide ores of common metals, and they are marketed as by-products. Their low melting points make them valuable in low-melting alloys. Gallium, Ga, melts at 30 °C. It will turn to liquid in the palm of the hand. Indium, In, melts at 157 °C, and thallium, Tl, melts at 303 °C. Indium is used as a protective coating for iron or steel, particularly in certain types of bearings.

Gallium arsenide, GaAs, is used in solar cells to convert sunlight into electricity. Thallium oxide, Tl_2O_3, has been used to make glass with a high index of refraction.

GERMANIUM, TIN, AND LEAD

Although none of these metals is very abundant, tin and lead occur concentrated in easily worked deposits, whereas germanium, which is at least as abundant as tin, is thinly distributed in Earth's crust and less well known. Germanium is usually found associated with the sulfide ores of the common metals. The chief ore of tin is cassiterite, SnO_2, which is found in Malaysia and Bolivia. Although lead is extracted from cerussite, $PbCO_3$; anglesite, $PbSO_4$; and crocoite, $PbCrO_4$, its chief source is galena, PbS.

Tin is easily reduced from its oxide ore by heating with carbon:

$$SnO_2(s) + 2\,C(s) \rightarrow 2\,CO(g) + Sn(s)$$

If the process is carried out in a sloping furnace, the molten reduced metal flows from the furnace into molds. This process of heating a metal oxide with some reducing agent such as coke (carbon) to produce the elemental metal is called *smelting*.

Lead is reduced from its sulfide ore by a process of partial roasting followed by reduction. The sulfide is first heated to a relatively low temperature in air, causing part of the sulfide to react as follows:

$$2\,PbS(s) + 3\,O_2(g) \rightarrow 2\,PbO(s) + 2\,SO_2(g)$$
$$PbS(s) + 2\,O_2(g) \rightarrow PbSO_4(s)$$

Then, the temperature is raised to the smelting temperature, at which the remaining lead sulfide reacts with the products formed above as follows:

$$PbS(s) + 2\,PbO(s) \rightarrow 3\,Pb(s) + SO_2(g)$$
$$PbS(s) + PbSO_4(s) \rightarrow 2\,Pb(s) + 2\,SO_2(g)$$

The molten lead is cast into molds.

Germanium may be extracted from the residues of refined lead or zinc electrolytically or by fractional crystallization, a process in which the melted residues can be carefully cooled, allowing the metal to slowly solidify. It is hard and brittle and is a relatively poor conductor of electricity; however, its crystals can rectify alternating current, allowing the current to pass in only one direction. Thus, alternating current (AC) can be turned into direct current (DC). For that reason, germanium crystals have been used as semiconductors in the electronics industry. Solar *photovoltaic cells* in the space program have also included germanium cells. They work by emitting electrons (electricity) when exposed to sunlight. Typical solar cells are only about 10% efficient, but incorporation of compounds of elements like germanium can dramatically increase the efficiency. This may allow solar cells to become economically competitive with fossil fuels in the near future.

Tin is a soft, white, low-melting metal. When exposed to cold weather for an extended time, it reverts to an allotropic form in a process known as *tin disease*. This allotrope, gray tin, is powdery and dull and is a poor conductor of electricity. Gray tin is not nearly as useful as the white form because of its powdery nature. Tin is only slightly above hydrogen in the activity series of metals and so is only moderately active chemically. It exhibits valences of 2 and 4 in its compounds. Tin does not tarnish in air, being protected by a thin oxide film. It is used chiefly as tinplate for the

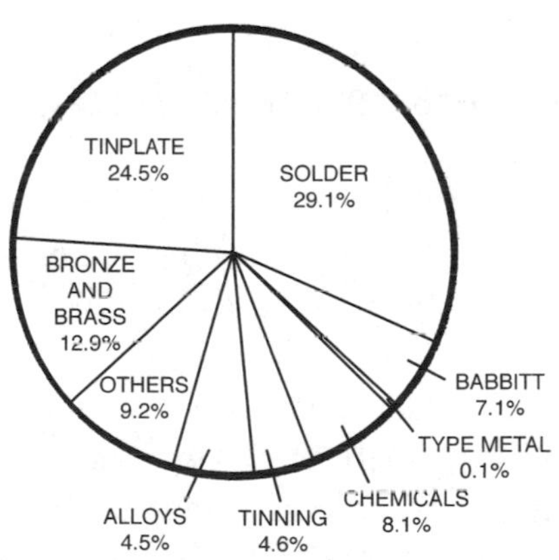

Figure 20.5 Uses of Tin

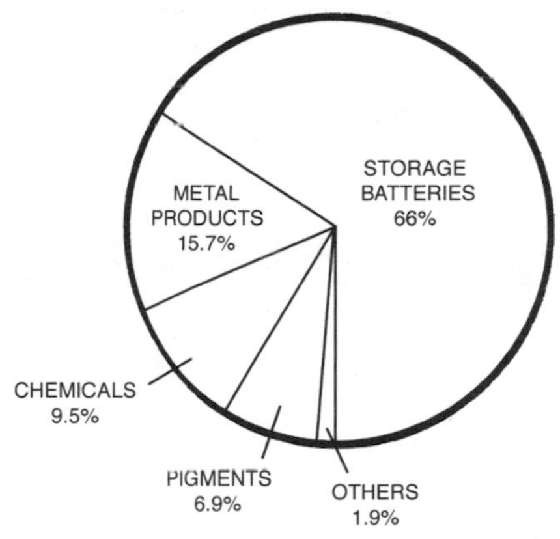

Figure 20.6 Uses of Lead

protection of iron and steel, and in many important nonferrous alloys (Table 20.1).

Lead is a soft, nonelastic, low-melting bluish white metal that turns dark gray in air from an oxide film. The film protects the metal from further oxidation. Lead is just above hydrogen in the activity series and is unaffected by acids at ordinary temperatures. Water containing dissolved oxygen slowly oxidizes and dissolves lead, thus making it unfit for drinking-water piping, because all lead salts are poisonous and lead accumulates in the body. Lead, like tin, exhibits valences of 2 and 4 in its compounds. Lead is used extensively for piping and as a coating around telephone and other electric cables to protect them from corrosion. It is a vital alloy metal. Table 20.1 describes some of the important lead and tin alloys.

Stannic chloride or tin (IV) chloride, $SnCl_4$, is an important intermediary in the recovery of tin from tin cans and other tin-plated objects. Chlorine gas is passed over the tin, forming the liquid chloride:

$$Sn(s) + 2\ Cl_2(g) \rightarrow SnCl_4(l)$$

Alloy	Composition (%)	Properties	Uses
babbitt metal	Sn 89, Sb 7.3, Cu 3.7	soft, low friction	bearings
battery plate	Pb 94, Sb 6	easily cast	batteries
bearing metal	Sn 75, Sb 12.5, Cu 12.5	soft, low friction	bearings
bell metal	Cu 75, Sn 25	clear tone when struck	bells
bronze	Cu 88, Sn 10, Zn 2	corrosion resistant, easily cast	bearings, castings, statues
fuse metal	Bi 50, Cd 10, Pb 26.6, Sn 13.3	low melting	fuses, safety fire sprinklers
lead foil	Pb 87, Sn 12, Cu 1	malleable, inert	wrappings for tops of wine/champagne bottles
pewter	Pb 75, Sn 25	soft, easily worked	ornamental ware
rose metal	Sn 22.9, Pb 27.1, Bi 50	low melting	safety fire sprinklers
shot	Pb 99.5, As 0.5	hard	shotgun shot
solder	Pb 67, Sn 33	low melting	soldering
silver solder	Sn 40, Ag 40, Cu 14, Zn 6	low melting	soldering
type metal	Pb 82, Sb 15, Sn 3	low melting, expands on cooling	printing type, castings
wood's metal	Bi 50, Pb 25, Sn 12.5, Cd 12.5	low melting	fuses, safety fire sprinklers

Table 20.1 Alloys of Lead and Tin

The chloride is then dissolved in water and boiled, causing stannic oxide to precipitate:

$$SnCl_4(aq) + 2\ H_2O(l) \rightarrow SnO_2(s)\downarrow + 4\ HCl(aq)$$

Metallic tin is then reduced from the oxide by heating with carbon.

Litharge, PbO; red lead, Pb_3O_4; and white lead, $Pb_3(OH)_2(CO_3)_2$, were once important paint compounds; however, lead-based paint was banned for residential use and sale in the United States in 1978 because of lead's extreme toxicity. Even today, however, old, peeling paint that is accidentally or inadvertently consumed by children and domestic animals remains a threat to health.

Lead dioxide, PbO_2, is a brown powder obtained by the electrolytic oxidation of lead. It is an essential part of lead storage batteries. A grating of battery plate impregnated with lead dioxide it forms the cathode of the cell. Another grating of battery plate impregnated with spongy lead, a porous, electrically deposited form of lead metal, forms the anode of the cell. The discharge reaction of the lead storage cell is

$$Pb(s) + PbO_2(s) + 2\ H_2SO_4(aq) \rightarrow 2\ PbSO_4(s) + 2\ H_2O(l)$$

Heat shortens the life of a storage battery by causing the lead dioxide to decompose:

$$2\ PbO_2(s) \rightarrow 2\ PbO(s) + O_2(g)$$

Tetraethyl lead, $(C_2H_5)_4Pb$, was for many years added to gasoline to prevent engine knock. Its use was discontinued in favor of unleaded gasoline, which does not discharge toxic lead into the environment. In place of tetraethylead, the gasoline octane enhancer methylcyclopentadienyl manganese tricarbonyl (MMT) is allowed as an additive. There is concern, though, that the use of manganese additives could increase inhalation exposure to manganese, which is a neurotoxin.

Alcohols and ethers are oxygen-containing fuel additives that can boost octane quality, enhance combustion, and reduce exhaust emissions.

SUMMARY

Copper is second to iron in importance among the metals. It is fairly abundant, occurring free and in several important ores. Its sulfide ores are concentrated, roasted, and then reduced. The impure copper is refined electrolytically. Copper is used as an electrical conductor and in piping. Its principal alloys are brasses, bronzes, and coinage metals.

Zinc and mercury are obtained from sulfide ores. Much zinc is used in galvanized iron and in alloys, particularly brass and zinc die casting. Mercury is a liquid used in thermometers, barometers, and electrical switches. Gallium is being used in photovoltaic cells, increasing their efficiency and perhaps will someday provide a clean, reasonably priced energy source.

Tin is reduced by carbon from its oxide ore. It is used primarily as tinplate and in low-melting and bearing alloys. Lead is obtained from its sulfide ore by partial roasting followed by smelting. It is used in piping, batteries, and low-melting alloys.

PROBLEM SET NO. 22

1. What is the green deposit that forms in a sink from a dripping faucet, and what is the origin of its constituents?

2. How does the electrolytic refining of copper using copper electrodes take full advantage of the electrochemical properties of metals and their ions?

3. Write equations for the reduction of zinc from sphalerite.

4. Give one application of each of the following metals that would make its availability critical in a national emergency: (a) copper (b) tin (c) mercury

5. Why is lead unfit for use as pipes for drinking water?

CHAPTER 21

THE NOBLE AND RARE METALS

KEY TERMS

noble metals, inner transition elements

Because they occur free in nature, silver, gold, and platinum are called *noble metals,* a term that describes any metal of low chemical activity that is found naturally in the free state. A metal found free is also called a *native* metal.

SILVER AND GOLD

Gold and silver are associated with wealth and luxury because of their use for coins and jewelry; but in truth, iron, copper, and other common metals have greater economic value, and some of the rare metals cost considerably more per pound.

Silver is found free and as the sulfide argentite, Ag_2S, usually associated with copper and lead sulfide ores. Most metallic silver is obtained from the anode slimes produced in the electrolytic refining of copper and lead. Silver is also obtained from lead produced by the smelting process. Zinc is stirred into molten lead, but the two metals are immiscible. Silver dissolves in the zinc, which floats on the lead and is skimmed off. The zinc is then distilled away from the silver. This series of reactions is known as the *Parkes process.*

Silver is extracted from silver ore by treating the crushed ore with sodium cyanide, which dissolves the silver sulfide. Metallic zinc is then added to reduce the silver from solution, the reaction being a direct application of the activity series of metals. Silver is refined electrolytically.

Gold is usually found free as nuggets in quartz rock formations or as grains or flakes in sands formed by the weathering of siliceous minerals. Gold-bearing rock is crushed and treated with sodium cyanide, which dissolves the gold just as it does silver. Again, zinc is added to cause the less active gold to precipitate from solution. The high density of gold (19.3 g/cm^3) is utilized in panning and sluicing operations, in which water washes away the less dense rock, leaving the heavy gold particles behind. Mercury is frequently used to absorb gold from sand as an amalgam. The amalgam is heated to drive off the mercury. The gold is then refined electrolytically.

Silver is a bright, lustrous metal that is soft and very malleable. It is easily worked into complex shapes and hence is much used for jewelry and tableware. It is extensively used as a plating metal, being deposited electrolytically from cyanide solutions. Its ability to reflect light is utilized in mirrors. Silver is attacked by nitric and concentrated sulfuric acids but is inert to alkalis.

Atmospheric gases do not corrode silver, but sulfur compounds tarnish it badly, forming black silver sulfide, Ag_2S. Tarnished silver objects may be cleaned by boiling them in water containing a spoonful of baking soda in an aluminum pan. An electrochemical reaction takes place in which aluminum metal forms ions, and the silver ions in the sulfide are redeposited on the surface of the object. The baking soda serves as an inert electrolyte. The overall reaction is

$$3\ AgS(s) + 2\ Al(s) + 6\ H_2O(l) \rightarrow 2\ Al(OH)_3(aq) + 3\ H_2S(aq) + 3\ Ag(s)$$

Buffing the object then restores the luster of the silver.

EXPERIMENT 29: Place a tarnished silver spoon in a clean aluminum pot and cover it with water to a depth of 1–2 inches. Add a teaspoon of baking soda. Boil for several minutes. Remove the silver spoon and buff it with a soft cloth.

Pure silver is too soft for use as coinage and tableware. Consequently, a little copper is alloyed with it to harden and strengthen it. Sterling silver contains 7.5 % copper. The amount of silver in an alloy is often indicated as *fineness*, which is the parts by weight of silver per 1000 parts of alloy. U.S. dimes, quarters, and half-dollars minted before 1965 were 90% silver or 900 fine. No silver coins have been made for general circulation in the United States since 1969.

The yellow color and the luster of gold are well known. It is a soft metal and is highly ductile and malleable. It can be drawn into extremely fine thread, and it can be beaten between sheets of parchment paper into gold leaf as thin as 0.00002 mm. Gold is very inert to ordinary reagents, although chlorine gas will corrode it. It dissolves in aqua regia, forming a solution of chlorauric acid, $HAuCl_4$.

Gold leaf, used in window signs, bookbinding, electrical instruments, and for various decorative purposes, is the chief application of pure gold. The gold used in coinage, jewelry, and dentistry is alloyed with other metals to give it hardness. Copper, nickel, and zinc are usually used. The gold content of these alloys is rated in carats, indicating the parts by weight of gold in 24 parts by weight of alloy. This means that a 10-carat (10K) gold ring is about 42% gold (10/24), and 24K gold is pure (actually 99.95%) gold. Some selected alloys of silver and gold are shown in Table 21.1.

PHOTOGRAPHY

Silver halides are sensitive to light energy. The precise nature of this sensitivity has not yet been fully explained, but when light strikes the silver ions in these salts, they become more susceptible to chemical reduction. This phenomenon is utilized in photographic film and paper.

Photographic film consists of a piece of plastic coated with a thin film of silver bromide and silver iodide suspended in gelatin. Four basic steps create a picture from photographic film: exposure, developing, fixing, and printing.

During exposure, light strikes the sensitive layer on photographic film. The silver ions struck by the light become susceptible to easy reduction to metallic silver *in direct proportion to the intensity of the light striking them.* Brief exposure causes no chemical or physical change in the ions, but longer exposures cause a reduction to fine specks of jet black, metallic silver known as *colloidal* silver.

In the developing process, the exposed film is immersed in a solution of a mild reducing agent such as pyrogallol, $C_6H_3(OH)_3$, or

Alloy	Composition (%)	Properties	Uses
sterling silver	Ag 92.5, Cu 7.5	corrosion resistant, easily worked	coinage, jewelry, tableware
U.S. commemorative silver coins	Ag 90, Cu 10	corrosion resistant, durable	coinage
U.S. commemorative gold coins	Au 90, Cu 10	corrosion resistant, durable	coinage
18K yellow gold	Au 75, Ag 12.5, Cu 12.5	corrosion resistant, easily worked	jewelry, accessories
18K yellow gold	Au 75, Cu 3.5, Ni 16.5, Zn 5	corrosion resistant, easily worked	jewelry, accessories
inlay	Au 90, Ag 1.5, Cu 7, Zn 0.5, Pt 1.0	corrosion resistant, easily worked	decoration
14K dental gold	50 Au, 33 Ag, 17 Cu	corrosion resistant, easily worked	dental fillings and braces

Table 21.1 Alloys of Noble Metals

hydroquinone, $C_6H_4(OH)_2$. These reducing agents cause a reduction of only the silver ions made sensitive by light during exposure. The developing solution contains Na_2CO_3, sodium carbonate, because the developing action takes place only in mildly alkaline solution; it also contains sodium sulfite, Na_2SO_3, which prevents attack on the developers by atmospheric oxygen.

The developed film is thus coated with black colloidal silver wherever white light originally struck it and is called a *negative*. When the developing is completed, the film is dipped into a solution of acetic acid that neutralizes the sodium carbonate and stops the reducing action of the developers.

In the fixing process, the unexposed silver salts are removed from the film. The fixing solution contains sodium thiosulfate, $Na_2S_2O_3$, also called *hypo*. It reacts with the silver salts still in the gelatin as follows:

$$2\,AgBr(s) + 3\,Na_2S_2O_3(aq) \rightarrow 2\,NaBr(aq) + Na_4Ag_2(S_2O_3)_3(aq)$$

Because both the sodium bromide and the sodium silver thiosulfate formed in this reaction are soluble, the fixing solution washes the silver salts from the film. The fixing solution also contains a little acetic acid and some alum, which hardens the gelatin film. When fixing is completed, the film is thoroughly washed with water to remove excess hypo, which would otherwise bleach the negative. The negative is then dried.

> Photography is possible because silver halides are sensitive to light energy. When light strikes the silver ions in these salts they become more susceptible to chemical reduction.

In the printing process, the developed negative is held against a sensitized film on the printing paper. The film on this paper usually contains silver chloride suspended in gelatin. Silver chloride is less sensitive to light than the other silver halides. Light is then permitted to pass through the negative to strike the silver ions on the printing paper. This exposes them just as the original film was exposed, except that this time the light strikes most heavily on areas that were dark when the original picture was taken. The black silver on the negative shields the bright areas in the original object. The exposed printing paper is then developed and fixed in the same manner as the film. The finished print is thus a reproduction of the original object and is known as a *positive*.

The film used in color photography has three sensitized layers, each containing a dye that is absorbed by metallic silver in the layer to produce the three primary colors—red, yellow, and blue—when the film is developed. When white light passes through the developed film, not only are the light and dark areas of the object reproduced but also all the hues and shades of its colors. The printing of colored film involves the use of special paper with several layers of silver salts containing dyes capable of reflecting the colors of the original object after developing.

THE PLATINUM METALS

The six metals below iron, cobalt, and nickel in the periodic table are known as the platinum metals. The first three, ruthenium, Ru; rhodium, Rh; and palladium, Pd, are the light platinum metals. The last three, osmium, Os; iridium, Ir; and platinum, Pt, are the heavy platinum metals. They are all very inert except toward alkalis, and all occur free in nature. They are both rare and very expensive. Osmium is the densest of all naturally occurring elements, at 22.5 g/cm^3. None of these metals have important compounds.

Platinum is the most important of these metals, being used especially in jewelry making and in dentistry, because it does not tarnish and is easily worked. It is used in the construction of electrical apparatus because its conductivity is excellent. Finely divided platinum is an important industrial catalyst, especially in the contact process of making sulfuric acid. The international standard of mass, the kilogram, is made of a platinum–iridium alloy.

THE TRANSITION METALS

The last remaining metals of the periodic table include scandium, Sc; yttrium, Y; and the 30 metals of the lanthanide and actinide series (the *inner transition elements*). These inner transition elements have electrons in their f-orbitals. These metals are rare, with uranium being the most abundant. The lanthanide metals were originally called the *rare earth* metals. Because the inner transition series metals differ structurally in the third outermost shell of electrons they are extremely difficult to separate and isolate. Thus, they have few uses. Uranium, U; thorium, Th; and plutonium, Pu, have isotopes that are sources of atomic energy (see Chapter 23). An alloy of cerium and iron, known as Auer metal, gives off sparks when struck, and is used in pocket lighters as "flints" and as tracer lights in tracer bullets. A mixture of cerium and thorium oxides forms the wick of gas mantle lamps. Cerium sulfate, $Ce(SO_4)_2$, is used in analytical chemistry. Europium oxide, Eu_2O_3, is a brilliant red phosphor used in TV receivers. The oxides of lanthanum, praseodymium, and neodymium

(La_2O_3, Pr_2O_3, Nd_2O_3) are used in sunglasses to absorb injurious ultraviolet radiation and to reduce the intensity of visible light. Uranium hexachloride, UF_6, was the gaseous compound used to separate the uranium isotopes in making the original atomic bomb.

SUMMARY

Noble metals—particularly silver, gold, and platinum—are very inert and occur free in nature. Silver is extracted from its ores by the cyanide process, but most silver is obtained as a by-product from the electrolytic refining of other metals. The Parkes process is a method of obtaining silver from lead in the impure state. Gold is found free. It is also separated from rock by the cyanide process.

Silverware may be cleaned of tarnish by boiling it in a solution of baking soda in an aluminum pan. The sensitivity of silver halides to light makes them useful in photography. Color photography combines dyes with the sensitive silver salts on the film and printing paper to produce the colored reproductions of objects.

The inner transition elements, the lanthanides and actinides, are rare and not very abundant. Because all these elements have electrons in their f orbitals, they are very similar and difficult to separate from each other. Economically, the most important member of these elements is uranium, which is used in nuclear power plants and bombs. The rest of the inner transition metals have limited uses.

PROBLEM SET NO. 23

1. Would a noble metal serve as an effective sacrificial anode (See Chapter 11)?

2. What is aqua regia?

3. To what do the terms *fineness* and *carats* refer?

4. Suggest a method of recovering silver from used fixing solutions.

5. Using the periodic table, list the names of the lanthanide and actinide series of metals.

CHAPTER 22

ORGANIC CHEMISTRY

KEY TERMS

organic chemistry, functional groups, isomer, aliphatics, aromatics, saturated, unsaturated, octane number, polymers

Compared with compounds of other elements, the compounds of carbon have a distinctive pattern of chemical behavior. In general, they react more slowly, and in reactions they tend to form equilibrium (see Chapter 9) mixtures instead of going to completion. Temperature and pressure have a greater effect on them, and frequently, carbon compounds of the same percentage composition have entirely different properties. Because carbon compounds were originally obtained from living things or from the remains of living things—such as coal and petroleum—their chemistry is called *organic chemistry.* Today, organic compounds are known to form from nonliving raw materials.

The field of organic chemistry is so vast that—despite there being thousands of organic compounds in daily use—only a tiny fraction of the possible compounds are used. Yet, organic chemicals play a vital role in the life of every person in terms of food, beverages, drugs, textiles, dyes, and plastics. Organic compounds are used as fuels, refrigerants, explosives, and adhesives.

The key element present in all organic chemicals is carbon. It has a valence of 4, for it has 4 electrons in its outermost shell, which form covalent bonds with atoms of hydrogen, oxygen, the halogens, nitrogen, sulfur, and other carbon atoms. Unlike inorganic compounds, in which electrovalent, ionic compounds predominate, the properties and chemistry of carbon compounds are intimately related to the nonionic, covalent bonds present in them. Single, double, and even triple covalent bonds form between a single pair of atoms, especially carbon atoms.

PROPERTIES

The physical properties and chemical behavior of carbon compounds are related to three factors. The first is the number of carbon atoms in the molecule. "Simple" organic molecules have been isolated that contain 60 or more carbon atoms, and complex molecules, like starch or rubber, consisting of endless chains of structural units, contain staggering numbers of carbon atoms. In general, an increase in number of carbon atoms tends to increase the melting and boiling point of the compounds and to reduce their chemical activity.

The second factor that affects the properties of carbon compounds is the type of bonding (single, double, or triple) present. A single covalent bond consists of a single pair of electrons shared between two atoms. Double bonds involve two shared pairs of electrons, and triple bonds involve three shared pairs. The bonding in organic compounds is usually indicated by dashes, as in the following examples:

Single	Double	Triple
H–C(H)(H)–H	$(H)_2C{=}C(H)_2$	H—C≡C—H
Methane—CH_4	Ethylene—C_2H_4	Acetylene—C_2H_2

In general, the single bond is the least active chemically. The double bond is more reactive, and compounds containing double bonds are somewhat more volatile than corresponding single-bonded compounds. The triple bond is the most active chemically, and compounds containing it are more volatile still.

The third major determinant of carbon compound chemistry is the types of functional groups present in the molecule. A *functional group* is the part of an organic molecule that gives that molecule its specific reactivity. It is similar to the radicals of inorganic chemistry, in that bonded groups of elements behave as a unit. They enter or leave an organic molecule together and impart particular properties to the molecules possessing them. Table 22.1 gives the names and formulas of the principal functional groups.

The dashes in the formulas of the functional groups in Table 22.1 indicate the location of the bonds available for attachment to basic

Name	Formula	Name	Formula
methyl	$—CH_3$	alcohol, or hydroxy	—OH
ethyl	$—C_2H_5$	aldehyde	—C(=O)H
propyl	$—C_3H_7$	ketone	—C(=O)—
butyl	$—C_4H_9$	acid	—C(=O)OH
phenyl	$—C_6H_5$	ether	—C(=O)O—
chloro	—Cl	ester	—O—
bromo	—Br	nitrile	—C≡N
iodo	—I	amino	$—NH_2$
fluoro	—F	nitro	$—NO_2$

Table 22.1 Organic Functional Groups

> Unlike inorganic compounds, in which electrovalent, ionic compounds predominate, the properties and chemistry of carbon compounds are intimately related to the nonionic, covalent bonds present in them. Single, double, and even triple covalent bonds form between a single pair of atoms, especially carbon atoms.

carbon atoms, to carbon compounds, or to other functional groups.

Because the chemistry of organic compounds is so closely related to the functional groups present in the molecules, the usual empirical formulas showing only the ratio of elements are inadequate. Instead, structural formulas are used. Structural formulas show not only the ratio of elements in a compound but also give information about how those elements are arranged. Consider ethyl alcohol as an example. Its empirical formula is C_2H_6O, a formula that reveals nothing about the functional groups present in the compound. Ethyl alcohol is actually composed of an ethyl group, $—C_2H_5$, combined with an alcohol group, —OH. Its structural formula shows how the atoms are arranged and bonded:

```
     H   H
     |   |
H—C—C—O—H (ethyl alcohol)
     |   |
     H   H
```

Frequently, to save space and time, a condensed structural formula is used to represent an organic compound. The condensed structural formula for ethyl alcohol is C_2H_5OH. That formula indicates the functional groups present without showing exactly the types of bonds and where they are located.

Many organic compounds have identical empirical formulas, but they differ both in their functional groups and in their properties. For example, CH_3OCH_3, dimethyl ether, has the same empirical formula as ethyl alcohol, but it has totally different properties because it possesses different functional groups. Different compounds possessing the same empirical formulas are known as *isomers*.

REACTIONS

There are two major types of organic reactions, depending on the bonding involved.

Substitution Reactions

A substitution reaction usually involves a single covalent bond and results in the replacement of one atom or functional group by another. For example, if the gas methane, CH_4, is treated with chlorine gas in the presence of ultraviolet light, the following series of reactions take place, each involving the substitution of a chlorine atom for a hydrogen atom:

$CH_4(g) + Cl_2(g) \rightarrow$
$HCl(g) + CH_3Cl(g)$ (methyl chloride)

$CH_3Cl(g) + Cl_2(g) \rightarrow$
$HCl(g) + CH_2Cl_2(g)$ (methylene chloride)

$CH_2Cl_2(g) + Cl_2(g) \rightarrow HCl(g) + CHCl_3(g)$ (chloroform)

$CHCl_3(g) + Cl_2(g) \rightarrow$
$HCl(g) + CCl_4(g)$ (carbon tetrachloride)

Methyl chloride and chloroform were used as anesthetics in the past; and methylene chloride, chloroform, and carbon tetrachloride were used as solvents. Halon, $CClBrF_2$, is a methane substitute that used to be widely used in portable fire extinguishers. It is a vaporizing liquid that can put out fires at much lower concentrations than carbon dioxide, CO_2 and is safe to use on electronic equipment. Because of the bromine content of Halon, other similar products are even more destructive to ozone than CFCs. Their use was banned by 1994. They and HCFCs are expected to be replaced by hydrofluorocarbons by 2020.

Addition Reactions

An addition reaction involves only double or triple bonds, resulting in the splitting of one of the bonds and the addition of a functional group to each of the original atoms in the multiple bond. For example, ethylene, $H_2C{=}CH_2$, adds chlorine gas readily with one chlorine atom attaching to each of the carbon atoms:

$$H_2C{=}CH_2(g) + Cl_2(g) \rightarrow CH_2ClCH_2Cl(g)$$
(dichloroethane)

COMPOUNDS

Organic compounds may be divided into two general classes, aromatic and aliphatic. *Aromatic* compounds all contain a molecular structure consisting of one or more benzene rings or a series of alternating single and double bonds. Benzene, C_6H_6, is a primary example of a compound of this class. Its six carbon atoms are arranged in a hexagon with alternating single and double bonds between them. One hydrogen atom is bonded to each carbon atom. The structure may be represented as follows:

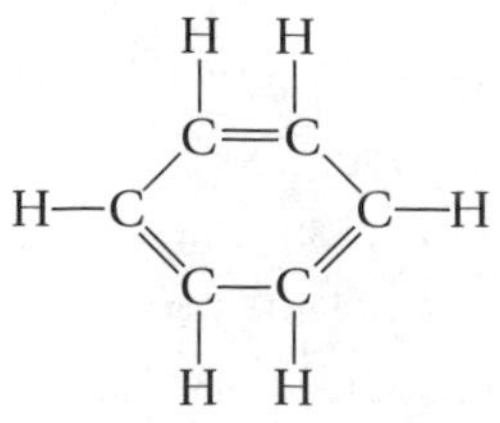

```
      H   H
      |   |
      C===C
     /     \
H---C       C---H
     \\    //
      C---C
      |   |
      H   H
```

Benzene

Other functional groups may then replace the hydrogen atoms to form an almost endless variety of aromatic compounds.

Aliphatic compounds include all organic compounds that do not possess a benzene ring or the series of alternating double and single bonds. Their basic structure involves chains of carbon atoms that may be puckered, coiled, or branched, such as in proteins. These aliphatic compounds may range from simple compounds containing one carbon (methane, CH_4) to compounds containing hundreds or thousands of carbon atoms. The carbon atoms may also be arranged in rings that are different from the benzene ring because they do not possess the alternating double and single bond structure. Such compounds are called *alicyclic* compounds. An almost endless variety of compounds of this class is possible.

The organic compounds that serve as raw materials for the preparation of the many specific synthetic compounds are obtained from three sources:

1. *Coal tar*. When coal is heated in the absence of air (destructive distillation) in huge ovens to form coke, all the volatile matter escapes from the coal and is collected and condensed to a black, tarry liquid called *coal tar*. This is the chief source of aromatic hydrocarbons.

2. *Petroleum and natural gas*. Petroleum is a mixture of principally aliphatic hydrocarbons. Many of them are liquid at ordinary temperatures, and they serve as solvents for both low molecular weight compounds that would otherwise be gases and high molecular weight compounds that would otherwise be solids. Natural gas is a mixture of light hydrocarbons that escaped from the petroleum solution.

3. *Plant life.* Plants are the principal source of the carbohydrates—sugar, starch and cellulose—as well as some special hydrocarbons such as turpentine and natural rubber.

From the point of view of composition, the most common organic compounds—whether aromatic or aliphatic—fall into six major classes: hydrocarbons, oxidation products, esters, ethers, derivatives (or hydrocarbons that have one or more hydrogen atoms replaced by functional groups), and carbohydrates. We'll look briefly at the first five here. The sixth, carbohydrates, is treated in Chapter 26.

HYDROCARBONS

TRIVIA: Ethylene gas, $H_2C{=}CH_2$, is produced by ripening fruit and stimulates nearby fruit to ripen also. Thus, one rotten apple can spoil the entire barrel.

Hydrocarbons are compounds of carbon and hydrogen. Several important families of hydrocarbons, or homologous series—each containing many compounds—are known. All the members of a homologous series have formulas that fit the general formula of the series, as summarized in Table 22.2.

Although the alkene and cycloparaffin series have the same formula, their chemical behavior differs. In the alkene series, a double bond between two carbon atoms greatly increases chemical reactivity. In the cycloparaffin series, the carbon atoms are joined to one another, forming a ring. Such a structure is relatively inert.

Both the paraffin and the cycloparaffin series are said to be *saturated* because they contain only single bonds. The other series are all *unsaturated* because they contain either double or triple bonds. You may be familiar with the terms *saturated* and *unsaturated* fats. Unsaturated fats contain a double bond in their structure. In the process of *hydrogenation*, that double bond is broken and hydrogen is added. This turns an unsaturated fat into a saturated one. This basic process converts vegetable oil into margarine.

In making margarine from vegetable oil, the fat is saturated with hydrogens. Double bonds between carbon atoms are broken, and single bonds form to hydrogen atoms. The process is hydrogenation and the result is a saturated fat.

Name	Formula	Familiar Members
alkane or paraffin	C_nH_{2n+2}	methane—CH_4; ethane—C_2H_6; propane—C_3H_8; butane—C_4H_{10}; octane—C_8H_{18}
alkene	C_nH_{2n}	ethylene—C_2H_4
alkyne	C_nH_{2n-2}	acetylene—C_2H_2
cycloparaffin or naphthene	C_nH_{2n}	cyclopentane—C_5H_{10}
benzene	C_nH_{2n-6}	benzene—C_6H_6; toluene—$C_6H_5CH_3$

Table 22.2 Homologous Series

Because of their widespread and important uses, some of the hydrocarbons deserve special mention.

Methane, CH_4. This gaseous compound is the chief ingredient of natural gas. As marsh gas it escapes from peat bogs and is responsible for the eerie will-o'-the-wisp observed when the gas ignites over swamps. Called fire damp, methane is the gas that escapes from coal that can cause disastrous mine explosions. Methane is highly combustible and an excellent fuel. Like all the other simple hydrocarbons, it is an important raw material for the manufacture of synthetic organic chemicals.

Ethane, C_2H_6. This gas makes up about 10% of natural gas. Its uses are similar to those of methane.

Propane, C_3H_8. This is one of the commonly used ingredients in bottled gas. It is also used as a high-pressure solvent.

Butane, C_4H_{10}. In addition to being useful as a bottled gas and as a solvent, butane is used in the production of high-test gasoline. Actually there are two butanes: normal butane, containing a straight chain of carbon atoms; and isobutane, which has a branched carbon chain. The prefix "normal" (*n*-) always refers to a straight chain of carbon atoms. The prefix "iso-" always refers to a chain of carbon atoms with a branch on the second carbon from one end. The structures of the isomers of butane are as follows:

```
   H  H  H  H           H  H  H
   |  |  |  |           |  |  |
H—C—C—C—C—H      H—C—C—C—H
   |  |  |  |           |  |  |
   H  H  H  H           H  |  H
                          H—C—H
                             |
                             H

   n-Butane             iso-butane
```

Octane, C_8H_{18}. This compound has 18 isomers, one of which, $(CH_3)_3CCH_2CH(CH_3)_2$, isooctane, is the reference compound for rating the octane number of gasoline. By definition, the *octane number* of a gasoline is the percentage of isooctane that must be added to *n*-heptane, C_7H_{16}, to cause a standard engine to operate with the same characteristics as the gasoline being tested. Pure isooctane has a rating of 100. High-test automotive gasolines now have octane numbers between 90 and 98. Aviation gasolines have octane numbers above 100, which simply means that they give better performance in the standard engine than pure isooctane.

Rubber, $(C_5H_8)_n$. Natural rubber consists of very long chains of C_5H_8 units. The long molecules are coiled and easily stretched. When rubber is vulcanized, sulfur atoms bond between carbon atoms of different chains, resulting in better wearing properties over a wider temperature range. The addition of fillers like zinc oxide and carbon black makes rubber less susceptible to oxidation.

Benzene, C_6H_6. This liquid is the fundamental hydrocarbon of the aromatic class. It is one of the principal ingredients of coal tar and is refined from it by fractional distillation. It is an important solvent and a fundamental raw material from which many other chemicals are made.

Toluene, $C_6H_5CH_3$. This compound is also called methylbenzene. It is obtained along with benzene from coal tar and is used both as a solvent and as a raw material for making other chemicals, including explosives like trinitrotoluene TNT, $C_6H_2CH_3(NO_2)_3$.

Naphthalene, $C_{10}H_8$. This white, crystalline solid has long been used as mothballs. It is obtained from coal tar and is a raw material for preparing other chemicals. It has a structure consisting of two benzene rings joined side by side. Paradichlorobenzene, $C_6H_4Cl_2$, is also used for mothballs.

Anthracene, $C_{14}H_{10}$. This crystalline, fluorescent solid is also obtained from coal tar and is used in the preparation of dyes. Its structure consists of three benzene rings joined side by side.

PETROLEUM REFINING

In the refining of petroleum, many useful mixtures of hydrocarbons are made available. The basic process in refining is fractional distillation, which separates the various products according to ranges of boiling points. The products formed by the initial distillation of crude petroleum are called *straight-run* products. Table 22.3 gives the main products formed, together with other pertinent information about them.

Because gasoline is only a small portion of the straight-run product, and since the demand for it is so great, two other steps are used in the refining process to increase the yield of gasoline:

1. *Re-forming*. In this process, molecules containing fewer than the desired number of atoms of carbon for gasoline are linked together. The key to the process is the use of certain catalysts that promote the formation of larger molecules.

2. *Cracking*. In this process, molecules larger than those of gasoline are broken apart, either by heat or by catalytic action.

Refined gasoline—but not straight-run gasoline—contains additives that promote complete combustion and raise the octane number. In 1922, $(C_2H_5)Pb$, tetraethyl lead, was first added to gasoline to prevent knocking, a condition created when the gasoline is ignited too soon in the engine. Because the combustion of tetraethyl lead caused deposits

Name	Boiling Range, °C	Number of Carbon Atoms Present
natural gas	below 32	1–4
gasoline	40–200	4–12
naphthas	50–200	7–12
kerosene	175–275	12–15
fuel oil	200–300	15–18
lubricating oil	above 300	16–20
wax	above 300	20–34
asphalt	residue	more than 34

Table 22.3 Petroleum Products

of metallic lead to form on spark plugs and in the engine cylinders, ethylene dibromide, $C_2H_4Br_2$, was then added to remove the lead as the volatile lead bromide, $PbBr_2$; however, this process then released significant amount of toxic lead into the environment and constituted a significant health hazard. The use of lead additives in gasoline was therefore discontinued in the 1980s in the United States.

Since 1975 automobiles sold in the United States have been equipped with catalytic converters as part of the exhaust system. These catalytic converters help reduce the levels of air pollution by aiding in the oxidation of unburned hydrocarbons and decomposition of oxides of nitrogen to less environmentally harmful compounds; however, the lead compounds that were once added to gasoline fouled these converters, and lead-free gasoline was developed for use in newer engines. Instead of tetraethyl lead, gasoline now contains other additives such as methylcyclopentadienyl manganese tricarbonyl (MMT) and alcohols and ethers, which increase the octane number and improve burning characteristics. The octane rating is also increased by increasing the ratio of branched-chain to straight-chain hydrocarbons during the refining process.

OXIDATION PRODUCTS

Hydrocarbons are highly combustible substances that burn when vigorously oxidized to form CO_2 and H_2O; however, if the oxidation of hydrocarbons is carried out gently, a number of important intermediate compounds form. The mild oxidation of hydrocarbons involves either the insertion of an oxygen atom into the molecule or the removal of two hydrogen atoms from the molecule. We may summarize the mild oxidation of methane as follows:

$$\underset{\text{Methane}}{CH_4} \xrightarrow{(+O)} \underset{\text{Methyl Alcohol}}{CH_3OH} \xrightarrow{(-H_2)} \underset{\text{Formaldehyde}}{HC(=O)H} \xrightarrow{(+O)}$$

$$\underset{\text{Formic Acid}}{HC(=O)OH} \xrightarrow{(+O)} \underset{\text{End Products}}{CO_2 + H_2O.}$$

It is sometimes necessary in practice to carry out the steps in the oxidation of hydrocarbons in an indirect fashion instead of in a direct way. For example, in making methyl alcohol from methane, the methane may be first treated with chlorine to form methyl chloride, which reacts with sodium hydroxide to produce the alcohol:

$$CH_4(g) + Cl_2(g) \rightarrow HCl(g) + CH_3Cl(g)$$

$$CH_3Cl(g) + NaOH(aq) \rightarrow NaCl + CH_3OH(aq)$$

In either case, between the hydrocarbon and the end products of its oxidation, three types of compounds appear: alcohols, aldehydes, and organic acids.

Alcohols

Methyl alcohol or methanol, CH_3OH. This volatile liquid is also known as *wood alcohol,* because it is an important by-product from the destructive distillation of wood to produce charcoal. Great quantities of it are also produced by the catalytic hydrogenation of carbon monoxide, in which hydrogen reacts with the carbon monoxide in the presence of a catalyst.

Methanol is poisonous and can cause blindness. Because it is absorbed through the skin, methanol is unfit for use as rubbing alcohol. It is an important solvent and is used both as an antifreeze and as a denaturant (poisonous additive) for ethyl alcohol, making the latter unfit for human consumption. It is the starting point in the production of many synthetic chemicals.

Ethyl alcohol or *ethanol*, C_2H_5OH. Also known as *grain alcohol*, this important compound is produced in large quantities from the fermentation of carbohydrate compounds contained in molasses, corn, rye, barley, and potatoes. Enzymes in yeast cause the fermentation. (*Fermentation* is the breakdown of sugar in the absence of oxygen, yielding alcohol.) Ethyl alcohol absorbs moisture from the atmosphere until it has a composition of 95% alcohol and 5% water. It is used extensively in the production of other chemicals and chemical products, particularly in industries preparing drugs, medicines, and cosmetics. It is present in a variety of beverages, causing them to be intoxicating. By law, the ethyl alcohol intended for human consumption must be produced by fermentation, but large amounts are produced synthetically for a variety of other purposes. Ethyl alcohol ranks next to water as an important solvent, and alcoholic solutions are known as *tinctures*. To prevent its illegal use in the manufacture of beverages, synthetic ethanol is often denatured by dissolving poisonous and nauseating compounds in it.

Isopropyl alcohol, $CH_3CHOHCH_3$. This liquid is extensively used as rubbing alcohol. It is prepared from the gases obtained from the cracking of petroleum.

Glycerin, $C_3H_5(OH)_3$. This sweet, syrupy liquid is obtained as a by-product from the manufacture of soap. It is used as an antifreeze and as a moistening agent in tobacco and cosmetics. When treated with concentrated nitric and sulfuric acids, it is converted to nitroglycerin, which, in turn, is absorbed in clay to form dynamite.

An important medical use of nitroglycerin is as a vasodilator to relieve the pain of the heart ailment called angina.

Phenol, C_6H_5OH. Known also as carbolic acid, this compound is obtained from coal tar. It is an important disinfectant, and great quantities of it are used in the preparation of plastics and other aromatic chemicals. Phenol may also be called monohydroxybenzene. Its related polyhydroxybenzenes such as hydroquinone, $C_6H_4(OH)_2$, and pyrogallol, $C_6H_3(OH)_3$, are important photographic developers.

The oxidation of an alcohol like isopropyl alcohol, $CH_3CHOHCH_3$, in which the alcohol group is in a nonterminal position in the carbon chain, produces a compound known as a ketone. In ketones the OH group of the alcohol is oxidized to an oxygen atom attached by a double bond. *Ketones* cannot be oxidized to acids, but vigorous oxidation (burning) yields CO_2 and H_2O.

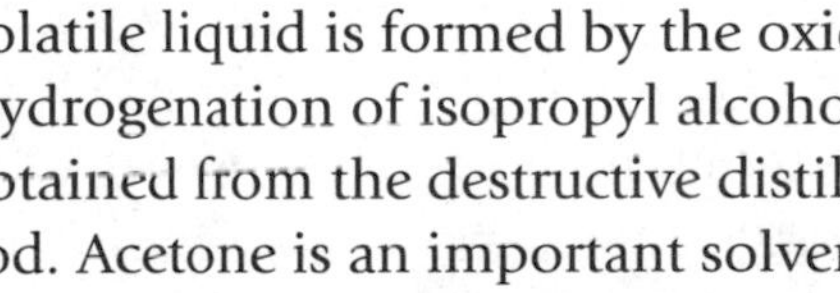

$$(H_3C)_2CH{-}OH \xrightarrow{(-H_2)} (H_3C)_2C{=}O.$$

Isopropyl Alcohol → Acetone

One important ketone is acetone, CH_3COCH_3. This volatile liquid is formed by the oxidation or dehydrogenation of isopropyl alcohol. It is also obtained from the destructive distillation of wood. Acetone is an important solvent as well as a raw material in the preparation of other organic compounds. It may be best known to the public as nail polish remover.

TRIVIA: The sweet properties of saccharin, the sugar substitute, were discovered in 1879 when a chemist noticed that a substance that he had spilled on his hand tasted sweet (no wonder chemists had a short life expectancy back then). Two other artificial sweeteners, cyclamate and aspartame, were also discovered in a similar fashion.

Aldehydes

Formaldehyde, HCHO. This gas is prepared by oxidizing methyl alcohol with hot copper oxide. A 40% solution of it is called *formalin.* Formaldehyde is poisonous and is an important germicidal agent and preservative. Great quantities of it are used with phenol in the preparation of plastics.

Acetaldehyde, CH_3CHO. Small amounts of this and other aldehydes are responsible for the flavor and scent of alcoholic beverages. Acetaldehyde is formed by the gradual oxidation of ethyl alcohol and is used as an intermediate in the preparation of other organic chemicals because it is very reactive chemically.

Carboxylic Acids

Organic acids are weak electrolytes in general. They form salts with bases just as inorganic acids do.

Acetic acid, CH_3COOH. Some acetic acid is obtained from the destructive distillation of wood, but most of it is produced by the fermentation of fruit juices. Vinegar is a dilute solution of this acid. Pure acetic acid is known as *glacial acetic acid.* It is an excellent solvent and is used in the preparation of photographic film and synthetic fibers.

Formic acid, HCOOH. Produced by the oxidation of formaldehyde, this liquid with an irritating odor is found in ants (from which it was originally prepared by distillation) and in stinging plants. It is a constituent of perspiration.

Citric acid, $H_3C_6H_5O_7$. This solid compound is contained in the juices of citrus fruits like lemons and limes. It was once obtained almost exclusively from these fruits. It is now prepared by the action of molds on molasses and sugar. Citric acid is used as an ingredient in drugs and soft drinks.

Oxalic acid, $H_2C_2O_4$. This solid acid is produced by the action of the same molds on molasses and sugar as used to produce citric acid. When the pH of the solution is 6–7, oxalic acid is formed. When the pH is reduced to 1–2, citric acid is produced. Oxalic acid is an important reducing agent and is used to bleach wood and rust stains in clothing.

ESTERS

Esters are compounds formed from the reaction of an alcohol with an organic acid. The general reaction is

$$\text{Alcohol} + \text{Acid} \underset{\text{hydrolysis}}{\overset{\text{esterification}}{\rightleftharpoons}} \text{Ester} + \text{Water}$$

$$R_1-\underset{\displaystyle OH}{\overset{|}{\underset{|}{C}}}- + R_2-C\begin{matrix}\nearrow O \\ \searrow OH\end{matrix} \rightleftharpoons R_1-C-O-\overset{\displaystyle O}{\overset{\|}{C}}-R_2 + H_2O$$

The reaction to the right, esterification, is favored by the presence of concentrated sulfuric acid, which is a powerful dehydrating agent and removes the water as it is formed. When water is added to an ester, hydrolysis (the reaction with water; see Chapter 9 for another type of hydrolysis) readily occurs, especially if either the alcohol or the acid can be removed from the system by precipitation or volatilization.

Esters tend to have pleasant scents and flavors. For example, the reaction of butanoic acid with ethyl alcohol produces ethyl butanoate, strawberry flavoring. These properties are used extensively in beverages, confections, medicines, cosmetics, and perfumes. Some of the more commonly known esters are amyl acetate (banana oil) and methyl salicylate (oil of wintergreen). Esters are also used as solvents.

ETHERS

Ethers are compounds formed by the dehydration of alcohols by concentrated sulfuric acid. The elements of a water molecule are removed from two alcohol molecules. For example, diethyl ether, $C_2H_5OC_2H_5$, which was once a commonly used anesthetic, is made from ethyl alcohol:

$$2\ C_2H_5OH \xrightarrow{(H_2SO_4)} C_2H_5OC_2H_5 + H_2O$$

desiccant (drying agent)

$$R\text{–}OH + R\text{–}OH \longrightarrow R\text{–}O\text{–}R + H_2O$$

The ether formed is extremely volatile and is not miscible with water. Thus, it is easily distilled from the water. Ethers are extremely flammable.

PLASTICS AND SYNTHETIC FIBERS

Under the right conditions, certain small organic molecules will link together to form long chains or complex two- and three-dimensional networks. That process is *polymerization*, and the products are called *polymers*. Polymerization takes place only in molecules containing double or triple bonds and is usually dependent on temperature, pressure, and the presence of a suitable catalyst.

Rubber is a natural polymer, but many synthetic polymers have been made for a variety of uses. For example, the gas ethylene, $H_2C{=}CH_2$, when heated under pressure, polymerizes to form a solid transparent plastic called polyethylene. The polymer consists of CH_2 units linked together in an endless chain: . . . $CH_2CH_2CH_2CH_2$. . . . Ethylene is said to be the *monomer*, or single unit, of polyethylene.

More commonly, two or more different monomers may be mixed and then copolymerized to form an almost endless variety of different plastic substances. Some plastics result from adding one monomer to another, similar to the formation of polyethylene. Others involve condensation, or the splitting out of water molecules as the monomers link together.

In general, under heating plastics behave in one of two ways. Either they are *thermosetting*—they set to a hard infusible solid on heating—or they are *thermoplastic*—they become soft and can be remolded or reshaped. Bakelite is an example of a thermosetting plastic, whereas while Lucite and Plexiglas are thermoplastic. The process of copolymerization yields some synthetic fibers. To make nylon, hydrogen from water and nitrogen from air are united to form ammonia, which is then converted in a series of steps to a compound called hexamethylenediamine. Phenol, extracted from coal tar, is converted by another series of steps to a compound called adipic acid. These two synthetic compounds are then condensed together, eliminating water and forming giant, fibrous, nylon molecules.

Rayon, in contrast, is pure cellulose. In its manufacture, purified cotton fibers (virtually pure cellulose) are converted to sodium cellulose and cellulose xanthate. The latter is then added to an acid solution that regenerates pure cellulose with characteristic silky properties. Acetate rayon is the compound cellulose acetate. These fibers are not related to plastics.

LIQUID CRYSTALS

Liquid crystals are jellylike, organic compounds that resemble liquids in some ways and crystals in other ways, such as their ability to scatter and reflect light. This ability to scatter and reflect light arises when an electrical field is varied around the liquid crystal. This causes the liquid crystal to change its molecular orientation. Numerical liquid crystal displays (LCDs) on some watches, calculators, clocks, and automobile dashboards depend on their use.

SUMMARY

Organic chemistry relates to the synthesis and chemistry of the compounds of carbon, of which many millions are known. Partly because the valence of carbon is 4, it can make numerous compounds based on carbon–carbon and carbon–hydrogen bonds, as well as many compounds based on carbon linkages with oxygen, nitrogen, sulfur, and with a variety of other elements and functional groups. Organic compounds may be saturated or unsaturated. They may also be aliphatic, aromatic, or alicyclic.

The sources of carbon for organic chemistry are coal, petroleum, natural gas, and plants. Organic chemical reactions proceed usually by addition or substitution. Important classes of organic compounds include alcohols, aldehydes, ketones, acids, esters, plastics, and synthetic fibers.

PROBLEM SET NO. 24

1. Write the structural formula for hydroxybenzene, or phenol.

2. Write the esterification–hydrolysis reaction between ethyl alcohol and acetic acid.

3. One liter of propylene weighs 1.88 g at STP. Its composition by weight is C, 85.6%; and H, 14.4%. What is the molecular formula of propylene?

4. Match the compound with its class name:

1. CH_3OCH_3	a. chlorobenzene
2. $CH_3\overset{\overset{O}{\|}}{C}CH_2CH_3$	b. polyethylene
3. $CH_3CH_2CH_2CH_2CH_2CH_2CH_3$	c. heptane
4. C_6H_5—Cl (benzene ring—Cl)	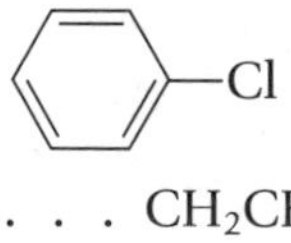 d. dimethyl ether
5. . . . $CH_2CH_2CH_2$.	e. methyl ethyl ketone

5. Given the structure of an *n*-propyl chloride,

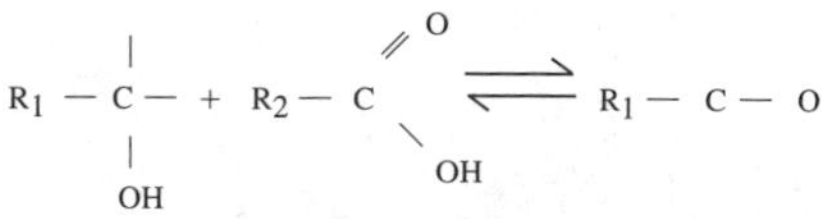

$$R_1 - \underset{\underset{OH}{|}}{\overset{|}{C}} - \; + \; R_2 - C\begin{matrix} /\!\!/ O \\ \diagdown OH \end{matrix} \rightleftharpoons R_1 - C - O$$

write the structure for isopropyl chloride.

CHAPTER 23

NUCLEAR CHEMISTRY

KEY TERMS

radioactivity, alpha rays, beta rays, gamma rays, half-life, fission, fusion, chain reaction

In 1896 the French scientist Henri Becquerel accidentally discovered a strange natural phenomenon. By chance, he placed a sample of a mineral called pitchblende in his desk on top of a covered photographic plate. A key happened to be lying between the mineral and the plate. When Becquerel later developed the plate, he found he had a photograph of the key. The only possible explanation was that some mysterious emanation from the mineral had penetrated the cover and exposed the plate.

Shortly thereafter, some European explorers in what was then the Belgian Congo of Africa came on a native village in which the tribal witch doctor cured his sick tribesmen by burying them to their neck in the ground near the village. It was later shown that the ground there gave off the same emanation as the pitchblende.

RADIOACTIVITY

The emanation coming from these substances is called *radioactivity*. It is the property that some elements or isotopes have of spontaneously emitting energetic particles by the disintegration of their atomic nuclei. Three different types of emanations, or "rays," come from radioactive substances (see Figure 23.1), namely, alpha and beta particles, and gamma rays, after the first three letters in the Greek alphabet.

1. *Alpha particles*, represented by ${}^{4}_{2}He$ or α, are identical with nuclei of helium atoms, consisting of 2 neutrons and 2 protons. Although they initially have a charge of +2, they quickly acquire 2 electrons. For this reason, they are commonly represented without the charge. They move at speeds ranging from 2000 to 20,000 miles per second, or 1–10% of the speed of light.

2. *Beta particles*, represented by ${}^{0}_{-1}e$ or β, are a stream of negatively charged electrons that move almost at the speed of light.

3. *Gamma rays*, represented by γ, are photons (small packets) of electromagnetic energy.

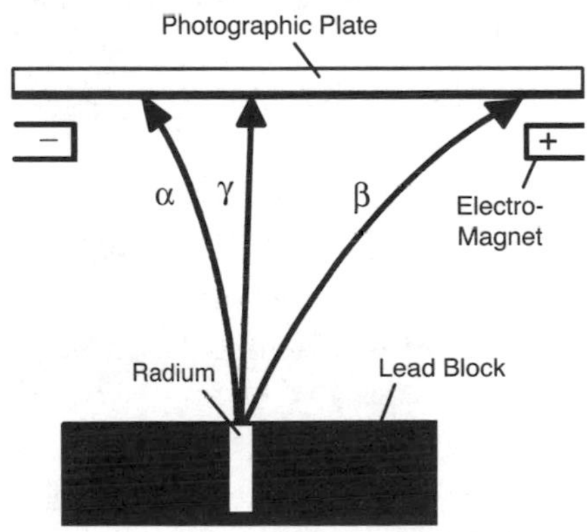

Figure 23.1 Type of Radiation

They are best described as bundles of energy that are more energetic and therefore more penetrating than x-rays. They move at the speed of light, but they have neither mass nor electrical properties. Often, gamma emission occurs simultaneously with alpha or beta emission.

Radioactive particles are emitted exclusively from the nuclei of atoms. Whether the atoms are in the free elemental state or whether they are chemically combined in compounds has no effect on their radioactivity, but because the nuclei of elements undergo change during alpha and beta emissions means that *one element is transmuted into another*. Such a change is not an ordinary chemical change, for it involves the nucleus rather than the electrons of the atom. All the elements of higher atomic number than lead have naturally radioactive isotopes.

> Radioactivity is the property that some elements or isotopes have of spontaneously emitting energetic particles by the disintegration of their atomic nuclei. The result is the change of one element into another.

When the nucleus of an atom loses an alpha particle, it loses two protons; therefore, the atomic number of the remaining nucleus is 2 less than the atomic number of the original atom. Thus, a uranium nucleus is transformed into a thorium nucleus as the result of the loss of an alpha particle. A radium nucleus is transformed into a radon nucleus by the same process.

The loss of a beta particle by a nucleus of an atom involves first the decomposition of a neutron. The neutron splits into a proton and an electron. The electron leaves the nucleus, and the remaining nucleus has one more proton than the original nucleus. Thus, the atomic number is increased by 1 during a beta emission. In this way, a uranium nucleus becomes a neptunium nucleus, or a neptunium nucleus becomes a plutonium nucleus.

Gamma radiation does not cause transmutation of the elements, for it does not involve change in either mass or electrical charge. It may accompany either alpha or beta radiation, and it accounts for the energy changes that accompany radioactivity.

Because radioactive atoms decay, they are said to be unstable. One method of expressing the stability of radioactive elements is known as the *half-life*, $t_{1/2}$, of the element. This is simply the time required for half of a given sample of a radioactive element to decay. The half-life for a given isotope may vary from a few billionths of a second to millions of years. Radium has a half-life of 1590 years, which means that from a 1-g sample of radium today, 0.50 g of radium will be left 1590 years from now. After another 1590 years, 0.25 g of the original radium will remain, and so on. After one half-life, 50% of the sample remains, 25% at the end of two half-lives, 12.5% at the end of three half-lives, and so on.

These half-lives are statistical in nature. They predict an average trend, not the future of an individual atom. If two atoms of radium are placed on a table, one may disintegrate immediately, whereas the other may remain intact for millions of years. In a similar fashion, two people may die today, or both may live to be 100. There is no way of knowing what will happen to a single person or to a single atom, but it is certain that of all the people born in the year 1950 , nearly all will be dead by the year 2050. Nevertheless, in any case, it is obvious that an element with a long half-life is more stable than one with a short half-life. Also, generally speaking, a radioactive sample will have decayed below detection limits after 10 half-lives.

Although early researchers learned to recognize the effects of radioactivity, it took many years of experimentation to learn what causes it. Today, in general, we know that

1. The principal nuclear particles, neutrons and protons (known collectively as nucleons), are held together by a "nuclear glue" called the strong nuclear force, which results from the tremendous density of these particles (hundreds of thousands of times the density of water) and their closeness to each other. This attractive force tends to cause nuclear particles to combine and build up increasingly larger nuclei.

2. Positively charged protons exert tremendous repulsive forces on one another. The neutrons serve to buffer the effect of these repulsive forces by maintaining a space interval between the protons, but the repulsive force cannot be discounted, and it causes some nuclei to fly apart and form smaller nuclei. This is radioactive decay. All elements with 84 or more protons are unstable and undergo decay.

3. Another factor affecting the stability of nuclei is the ratio of the number of neutrons to the number of protons in the nucleus. For the very light atoms, the ratio is 1; for atoms in the middle of the periodic table, it is about 1.3; for the heavy elements, it is about 1.6. If there is any marked variation from the stable ratio, the atom will be radioactive. For example, carbon-12, $^{12}_{6}C$, contains 6 protons and 6 neutrons in its nucleus. Its n/p ratio is 1. It is not radioactive. But carbon-14, $^{14}_{6}C$, has 8 neutrons and only 6 protons. Its n/p ratio is 1.4. That atom is radioactive because its ratio differs from the stable ratio. Thus, radioactivity is spontaneous and is independent of temperature, pressure, catalysts, or any other external conditions.

Both *fission*, the decay of a nucleus by breaking apart into two or more lighter isotopes, and *fusion*, the combining of two light isotopes into a heavier one, are different from ordinary radioactivity. Before heavy nuclei will split to form fragment nuclei of about equal size (fission), and before light nuclei will meld to form heavy nuclei (fusion), an input of energy, called *activation energy*, is required. In the case of fusion, the activation energy can be supplied by extremely high temperatures such as those of the Sun or other stars, or by the energy associated with extremely high velocity particles at elevated temperature. In the case of fission, most nuclei also require very high activation energy; however, naturally occurring uranium-235 has an activation energy that can be attained by bombarding it with neutrons.

NUCLEAR REACTIONS

Bombarding the nuclei of atoms with atomic particles can cause the artificial transmutation of elements. In general, such reactions lead to the production of artificially radioactive isotopes of the various elements. Successful transmutation has been accomplished by using the following as "bullets":

- neutrons—$^{1}_{0}n$
- protons (hydrogen nuclei)—$^{1}_{1}H$

- deuterons (deuterium nuclei)—$^{2}_{1}H$
- alpha particles (helium nuclei)—$^{4}_{2}He$
- gamma rays (pure energy)—γ.

Thus, five different types of transmutation reactions are possible. Notice that these nuclear reactions are balanced not in the same fashion as ordinary chemical reactions but by making sure that the sum of all atomic numbers of the reactants equals the sum of all atomic numbers of the products. Likewise, the sum of all mass numbers on the left has to equal the sum of all mass numbers on the right. The five types of nuclear reactions are as follows:

1. *Neutron-induced reactions.* These result in the formation of a heavier isotope of the bombarded element, or they may cause the emission of protons or alpha particles. Examples are

$$^{107}_{47}Ag + ^{1}_{0}n \rightarrow ^{108}_{47}Ag$$
$$^{10}_{5}B + ^{1}_{0}n \rightarrow ^{7}_{3}Li + ^{4}_{2}He$$
$$^{14}_{7}N + ^{1}_{0}n \rightarrow ^{14}_{6}C + ^{1}_{1}H$$

The last reaction is a side reaction of the explosion of an atomic fission bomb. Carbon-14 is radioactive, has a half-life of 5730 years, and is formed in such blasts in large quantities.

2. *Proton-induced reactions.* In these reactions, the atomic number of the element involved increases by 1, resulting in the formation of the next higher element. Neutrons may or may not be emitted during these reactions. Examples are

$$^{12}_{6}C + ^{1}_{1}H \rightarrow ^{13}_{7}C$$
$$^{18}_{8}O + ^{1}_{1}H \rightarrow ^{18}_{9}F + ^{1}_{0}n$$

The latter reaction may be used as a source of neutrons for other nuclear reactions.

3. *Deuteron-induced reactions.* These result in the formation of protons, neutrons, or alpha particles. This type is commonly employed in the manufacture of artificially radioactive elements. Examples are

$$^{9}_{4}Be + ^{2}_{1}H \rightarrow ^{10}_{5}B + ^{1}_{0}n$$
$$^{23}_{11}Na + ^{2}_{1}H \rightarrow ^{24}_{11}Na + ^{1}_{1}H$$
$$^{20}_{10}Ne + ^{2}_{1}H \rightarrow ^{18}_{9}F + ^{4}_{2}He$$

The first reaction is commonly employed as a source of neutrons.

4. *Alpha particle-induced reactions.* These result in the emission of either protons or neutrons. Examples are

$$^{14}_{7}N + ^{4}_{2}He \rightarrow ^{17}_{8}O + ^{1}_{1}H$$
$$^{10}_{5}B + ^{4}_{2}He \rightarrow ^{13}_{7}N + ^{1}_{0}n$$
$$^{9}_{4}Be + ^{4}_{2}He \rightarrow ^{12}_{6}C + ^{1}_{0}n$$

The first reaction was the first artificial transmutation ever carried out experimentally. The last one produces nonradioactive carbon. It was the original source of neutrons for experimental purposes.

5. *Gamma ray-induced reactions.* Again, these may produce either protons or neutrons. Examples are

$$^{26}_{12}Mg + \gamma \rightarrow ^{25}_{11}Na + ^{1}_{1}H$$
$$^{9}_{4}Be + \gamma \rightarrow ^{8}_{4}Be + ^{1}_{0}n$$

The particles used for bombardment in nuclear reactions are usually energized to a definite energy level before they are permitted to strike the target nuclei. This energizing is accomplished in various types of accelerating

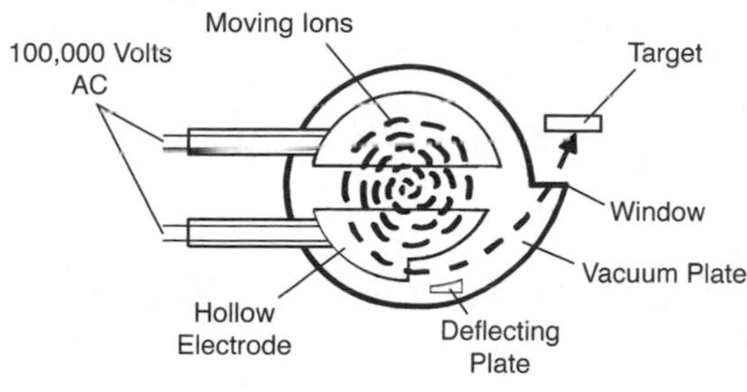

Figure 23.2 The Cyclotron

devices known as cyclotrons (Figure 23.2), betatrons, synchrotrons, linear accelerators, and various other names. In general, these devices employ the energy of magnetic and electrical fields to speed up the bombarding particle until it has enough energy to carry out a desired nuclear reaction.

The preceding examples were types of artificial transmutation, the deliberate conversion of one element into another by bombardment of its nucleus. The following are types of decay of naturally occurring isotopes (see Chapter 3).

1. *Alpha decay.* In this decay mode, an alpha particle (2 protons + 2 neutrons) is emitted from the nucleus. For example,

$$^{226}_{88}\text{Ra} + \rightarrow {}^{222}_{86}\text{Rn} + {}^{4}_{2}\text{He}$$

   Heavy, large isotopes tend to undergo alpha decay, ridding the unstable nucleus of two protons.

2. *Beta decay.* In beta decay, an electron is emitted from the nucleus. The electron results from the decay of a neutron into a proton and an electron (beta particle).

$$^{14}_{6}\text{C} + \rightarrow {}^{14}_{7}\text{N} + {}^{0}_{-1}\text{e}$$

   Isotopes that have a high ratio of neutrons to protons commonly undergo beta emission, since a neutron is lost, and a proton is formed.

3. *Gamma radiation emission.* This short-wavelength, high-energy radiation commonly accompanies other radioactive decays, such as

$$^{238}_{92}\text{U} + \rightarrow {}^{234}_{90}\text{Th} + {}^{4}_{2}\text{He} + 2\gamma$$

4. *Positron emission.* A positron is a particle of the same mass as an electron, but with a positive charge.

$$^{40}_{19}\text{K} + \rightarrow {}^{40}_{18}\text{Ar} + {}^{0}_{+1}\text{e}$$

   The decay of a proton in the nucleus yields a neutron and a positron. Positron emission is common in elements that have a low ratio of neutrons to protons, since a proton is lost, and a neutron is formed.

   A positron can combine with an electron to form a cosmic ray. In this reaction, the mass of both particles is transformed completely into pure energy. Positrons may also combine with neutrons to form protons.

5. *Electron capture.* In electron capture, an electron from the orbital closest to the nucleus (the 1s) is captured by the nucleus and combines with a proton to form a neutron. As electrons from higher energy levels drop down to fill the vacancy, energy in the x-ray region of the electromagnetic spectrum is emitted:

$$^{55}_{26}\text{Fe} + {}^{0}_{-1}\text{e} \rightarrow {}^{55}_{25}\text{Mn} + \text{x-rays}$$

   Again, elements with a low ratio of neutrons to protons are candidates for electron capture.

Just as in ordinary chemical reactions, energy is either absorbed or released during nuclear reactions. The amount of energy associated with nuclear reactions is, in general, higher than that associated with chemical change. In the late 1930s Otto Hahn, Fritz Strassmann, Lisa Meitner, and Enrico Fermi discovered that bombarding uranium with neutrons liberated unexpectedly huge quantities of energy and that the products of the reaction consisted of two atoms each about half the atomic mass of uranium. This reaction was the splitting, or *fission*, of uranium atoms into two alomst equal parts.

This fission of uranium was interesting for two reasons. First, its energy output was enormous. Second, additional neutrons were liberated during the reaction that could continue the fission reaction with neighboring atoms. The neutrons produced made possible a *chain reaction*, or one in which one of the agents necessary to the reaction is itself produced by the reaction, thus causing further like reactions. The reaction liberated previously unheard-of quantities of energy from relatively small amounts of matter, just as Einstein had predicted. It was also found that certain isotopes would not undergo chain–reaction because they did not produce an excess of neutrons in the initial fission reaction. Uranium-235 is fissionable, whereas uranium-238 is not.

The first practical use of atomic energy was in the atomic bombs that brought World War II to an end. They used the fission chain reactions of uranium and plutonium. These same reactions are employed for peaceful purposes inside the carefully controlled environment of a nuclear reactor (see Figure 23.3). The heat generated in a nuclear power plant supplies steam for the generation of electrical energy. Nuclear reactors are also a source of radioactive materials used in medical and scientific research.

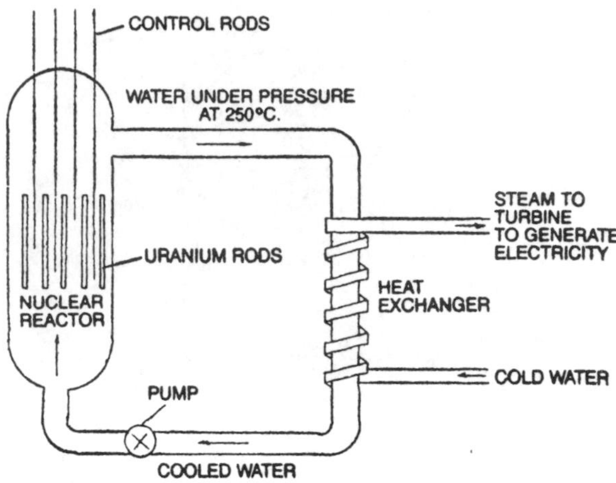

Figure 23.3 Nuclear Reactor to Make Steam to Make Electricity

About 20% of the electricity produced in the United States comes from nuclear energy. In a typical fission reaction in a nuclear reactor, uranium-235 is bombarded with neutrons. It splits or fissions into barium, krypton, and additional neutrons with an enormous release of energy. The energy is transferred by heat exchange to cold water, which is converted into steam. The steam runs a turbine that drives a generator to produce electricity. Such turbines also power nuclear submarines.

Another important use of radioactivity results from the fact that radioactive atoms behave chemically in exactly the same manner as the ir nonradioactive natural isotopes. For this reason, radioactive isotopes can be used as tracers. Following the path of the radioactive emission traces the path of the nonradioactive isotope. Isotopes can therefore reveal the intricacies of complex chemical processes that occur in the human body, in the growth of plant and animal life, and in industrial processes.

FUSION REACTIONS

The energy we receive from the Sun comes from the nuclear reactions that take place there. They are not fission reactions that split atoms apart; they are fusion reactions that combine light elements into heavier ones. Scientists generally agree that the following fusion reactions called the proton–proton chain produce most of the Sun's energy in its core:

(I) ${}^{1}_{1}H + {}^{1}_{1}H \rightarrow {}^{2}_{1}H + {}^{0}_{1}e$ positron + neutrino

(II) ${}^{1}_{1}H + {}^{2}_{1}H \rightarrow {}^{3}_{2}He$ + energy

(III) ${}^{3}_{2}He + {}^{3}_{2}H \rightarrow {}^{4}_{2}He + 2{}^{1}_{1}H$ + energy

Net reaction:

$$4{}^{1}H \rightarrow {}^{4}He + \text{energy} + 2 \text{ neutrinos}$$

The positrons produced in step I collide with electrons and are annihilated, releasing gamma rays.

Fusion reactions are also known as *thermonuclear* reactions. They require exceedingly high temperatures to take place. The energy liberated by this type of reaction is far greater than that of fission reactions. The thermonuclear reaction is the basis of the hydrogen bomb, which releases tremendous amounts of energy when hydrogen nuclei collide at high temperatures to form helium.

Because the thermonuclear reaction is difficult to control once it begins, it has not been harnessed for power generation, as the fission reaction has. Experiments have been performed in the past, however—such as in the Tokamak reactor at Princeton, which was shut down in 2002—and are planned for the future—in the International Thermonuclear Experimental Reactor scheduled for operation in 2014. The effort is being made to fuse small nuclei on a sustainable, controlled basis to create energy for electricity and other purposes.

One of the fusion reactions that scientists would like to control for peaceful purposes is the fusion of two deuterium nuclei to form an alpha particle, helium-4:

$${}^{2}_{1}H + {}^{2}_{1}H \rightarrow {}^{4}_{2}He$$

About 1×10^8 kilocalories (kcal) of energy are released per gram of deuterium. (One kilocalorie equals 1000 calories.) This is five times the energy released in the combustion of 1 g of petroleum. The ocean contains 2×10^{23} g of deuterium, enough to support controlled fusion reactions for hundreds of years.

In Chapter 5 you read about the law of conservation of matter. In an ordinary chemical reaction, the loss or gain in mass is too minute to be detected. Fission and fusion illustrate vividly, however, that it is mass–energy that is neither created nor destroyed in any chemical transmutation. In fission and fusion, the loss in mass is sufficient to create an enormous release of energy. The relation between mass and energy is Einstein's famous equation

$$E = mc^2$$

where

E = energy in ergs (4.184×10^7 ergs = 1 calorie)

m = mass in grams

$c = 3 \times 10^{10}$ cm/s, the speed of light

If 2 mol of deuterium fuse to make 1 mol of helium, then 2 (2.016) = 4.032 g of deuterium yield 4.00 g helium. That equation does not balance. A mass of 0.032 g is lost. The explanation is that this mass has been transformed into energy. Let's calculate its value:

$$\begin{aligned} E &= \frac{0.032}{4.032} \text{ g lost per gram of deuterium} \\ &\quad \times (3 \times 10^{10} \text{ cm/s})^2 = 0.0714 \times 10^{20} \text{ ergs} \\ &= 1.71 \times 10^{8} \text{ kilocalories per gram of deuterium reacted} \end{aligned}$$

That is about 10 million times the energy released in burning 1 g of carbon, methane, or hydrogen (see Table 24.2 in Chapter 24).

Fusion reactions are used to create the transuranium elements, those with atomic numbers greater than 92. Element 109, meitnerium, was created at the Institute for Heavy Ion Research in Darmstadt, Germany, in 1982 by the bombardment of a bismuth nucleus with iron ions. Its half-life was 3.4 milliseconds. Some scientists believe that only a few more superheavy elements remain to be created experimentally by fusion. Superheavy elements of atomic numbers in the region of 114 to 118 have been postulated.

SUMMARY

Nuclear chemistry involves fission or disintegration of the nuclei of the atoms of elements naturally or by artificial means. Natural fission is called radioactivity. Radioactive elements give off α and β particles, and γ rays. Their stability is measured by their half-lives. Bombarding their nuclei with neutrons, protons, deuterons, alpha particles, and gamma rays can effect artificial transmutation of the elements. Useful applications of nuclear fission include nuclear power plants, in which fission releases enough heat to generate electricity. Nuclear chemistry also involves the fusion of the nuclei of light atoms such as hydrogen, with the resultant release of huge amounts of energy.

PROBLEM SET NO. 25

1. Complete the following nuclear reaction:

$$^{14}_{7}\text{N} + {}^{4}_{2}\text{He} \rightarrow {}^{17}_{8}\text{O} + x$$

2. How does a fusion reaction differ from a fission reaction and from an ordinary chemical reaction?

3. Do the emissions of radioactivity, α and β particles, and γ rays, come form (a) the nucleus, (b) the electrons surrounding the nucleus, or (c) both nucleus and surrounding electrons?

4. Do neutron-induced reactions create heavier or lighter elements?

5. Is the energy from the Sun produced by fusion or fission reactions?

CHAPTER 24

THE CHEMISTRY OF ENERGY

KEY TERMS

fossil fuels, renewable, nonrenewable, BTU

Chemistry is committed to developing more abundant energy resources at a time when world population is rapidly increasing, people are demanding higher living standards, and the most depended-on source of energy, petroleum or oil, is being depleted rapidly.

SOURCES OF ENERGY

Energy can be harnessed for human use by physical or chemical means. Hydroelectric energy, generated from the power of moving water, is physical. The rotation of the vanes of a windmill to create wind energy is also physical, as is geothermal energy, which taps underground beds of steam to produce electrical energy.

Most forms of energy, however, are chemical; they employ the chemical reaction oxidation, which releases energy as heat. Important sources of chemical energy are petroleum, coal, natural gas, and, to a lesser extent, wood, methane, and alcohol.

Fossil fuels, including coal and petroleum, formed millions of years ago from the remains of plants and animals. The once-living organic compounds were transformed under conditions of great heat and pressure into the concentrated energy sources we pump, extract, and mine today. Petroleum contains the organic compounds alkanes, which can react with oxygen as follows:

$$2\ C_nH_{2n+2}(g) + (3n+1)\ O_2(g) \rightarrow 2nCO_2(g) + 2(n+1)\ H_2O(g) + \text{heat energy}$$

The carbon in coal can be burned directly:

$$C(s) + O_2(g) \rightarrow CO_2(g) + \text{heat energy}$$

or it can be reacted in the water gas reaction, with steam:

$$C(s) + H_2O(g) \rightarrow CO(g) + H_2(g) + \text{heat energy}$$

The mixture of carbon monoxide, CO, and hydrogen, H_2, is called water gas. It can also be used as a fuel or, by adjusting the CO/H_2 ratio, as a reactant to form synthesis gas, which is used to make liquid fuels comparable to gasoline, where C_6H_{14} is a form of gasoline:

$$6\ CO(g) + 13\ H_2(g) \rightarrow C_6H_{14}(l) + 6\ H_2O(l)$$

Natural gas is 85% or more methane, CH_4; with some ethane, C_2H_6; propane, C_3H_8; and trace amounts of higher alkanes and carbon dioxide. The methane reacts with oxygen to release heat:

$$CH_4(g) + 2\ O_2(g) \rightarrow CO_2(g) + 2\ H_2O(g) + \text{heat energy}$$

Wood is a complex mixture of 70–90% cellulose, 10–30% lignin, and a small percentage of wood ash. Wood ash contains inorganic oxides left after combustion. Lignin is a complex phenolic polymer that gives seasoned wood its color, hardness, and mass. Cellulose is a polymer of repeating units of composition $C_6H_{10}O_5$ and is the most abundant organic compound on Earth. Wood contains some oxygen, so pound for pound its heating value is less than that of petroleum, coal, and natural gas. Municipal garbage and plant stalks, which also contain cellulose, lignin, and other combustibles, may also be burned to yield useful heat energy.

Methanol, CH_3OH, and ethanol, CH_3CH_2OH, also release energy as a result of chemical reactions:

$$2\ CH_3OH(l) + 3\ O_2(g) \rightarrow 2\ CO_2(g) + 4\ H_2O(g) + \text{heat energy}$$

$$C_2H_5OH(l) + 3\ O_2(g) \rightarrow 2\ CO_2(g) + 3\ H_2O(g) + \text{heat energy}$$

Methanol can be made by destructive distillation (heating in the absence of air), or it can be made from a mixture of CO and H_2 from coal or natural gas:

$$CO(g) + 2\ H_2(g) \rightarrow CH_3OH(l)$$

Ethanol can be made by the fermentation of carbohydrates from grain, corn, sugarcane, cassava roots, and other high-starch-producing plants. Starch is a complex of repeating sugar units. In the process of fermentation, starch is first broken down into molecules of the sugar glucose, $C_6H_{12}O_6$. Then, glucose is converted to ethanol:

$$C_6H_{12}O_6(s) \rightarrow 2\ C_2H_5OH(l) + 2\ CO_2(g)$$

Photovoltaic cells have been developed to convert solar energy to electrical energy. The most popular photovoltaic material is ultrapure silicon. GaAs, CdS, and CdSe can also be used. The Sun's rays strike an atom of silicon and knock an electron from the outer shell, starting an electrical current. Solar cells are used to power buoys, irrigation systems, communications equipment in remote areas, calculators, watches, and—when mounted on rooftops—to provide electricity for the home. If solar cells could be used to provide the electricity for the electrolysis of water to hydrogen, H_2, and oxygen, O_2, the hydrogen could be made available as a fuel:

$$2\ H_2(g) + O_2(g) \rightarrow 2\ H_2O(l) + \text{heat energy}$$

Nuclear fission and nuclear fusion were discussed in Chapter 23. Nuclear fission involves splitting apart the nucleus of heavy atoms by bombardment with particles; nuclear fusion involves fusing together the nuclei of light atoms. Both processes occur with an enormous release of energy. These are not chemical reactions in the ordinary sense but energy-releasing chemical reactions involving nuclear transmutations.

RENEWABLE AND NONRENEWABLE ENERGY

An energy source is either renewable or nonrenewable. It is *renewable* if it can be replenished. Petroleum, natural gas, and uranium for nuclear fission are *nonrenewable*; sooner or later, their supply will be exhausted. Wood, municipal waste, plant stalks, methanol from wood, and ethanol from plant sources are renewable sources of energy. New trees and plants can be grown and consumed year after year. Wind and hydroelectric energy are renewable, as is geothermal energy derived from the heat of Earth's inner core.

Fossil fuels, including coal and petroleum, formed millions of years ago from the remains of plants and animals. They are a nonrenewable energy source. Their supply will run out.

Solar energy is also renewable. Photovoltaic cells can electrolyze water indefinitely because the product of combustion is water itself! The cells themselves are made of silicon obtained by reduction of silica sand, an inexpensive source available in superabundant quantities. Nuclear fusion would provide virtually unlimited energy, because there is enough heavy hydrogen, or deuterium, in the world's oceans to support fusion reactors for hundreds of years.

CONSUMPTION

A measure of energy consumption is the *quad*. One quad equals 10^{15}, or one quadrillion, *British thermal units* (BTU). A BTU is a unit equal to the amount of heat required to raise the temperature of 1 lb of water 1°F at 1 atm pressure.

The United States consumes about 94.3 quads per year. This rate of consumption is based on 1998 data from the U. S. Geological Survey and is expected to increase gradually. The sources of energy that supply the 94.3 quads are given in Table 24.1. The heat energy released in typical combustion reactions is listed in Table 24.2.

SUMMARY

Most energy for human use today is produced by the combustion of fossil fuels (oil, natural gas, coal). Alcohol, wood, and plant materials are also combusted. Nuclear fission is chemical energy in the sense that a reaction takes place, disintegrating the nucleus of radioactive materials. Solar cells are chemical in the sense that electrons flow into and from the outer valence shell of silicon under the influence of sunlight. Other sources of energy are physical: hydropower, geothermal, wind. Coal and nuclear energy will become more important energy sources in the future.

Energy sources are renewable or nonrenewable depending on whether they can be replenished. If the energy of nuclear fusion

Energy Source	1998	Projection for 2020
oil	36.6	20
natural gas	21.8	15
coal	21.6	71
nuclear fission	7.2	18
other (hydropower, solar, geothermal, wood, biomass, and wind power)	7.1	19
Total	94.3	143

Table 24.1 U.S. Energy Consumption in Quads

Fuel	Energy Released, kcal/g of fuel (Note: 1 BTU = 0.252 kcal.)
natural gas	11.6
petroleum or oil	11.3
coal (anthracite)	7.3
coal (bituminous)	7.0
ethanol (C_2H_5OH)	7.1
methanol (CH_3OH)	5.4
wood	4.5
hydrogen (H_2)	34.2
carbon monoxide (CO)	2.4
carbon (C)	7.8
methane (CH_4)	13.3
deuterium fusion ($2\ {}^{2}_{1}H \rightarrow {}^{4}_{2}He$)	1×10^8
${}^{235}_{92}U$ (uranium) fission	2×10^7

Table 24.2 Heat Liberated by Various Fuels (by combustion with oxygen, with the exception of fission and fusion)

can be harnessed, it will become significant, since there is enough deuterium in the oceans to last for centuries. Convenient units of measurement of energy are the quad and the British thermal unit (BTU).

PROBLEM SET NO. 26

1. Which of the following energy sources are renewable?
   (a) hydroelectric
   (b) coal
   (c) wood
   (d) nuclear fission
   (e) natural gas

2. The complete combustion of oil, natural gas, coal, and wood yields CO_2 and H_2O. True or false?

3. In what way can solar energy provide hydrogen (H_2) fuel?

4. What percentage of energy consumption in the United States is now provided by coal and natural gas? What is the projected percent in 2020? Why?

5. If 1 g of hydrogen (H_2) burns with oxygen to release 34.2 kcal /g hydrogen of energy, how much energy, in BTU per mole of hydrogen, is released?

CHAPTER 25

THE CHEMISTRY OF THE ENVIRONMENT

KEY TERMS

environment, pollution, acid rain, greenhouse effect, biological oxygen demand (BOD)

The *environment* is everything that surrounds and affects living things on Earth. The physical environment consists of the air or atmosphere; the water in lakes, rivers, underground wells, streams, and oceans; and the solid land, or ground. Our environment is now being assaulted by the products and by-products of the chemical and nuclear power industries, many of which are toxic or injurious to the health of plants and animals, including humans. Before the Industrial Revolution, which began about 1750, the chemical industry was virtually nonexistent. The chemistry of the environment operated through the same natural processes that had been at work for millions of years.

Pollution of the air, water, or land adds hazardous waste products to the environment. Toxic chemicals produced today react in chemical ways to impair the well-being of human, animal, and plant life. The safe or safest possible disposal of such chemicals involves chemical or physical principles.

AIR POLLUTION

The contamination of air with hazardous and toxic substances is air pollution. The most visible form is the cloud of photochemical smog that can be seen floating above some cities, but not all air pollution is so obvious.

The burning of fossil fuels releases oxides of nitrogen and sulfur into the air. Coal contains free or combined sulfur. The sulfur oxide gases, SO_2 and SO_3, produced by its combustion react with water vapor in the air and clouds to form sulfuric acid:

$$SO_3(g) + H_2O(g) \rightarrow H_2SO_4(aq)$$

Nitrogen monoxide (older name, nitric oxide), NO, is formed during lightning discharges and is released into the air from cars, trucks, buses, trains, electric generating plants, and industrial combustion processes. NO is in turn oxidized by oxygen to nitrogen dioxide, NO_2, a yellow to reddish brown gas that is one of the contributors of smog.

$$2NO(g) + O_2(g) \rightarrow 2NO_2$$

Nitrogen dioxide also reacts with water in the air in the same way as the sulfur oxides do, forming nitric acid:

$$3NO_2(g) + H_2O(l) \rightarrow 2H_2NO_3(aq) + NO(g)$$

The sulfuric and nitric acids dissolve in rainfall, lowering its pH to as low as 2 or 3 (highly acidic), producing *acid rain*. Acid rain attacks metals, marble surfaces, statues, and plant and animal life. Metals corrode, marble disintegrates, and plants and trees lose their leaves. Plant roots are attacked by acid rain leaching into the soil, and fish die in lakes and streams acidified by the rain.

The chemical reaction of sulfuric acid with marble shows the destructive effect of acid rain:

$$CaCO_3(s) + H_2SO_4(aq) \rightarrow CaSO_4(aq) + CO_2(g) + H_2O(l)$$

In an attempt to reduce acid rain, scrubbing systems containing alkaline materials such as $Mg(OH)_2$ are used in coal-burning electric plants to remove sulfur oxides from stack emissions:

$$Mg(OH)_2(aq) + SO_2(g) \rightarrow MgSO_3(s) + H_2O(l)$$

Laws limiting emissions from vehicles require controls on the amounts of nitrogen oxides that can be released. That is one of the purposes of the automobile catalytic converter. It helps decompose NO to nitrogen gas, N_2, and oxygen, O_2.

Carbon monoxide, CO, is a toxic gas that enters the atmosphere principally from tailpipes of cars from the incomplete combustion of gasoline:

$$2C_8H_{18}(l) + 17\ O_2(g) \rightarrow 16\ CO(g) + 18\ H_2O(g)$$

Since 1975, U.S.-made cars have been equipped with catalytic converters containing catalyst pellets coated with platinum or palladium to oxidize the CO in the tailpipe:

$$2\ CO(g) + O_2(g) \xrightarrow{Pt,\ Pd} 2\ CO_2(g).$$

Oxygen atoms, O, that are produced by the decomposition of NO_2 combine with diatomic oxygen to produce ozone, O_3, a major constituent of smog. Ozone causes eye irritation and breathing difficulties. It is produced as follows:

$$NO_2(g) \rightarrow NO(g) + O(g)$$
$$O_2(g) + O(g) \rightarrow O_3(g)$$

Although ozone near the ground is a pollutant that contributes to photochemical smog, in the upper atmosphere ozone protects plant and animal life from the harmful ultraviolet (UV) radiation that causes cancer. For that reason, worldwide concern was triggered in the 1970s when it was discovered that the ozone layer was being depleted at an alarming rate. Many scientists felt that the ozone depletion was at least partly linked to the release of chlorofluorocarbons (CFCs) into the atmosphere. CFCs were commonly used as refrigerants (especially in automobile air-conditioning systems), as propellants in aerosol cans, and in the production of certain plastics, such as Styrofoam. Laws were passed to restrict the use and release of CFCs, but it will take many years for the ozone layer to repair itself.

Since 1975, U.S.-made cars have been equipped with catalytic converters containing catalyst pellets coated with platinum or palladium. They oxidize carbon monoxide in the exhaust gases into carbon dioxide that is released through the tailpipe into the air. Carbon dioxide produced by the burning of gasoline and other fossil fuels is a major contributor to the greenhouse effect and resultant global warming.

Another problem is the *greenhouse effect,* which is the warming that results when solar radiation is trapped by the atmosphere. Heat from the Sun travels through our atmosphere and strikes the surface. Some of the energy is absorbed and some is reflected back into the atmosphere. Certain gases in the atmosphere (carbon dioxide, water vapor, methane, and others) absorb this heat. This process is called the greenhouse effect, and it helps moderate Earth's temperature and keep it relatively constant; however, since the Industrial Revolution began, massive amounts of carbon dioxide have been released into the atmosphere, increasing the amount of heat absorption. As a result, the average temperature of the atmosphere has been slowly increasing. This global warming has drawn the attention of many scientists and concerned citizens, because of the potential detrimental effect on climate and worldwide ecosystems. Steps are currently being taken by nations to try to reduce the amount of carbon dioxide released into the atmosphere by human activities.

CHEMISTRY OF TRACE METALS

Trace metals are present in chemical products and in nature. They find their way into the water and the soil. Some trace metals are found in food; derived from water, plants, and animals, they are essential to health. Other trace metals in the environment may be toxic; they include beryllium, cadmium, mercury, and lead. Beryllium can cause lung disease; cadmium can cause high blood pressure and affect the heart.

If mercury is discarded in or near the ocean, a lake, or a river, microorganisms convert the mercury into the extremely toxic dimethyl mercury, $(CH_3)_2Hg$, which finds its way into fish and up the food chain into people who eat the fish. The mercury damages the brain and nervous system, especially in the fetus, in infants, and in young children, in whom the nervous system is still developing. For that reason, children and pregnant women are advised to limit their intake of many kinds of fish, since most of the world's bodies of water are now polluted with mercury to a greater or lesser degree.

Tetraethyl lead, $(C_2H_5)_4Pb$, along with dichloroethane, $C_2H_4Cl_2$, and dibromoethane, $C_2H_4Br_2$, was for many years added to some gasolines to prevent knocking. The auto exhaust from leaded gasoline emits particles of PbClBr and $NH_4Cl \cdot 2\ PbClBr$, which are small enough to be breathed and retained in the lungs. The use of $(C_2H_5)_4Pb$ has been stopped by law. Cars are now designed to run on lead-free gasoline. Lead pigments in paints such as $Pb(OH)_2$, $PbCO_3$, and Pb_3O_4 are used less in paints than they once were because of their toxicity. Their residential use was banned in 1978.

Concentrations of trace metals in the ground, water, plants, and animals are expressed in parts per million (ppm) which means 1 gram of metal in 10^6 grams of host material. Fish normally contain 0.2 ppm or less of mercury, or

$$0.2 \times 10^{-6}\ \text{g} = 2 \times 10^{-7}\ \text{g mercury per gram of fish}$$

Higher levels are considered hazardous.

HAZARDOUS WASTES IN THE GROUND AND WATER

More than 40 million tons of hazardous waste are generated each year in the United States. Sources of waste as a percentage of the total are given in Table 25.1. Waste disposal methods are listed in Table 25.2.

Source	Percentage
organic chemicals	34.2
primary metals	38.2
electroplating	11.8
inorganic chemicals	11.8
textile, oil refinery, rubbers, and plastics	4.0
Total	100

Table 25.1 Hazardous Wastes Generated Annually by Industry

Method	Million Tons
wastewater treatment	171
land disposal	
landfill	2.4
deep-well injection	26.5
thermal treatments	
recovery as fuel	2.7
incineration	3.0
recovery	8.1
Total	213.7

Table 25.2 Annual Disposal of Hazardous Wastes

There is growing concern about the use of landfills for the disposal of hazardous wastes. Even the most secure landfills will eventually leak, oozing their contents into underground streams or into the surrounding surface environment. The following are alternatives to sanitary landfills:

- Reduce generation of waste products by industry.
- Recover and recycle waste materials.
- Incinerate waste properly on land or at sea.
- Reduce the volume and toxicity of wastes by physical, chemical, and biological treatments.
- Inject waste into deep wells or store it underground caverns.
- Provide interim storage in surface tanks.
- Prohibit products or processes that create unmanageable wastes.

JOKE: Why are chemists good at solving problems?—They have all the solutions.

Hazardous wastes include pesticide residues, halogenated hydrocarbons such as polychlorobiphenyls (PCBs), asbestos, acids and bases, and a host of other chemicals. Their disposal in any body of water is improper. Freshwater is eventually used by plants and animals, including humans. Ocean water is a reservoir for marine life, some of which becomes food for humans.

Solid wastes include any garbage, refuse, sludge, or discarded material generated by municipal, industrial, mineral, and agricultural activities. The last three are disposed of by their producers. Municipal wastes are disposed of in landfills, by incineration with air at 1050 K–1250 K, or by *pyrolysis*, the decomposition by heat in the absence of air. Solid waste may also be reclaimed for a use different from its original use; recycled, as scrap iron is converted to steel; or reused, as are glass beverage bottles.

Municipal and industrial wastewaters must be treated physically and chemically. In municipal wastewater, suspended solids are first allowed to settle, then microorganisms decompose organic matter in the water. *Biological oxygen demand*, or BOD, is a measure of the amount of organic material in the wastewater. It is the amount of oxygen consumed in a given time period by the biological processes that break down organic waste—the cleaner the water, the lower the BOD.

THERMAL POLLUTION

Thermal pollution is polluting water with heat. Power plants, whether nuclear or fossil fuel, generate large amounts of waste heat, which is usually discharged as hot water into nearby bodies of water, either lakes or streams. The consequent increase in water temperature decreases the solubility of dissolved gases such as oxygen. Thus, thermal pollution can have a disastrous effect on aquatic life, and many large fish kills have been attributed to it. Holding tanks and cooling towers help reduce thermal pollution.

RADIOACTIVE WASTES

Radioactive wastes are generated by nuclear power plants, military applications, and mining, industrial, and medical sources. In 2000, approximately 600,000 m^3 of difficult types of radioactive waste were generated, and approximately 700,000 m^3 was in storage awaiting disposal. It is difficult to dispose of them because radioactivity can persist for months, years, or centuries. At one time, the long-lived radioactive wastes of power plants were stored in ponds of water, but the Nuclear Waste Policy Act of 1982 allowed for the storage of radioactive waste in metal canisters underground. Stored 2000 to 4000 feet below the surface in solid-rock cavern formations, the wastes should remain sufficiently dry to remain safe for up to 10,000 years; however, the storage and disposal of radioactive wastes continues to be a major problem associated with nuclear power and other uses of radioactive materials.

SUMMARY

The environment consists of the air, water, and solid land at Earth's surface. Industrial chemical and by-products may be toxic or hazardous to living things if they get into the air, water, or soil. Toxicity of chemicals is also chemical in nature, involving reactions with different areas of the body, causing injury.

The burning of fossil fuels for power, and the use of gasoline in internal combustion engines, releases oxides of carbon, nitrogen, and sulfur into the air. These substances contribute to environmental problems such as acid rain, photochemical smog, and the greenhouse effect. The release of CFCs into the atmosphere may have caused a reduction in atmospheric ozone. This ozone protects the inhabitants of Earth from harmful ultraviolet radiation.

Some of Earth's water has become contaminated with heavy metals such as mercury and lead. The metals are extremely toxic to animals, including humans. Hazardous wastes from landfills have found their way into the water and soil. Thermal pollution arising from the release of heated water into lakes and streams has been linked to fish kills; and long-lived radioactive wastes must be stored safely to protect the environment.

It is becoming increasingly obvious that toxic chemicals, radioactive wastes, and even heat must be properly disposed of. Methods of disposal involve both physical and chemical means.

PROBLEM SET NO. 27

1. The highest concentration of lead allowed in water is 0.05 ppm. How many grams of lead is that per gram of water?

2. Smog comes mostly from (a) mining operations, (b) automobiles, (c) chemical plants, (d) none of these. Explain.

3. What does a pH value for rain of 4.5 signify?

4. Landfills may not be the best way to dispose of all solid wastes. Why?

5. Lime, CaO, can remove SO_2 and SO_3 from stack gases in coal-fired electric plants. True or false?

CHAPTER 26

BIOCHEMISTRY

KEY TERMS

carbohydrate, fat, protein, nucleic acid, DNA, biotechnology

Biochemistry is the organic chemistry of life. The major chemical building blocks or substances of life are carbohydrates, fats, proteins, and nucleic acids.

CARBOHYDRATES

Carbohydrates are compounds of carbon, hydrogen, and oxygen that usually have a hydrogen-to-oxygen atomic ratio of 2 to 1—the same as in water. Carbohydrates are also known as *saccharides*. The chief members of this family of compounds are cellulose, starches, and sugars. Both cellulose and starch are formed in plants from carbon dioxide and water by photosynthesis. Both may be represented by the following formula: $(C_6H_{10}O_5)_x$. This formula indicates that the molecules of these compounds consist of a series of units, each containing the basic carbon–hydrogen–oxygen ratio indicated, joined to form a long chain. The actual number of such units present in a single molecule varies. Cellulose contains more of the basic molecular units than does starch, and the units differ in their spatial arrangement, yet both are called *polysaccharides* because they contain many of the simple units. Starch is obtained chiefly from cereal grains and potatoes. Cellulose is the basic constituent of plant cell walls. Cotton and linen are almost pure cellulose, whereas wood is cellulose mixed with compounds called lignins. Our bodies have the enzymes necessary to hydrolyze starch to glucose but lack the enzymes required to digest cellulose and lignin.

Polysaccharides hydrolyze (react with water) in the presence of acid and enzymes to form simple sugars of the formula $C_6H_{12}O_6$; however, many different sugars have this same formula. They differ from one another again in the arrangement of the atoms in their molecules, and thus they have different properties. The two basic types of simple sugars are *glucose* (also called *dextrose*) and *fructose* (also called *levulose*). Note the structural formulas of these two compounds:

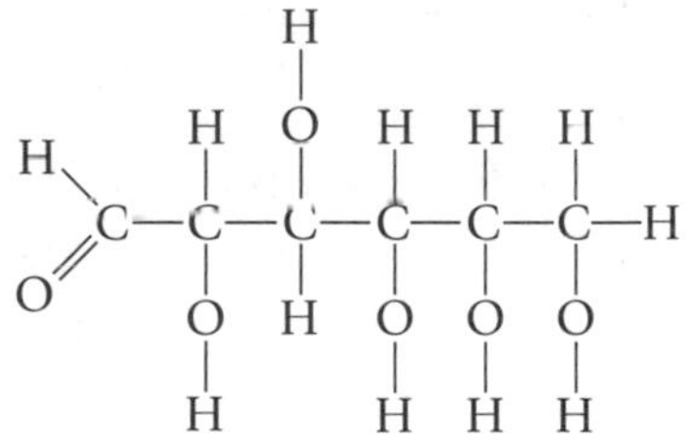

Glucose

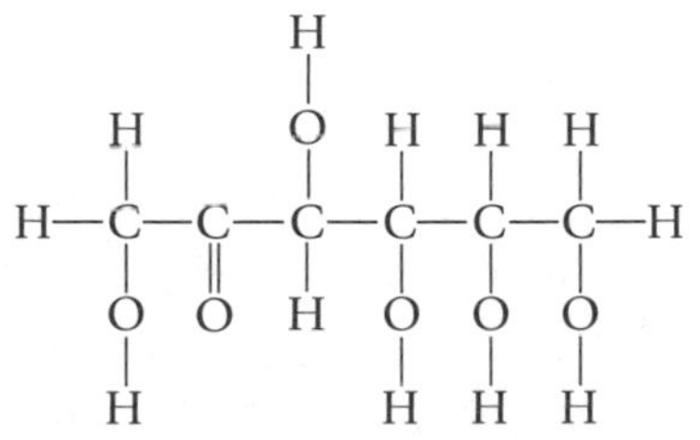

Fructose

Simple sugars such as these are called *monosaccharides*.

Ordinary cane sugar is *sucrose*, obtained from sugarcane or sugar beets. It is a disaccharide, consisting of two combined basic units, with formula $C_{12}H_{22}O_{11}$. In hydrolysis reactions, sucrose adds one molecule of water to form one molecule of glucose and one molecule of fructose. This equimolar mixture is known as *invert sugar*. Honey is essentially invert sugar.

FATS

A *fat* is an ester of glycerin and an organic acid. Butterfat is a mixture of fats with attached acid radicals that contain four or more carbon atoms. The chief compound in beef fat is glycerin tristearate, $(C_{17}H_{35}COO)_3C_3H_5$. Vegetable oils used in cooking, such as cottonseed oil, olive oil, soybean oil, and coconut oil, all contain unsaturated esters of glycerin that, when hydrogenated (reacted with hydrogen) in the presence of a nickel catalyst, yield solid fats. Fats are widely used in cooking and in the making of candles.

Fats also form the raw material from which soap is made. When a fat is heated with a solution of sodium hydroxide, hydrolysis takes place and the fat ester is converted to glycerin and the sodium salt of the acid. That salt is *soap*. The general reaction is

$$\text{Fat} + \text{Alkali solution} \rightarrow \text{Glycerin} + \text{Soap}$$

Because soap is a sodium salt, the soap is separated from the glycerin solution by the addition of sodium chloride which precipitates the soap by the common-ion effect. The glycerin is collected as a valuable by-product by distillation of the remaining solution. Ordinary cleansing soap is primarily sodium stearate, $C_{17}H_{35}COONa$. In facial soaps, the excess alkali is removed; but in laundry soaps, more alkali is added to increase grease-cutting action. Floating soaps are aerated to lower the specific gravity below 1. Shaving soaps and creams are normally made with potassium hydroxide rather than sodium hydroxide. Soaps may also contain perfumes and other germicidal or cleansing additives.

EXPERIMENT 30: Prepare various solutions of soap and water by shaking different kinds of soap with about 1 cup of water in each case. Be sure to include both face soap and laundry soap. Add a few drops of the purple cabbage indicator solution (prepared in Experiment 5) to each soap. Facial soaps should contain less free alkali than laundry soaps. Do your results concur?

In addition to soaps, there are detergents. The name comes from the Latin *detergere*, "to cleanse." Like soaps, detergents contain a long-chain aliphatic group attached to a sodium salt of an acidic group. Detergents have an advantage of over soap in hard water, which contains Mg^{2+} and Ca^{2+} ions. These ions form insoluble salts with soap, such as insoluble calcium stearate, $Ca(C_{17}H_{35}O_2)_2$. This waxy precipitate is commonly called "bathtub ring." The calcium and magnesium salts that might form with detergents, however, are soluble.

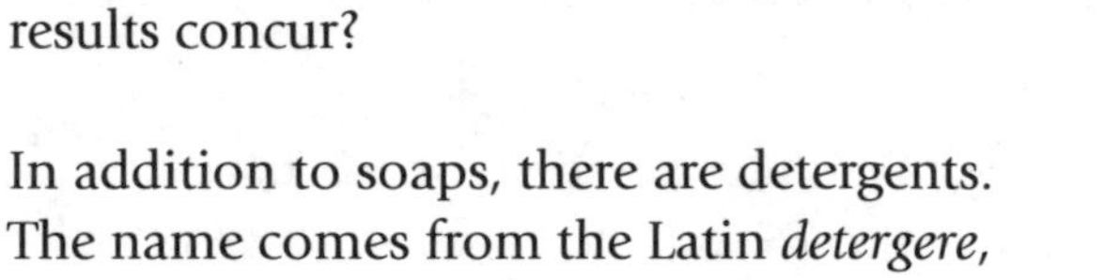

$CH_3CH_2CH_2CH_2CH_2CH_2CH_2CH_2CH_2CHCH_2CH_3$

SO_3Na—(benzene ring)

In 1916 the first synthetic detergent was developed in Germany in response to a World War I–related shortage of fats for making soap. The development of detergents began in earnest in the 1930s, and today detergents derived from petroleum have all but replaced soaps for residential and commercial laundry and dishwashing.

PROTEINS

TRIVIA: When an egg is cooked, the protein chains present unfold and link together in a framework trapping water. The longer the egg is cooked, the stronger the framework.

Proteins are complex polymer molecules that are the building blocks of all plant and animal tissues and organs. Hair, skin, nails, and feathers are mostly protein. When hydrolyzed, proteins break apart into the individual amino acid units from which they are made. An amino acid has the general structure

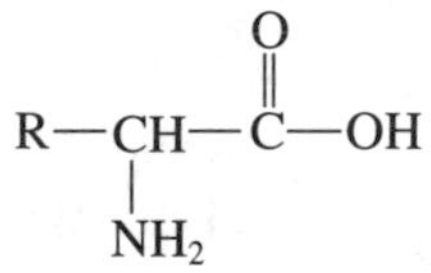

$$\mathrm{R{-}CH(NH_2){-}C({=}O){-}OH}$$

where R is a hydrocarbon group that may contain oxygen or sulfur atoms, NH_2 is the amine group, and COOH is the carboxyl group. The amine and carboxyl groups are the same for all amino acids, but the R groups vary. There are 20 "most common" amino acids. Glycine is the simplest. Its formula is

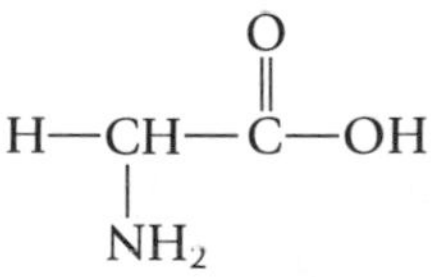

$$\mathrm{H{-}CH(NH_2){-}C({=}O){-}OH}$$

Among the most important (essential) amino acids are those that are required by humans for health and growth but which the human body itself cannot make. They must come from the foods we eat.

The linkage in the polymers of the amino acids is the peptide bond. The bond forms when one OH from the carboxyl group of one amino acid unites with one H from the amine group of another amino acid. A water molecule forms, leaving the two amino acids joined together.

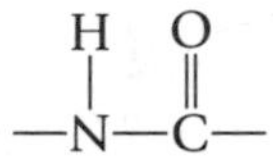

$$\mathrm{{-}N(H){-}C({=}O){-}}$$

In this manner, long chains of amino acids can form. A protein that contains many peptide bonds is a *polypeptide*. Three typical protein units of a polypeptide are shown here.

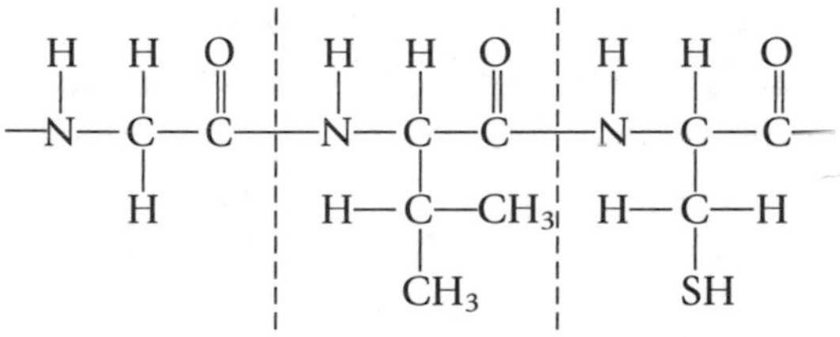

$$\mathrm{{-}N(H){-}C(H)(H){-}C({=}O){-}\;|\;{-}N(H){-}C(H)(CH(CH_3)CH_3){-}C({=}O){-}\;|\;{-}N(H){-}C(H)(CH_2SH){-}C({=}O){-}}$$

Polypeptides join, fold, and twist into the three-dimensional structures that are proteins.

NUCLEIC ACIDS

Nucleic acids are a group of polymers found in all living cells, both plant and animal, made of subunits containing a nitrogenous base (adenine, thymine, cytosine, and guanine in DNA), a phosphate molecule, and a sugar molecule. The collective name for this subunit is *nucleotide*. Important nucleic acid polymers are deoxyribonucleic acid, *DNA*, and ribonucleic acid, RNA. DNA is found mainly in the nucleus of all cells. RNA is found mainly in the material outside the nucleus. DNA carries the genetic information that allows a species to reproduce its own kind. It is DNA that determines whether a living thing is a robin or a bluefish, a tree or a human. This happens because DNA, the template, and RNA, the messenger, determine which proteins the cells of different species will make.

> DNA carries the genetic information that allows a species to reproduce its own kind and makes each human being unique. Genetic information is encoded in the sequence of bases that make up part of DNA's structure. DNA, the template, and RNA, the messenger, determine which proteins a cell will make.

DNA is formed in the shape of a double helix, as shown in Figure 26.1.

The two strands or chains are composed of two pentose sugar–phosphate chains that form the helix. They are held together by hydrogen bonding between the base pairs adenine–thymine and guanine–cytosine. Each of the ladder rungs is a pair of complementary nucleotides.

The order of bases in the chain is the basis for the genetic code that determines the proteins cells make. The code is "read" in "words" of three bases called *codons*. Each codon directs the addition of one of 20 amino acids to a polypeptitde chain, when proteins are made in the cell. A few hundred to several thousand codons make up a gene. Humans have about 30,000 genes. They are arranged into complexes of DNA with protein known as *chromosomes*. The constituents of DNA are shown in Figure 26.2.

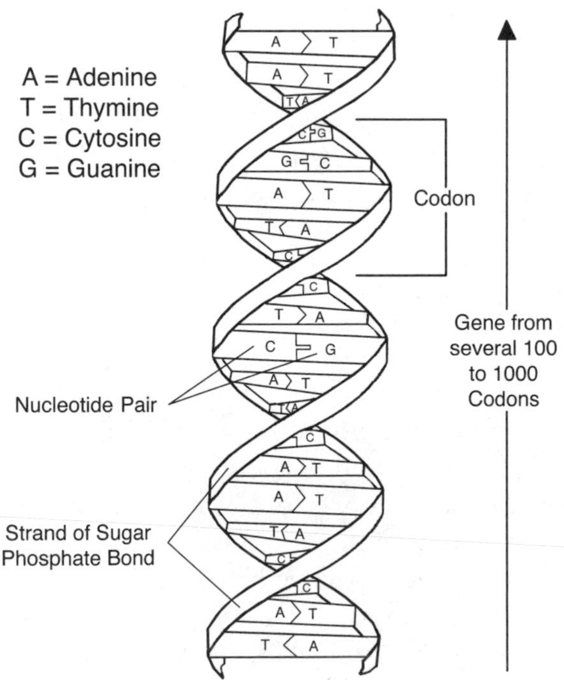

The DNA Double Helix

Figure 26.1

BIOTECHNOLOGY

Biotechnology is the newest area of biochemistry. The goal of biotechnology (biotech) is to use molecular engineering to modify or create parts or products of living organisms. One tool of biotechnology is *recombinant DNA technology*, in which a gene taken from one kind of living thing is transplanted and caused to operate in another. In 1978 the gene that causes human cells to manufacture insulin was engineered into bacterial cells, which caused the bacteria to produce human insulin. Insulin is the enzyme that people with type I diabetes do not manufacture in sufficient quantities. Pharmaceutical factories non produce human insulin via bacteria to treat people afflicted with diabetes. This process represents a new, inexhaustible source of insulin, which was previously extracted from the pancreas glands of cattle and swine.

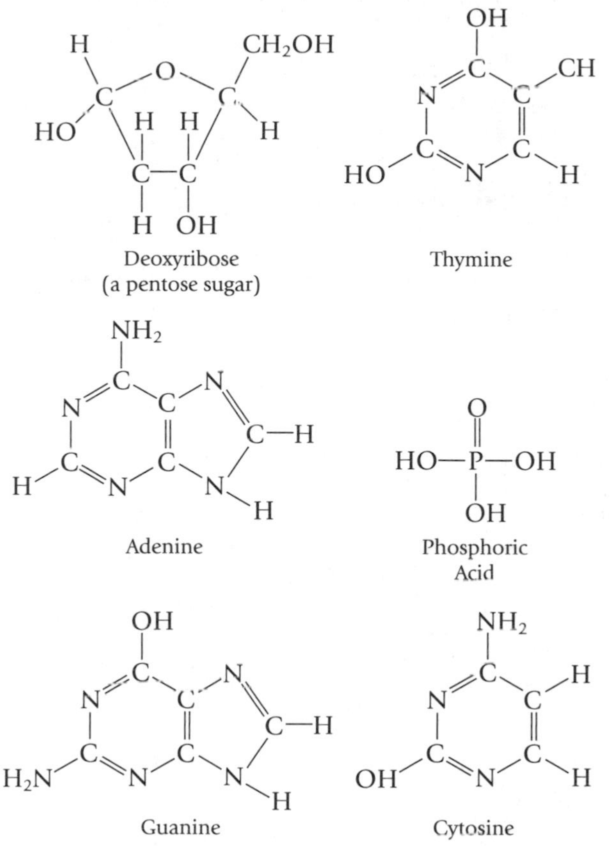

Deoxyribose
(a pentose sugar)

Thymine

Adenine

Phosphoric Acid

Guanine

Cytosine

Figure 26.2

Recombinant DNA technology is expected to produce many more useful pharmaceuticals, including antibiotics. It is also being used to create crop plants (grain, corn, soybeans, and more) and farm animals with improved nutritive components, growth yields, and drought and disease resistance.

TRIVIA: Minoxidil, commonly used in the treatment of baldness, was originally used in the treatment of high blood pressure. A bald patient experienced hair growth as a side effect of this drug treatment, and a new industry was born.

SUMMARY

Carbohydrates (saccharides) are compounds of carbon hydrogen and oxygen. Important members of the carbohydrates are sugars, starches, and cellulose. Important sugars are glucose and fructose (monosaccharides) along with sucrose, a disaccharide.

Fats are esters of glycerine and an organic acid. They are used extensively in cooking, in the making of candles, and in the preparation of soaps. Soaps form a waxy precipitate in hard water, that is, water containing calcium and magnesium ions. Detergents are synthetic cleaners that do not form this wax precipitate.

Proteins are complex biochemical polymers that are the building blocks of all animal tissue. They are composed of amino acid units. Nucleic acids are also biochemical polymers that are found in all living cells. They include such important compounds as DNA and RNA.

Biotechnology uses molecular engineering to modify or create parts of living organisms. It is used to create synthetic insulin and antibiotics, and holds great promise for the future in creating improved varieties of crops.

PROBLEM SET NO. 28

1. Identify each of the following as carbohydrate, fat, or protein.
   (a) glucose
   (b) safflower oil
   (c) hair and skin
   (d) butter
   (e) lard

2. Which of the following are monomers and which are polymers?

(a) DNA
(b) glucose
(c) fat
(d) protein
(e) adenine

3. Write the reaction between sodium stearate and the calcium ion, Ca^{2+}, and explain why the product forms a ring around the bathtub.

4. Fats are found in

(a) animals
(b) plants
(c) animals and plants

Explain your answer.

5. What are the functional groups of an amino acid?

ANSWERS TO PROBLEM SETS

Problem Set No. 1

1. From Table 1.1, 1 fl oz = 29.573 mL. Therefore, 3 fl oz = 88.719 mL.
2. From Table 1.1, 1 lb = 453.6 g; 1 oz = 28.35 g. Finding sum of 1 lb as grams and 3 oz as grams, and moving decimal three places to left we have: 1 lb 3 oz = 0.53865 kg.
3. $\frac{528 \text{ in.}^2}{1} \times \frac{(2.54 \text{ cm})^2}{(1 \text{ in.})^2} \times \frac{(1 \text{ m})^2}{(100 \text{ cm})^2} = 0.341 \text{ m}^2$
4. $\frac{22{,}400 \text{ cm}^3}{1} \times \frac{(1 \text{ in.})^3}{(2.54 \text{ cm})^3} \times \frac{(1 \text{ ft})^3}{(12 \text{ in.})^3} = 0.791 \text{ ft}^3$
5. From Table 1.1, 1 g = 0.0353 oz. Therefore, 250 g = 8.825 oz
6. $\frac{25 \text{ yd}^2}{1} \times \frac{(3 \text{ ft})^2}{(1 \text{ yd})^2} \times \frac{(12 \text{ in.})^2}{(1 \text{ ft})^2} \times \frac{(2.54 \text{ cm})^2}{(1 \text{ in.})^2}$

   $= 2.1 \times 10^5 \text{ cm}^2$
7. $\frac{1.00 \text{ cup}}{1} \times \frac{8 \text{ oz}}{1 \text{ cup}} \times \frac{29.573 \text{ mL}}{1 \text{ oz}} \times \frac{1\text{L}}{1000 \text{ mL}}$

   $= 0.237 \text{ L}$
8. From Table 1.1, 1 fl oz = 29.573 mL. Therefore, 1 tablespoon = 0.5 fl oz = 14.787 mL.
9. From Table 1.1, 1 gal = 3.7853 L. Therefore, 18 gal = 68.135L.
10. Multiplying the dimensions, we have 58.44 cu in. in the book. Then set up the conversion:

$$\frac{58.44 \text{ in.}^3}{1} \times \frac{(2.54 \text{ cm})^2}{(1 \text{ in.})^3} \times \frac{(1 \text{ dm})^3}{(10 \text{ cm})^3} = 0.9577 \text{ dm}^3$$

Problem Set No. 2

1. (a) intensive; (b) extensive; (c) intensive.
2. From Table 2.1, density of aluminum is 2.7 g/cm^3.

   Volume of 5.4 g of aluminum =

   $$\frac{5.4 \text{ g}}{1} \times \frac{1 \text{ cc}}{2.7 \text{ g}} = 2.0 \text{ cc}$$
3. From Table 2.1, 1 cm^3 of gold weighs 19.3 g; density of cork is 0.22 g/cm^3. Therefore, 19.3 g of cork would occupy

   $$\frac{19.3 \text{ g}}{0.22 \text{ g/cc}} = 87.7 \text{ cc}$$
4. Specific gravity = density in g/cm^3. Setting up the problem:

   $$\frac{1.28 \text{ g}}{\text{cm}^3} \times \frac{(2.54 \text{ cm})^3}{(1 \text{ in.})^3} \times \frac{(12 \text{ in.})^3}{(1 \text{ ft})^3} \times \frac{1\text{lb}}{454 \text{ g}} = \frac{79.8 \text{ lb}}{\text{ft}^3}$$
5. (a) Those with less density: cork, ice
   (b) Those with greater density: gold
6. Carbon, hydrogen, oxygen, phosphorus, potassium, iodine, nitrogen, sulfur, calcium, iron, sodium, chlorine
7. (a) mixture; (b) substance; (c) mixture; (d) mixture; (e) substance
8. (a) physical; (b) chemical; (c) chemical; (d) physical; (e) physical
9. (a) combination; (b) decomposition; (c) single displacement; (d) double displacement.

Problem Set No. 3

1. (a) carbon $1s^2 2s^2 2p^2$ or $[He]2s^2 2p^2$
   (b) silicon $1s^2 2s^2 2p^6 3s^2 3p^2$ or $[Ne] 3s^2 3p^2$
   (c) bromine $1s^2 2s^2 2p^6 3s^2 3p^6 3d^{10} 4s^2 4p^5$ or $[Ar] 3d^{10} 4s^2 4p^5$
2. (a) $^{11}_{5}B$ (b) $^{24}_{12}Mg$ (c) $^{40}_{18}Ar$
   (d) $^{79}_{34}Se$ (e) $^{93}_{50}Sn$

Problem Set No. 4

1. Each oxygen has accepted an electron from a sodium atom, so each of the oxygen atoms has a −1 charge and the sodium atoms have a 1+ charge. The compound would be *sodium peroxide.*
2. Al, a group III element, would tend to lose three electrons to give a cation with a charge of +3. S, a group VI element, would tend to gain two electrons to achieve noble gas configuration and have a charge of −2. Crisscrossing and writing as subscripts, we have: Al_2S_3. Its name is *aluminum sulfide.*
3. Mg, a group II element, would lose two electrons to have a +2 charge. N, a group V element, would gain three electrons to form an ion of −3 charge. Crisscrossing and writing as subscripts, we have: Mg_3N_2. Its name is *magnesium nitride.*
4. Each Na atom is +1. Total: +2. Each O atom is −2. Total: −6. Therefore, the C atom must be: +4.
5. The Ca atom is +2. Total: +2. The NO_3 ion is −1. Total: −2. The problem here, then, is to show how the nitrate ion has a net oxidation number of −1. Each O atom is −2. Total: −6. Therefore the N atom must be +5 if the polyatomic ion is to have 1 excess negative charge.
6. The NH_4 ion is +1. The SO_4 ion is −2. Crisscrossing and writing as subscripts, we have: $(NH_4)_2SO_4$.
7. Mg has a valence of +2. The PO_4 ion is −3. Crisscrossing and writing as subscripts, we have: $Mg_3(PO_4)_2$.
8. Atomic mass of C is 12.0 amu (× 22) = 144.0 amu
   Atomic mass of H is 1.0 amu (× 22) = 22.0 amu
   Atomic mass of O is 16.0 amu (× 11) = 176.0 amu
   Formula mass of sugar amu = 342.0 amu
9. Atomic mass of Al is 27.0 amu (× 2) = 54.0 amu
   Atomic mass of S is 32.1 amu (× 3) = 96.3 amu
   Atomic mass of O is 16.0 amu (× 12) = 192.0 amu
   Formula mass of aluminum sulfate amu = 342.3 amu
10. Formula wt. of NaOH is 23.0 g + 16.0 g + 1.0 g = 40.0 g/mol

    Thus, $\frac{200 \text{ g}}{1} \times \frac{1 \text{ mol}}{40.0 \text{ g}} = 5 \text{ mol}$
11. Carbon is the central atom, so O C O ; then, distributing the electrons so that all atoms have 8 valence electrons each gives $\ddot{O}=C=\ddot{O}$

Problem Set No. 5

1. By Avogadro's hypothesis, the equal volumes contain the same number of molecules. Therefore the nitrogen molecule weighs 14 times as much as the hydrogen molecule. Since hydrogen, H_2, has a molecular mass of 2 g/mol, the molecular weight of nitrogen must be:

   $$2 \text{ g/mol} \times 14 = 28 \text{ g/mol}$$

   Since the atomic mass of nitrogen is 14 g/mol, the number of atoms in the nitrogen molecule must be:

   $$\frac{28 \text{ g/mol}}{14 \text{ g/mol}} = 2$$

   Therefore the formula of nitrogen gas is N_2.

2.

	Number of Atoms	*Atomic Weight*	*Total Weight*
Carbon:	12	12 amu	144 amu
Hydrogen:	22	1 amu	22 amu
Oxygen:	11	16 amu	176 amu
Molecular weight of sugar:			342 amu

$$\% \text{ C} = \frac{144 \text{ amu}}{342 \text{ amu}} \times 100 = 42.10\%$$

$$\% \text{ H} = \frac{22 \text{ amu}}{342 \text{ amu}} \times 100 = 6.44\%$$

$$\% \text{ O} = \frac{176 \text{ amu}}{342 \text{ amu}} \times 100 = 51.46\%$$

3. The formula weight of $CuSO_4 \cdot 5H_2O$ is 249.6 g/mol.

   $$\% \text{ Cu} = \frac{63.5 \text{ g/mol}}{342 \text{ g/mol}} \times 100 = 25.44\%$$

4. First, find % Cl in pure NaCl.

   $$\% \text{ Cl} = \frac{35.5 \text{ g/mol}}{58.5 \text{ g/mol}} \times 100 = 60.68\%$$

   $$\% \text{ purity} = \frac{58\%}{60.65\%} \times 100 = 95.58\% \text{ pure NaCl}$$

5. Aluminum:

$$\frac{52.9\text{ g}}{27.0\text{ g/mol}} = 1.96\text{ mol}; \quad \frac{1.96\text{ mol}}{1.96\text{ mol}} = 1$$

Oxygen:

$$\frac{47.1\text{ g}}{16.0\text{ g/mol}} = 2.94\text{ mol}; \quad \frac{2.94\text{ mol}}{1.96\text{ mol}} = 1.5$$

To produce whole numbers, we multiply each result by 2. The formula of the compound is therefore Al_2O_3.

6. Copper: $\frac{50.88\text{ g}}{63.6\text{ g/mol}} = 0.8\text{ mol}; \quad \frac{0.8}{0.4} = 1$

Sulfur: $\frac{12.84\text{ g}}{32.1\text{ g/mol}} = 0.4\text{ mol}; \quad \frac{0.4}{0.4} = 1.5.$

Formula is Cu_2S

7. (a) $2\ NaCl + H_2SO_4 \rightarrow Na_2SO_4 + 2\ HCl$
(b) $4\ NH_3 + 3\ O_2 \rightarrow 2\ N_2 + 6\ H_2O$
(c) $2\ ZnS + 3\ O_2 \rightarrow 2\ ZnO + 2\ SO_2$
(d) $C_3H_8 + 5\ O_2 \rightarrow 3\ CO_2 + 4\ H_2O$
(e) $Ca_3(PO_4)_2 + 3\ SiO_2 + 5\ C \rightarrow 3\ CaSiO_3 + 5\ CO + 2\ P$

8. $CaCO_3 \rightarrow CaO + CO_2$

$$\frac{500\text{ lb}}{1} \times \frac{454}{1\text{ lb}} \times \frac{1\text{ mol }CaCO_3}{100.1\text{ g}} \times$$

$$\frac{1\text{ mol CaO}}{1\text{ mol }CaCo_3} \times \frac{56.1\text{ g}}{1\text{ mol CaO}} \times \frac{1\text{ lb}}{454\text{ g}} = 280\text{ lb}$$

9. (a) $2\ KNO_3 \rightarrow 2\ KNO_2 + O_2$
(b) 2 mol KNO_3 produce 1 mol O_2; therefore, 12 mol KNO_3 will produce 6 mol O_2
(c) Potassium nitrite
(d) 12 mol KNO_3 produce 12 mol KNO_2. One mole of KNO_2 is 39.1 g/mol + 14.0 g/mol + 32.0 g/mol = 85.1 g/mol of KNO_2; therefore, 12 × 85.1 g/mol = 1.02×10^3 g of KNO_2 produced

10. (a) Moles of N_2 admitted: $\frac{280\text{ g}}{28\text{ g/mol}} = 10\text{ mol}$

Moles of H_2 admitted: $\frac{100\text{ g}}{2\text{ g/mol}} = 50\text{ mol}$

Ratio of moles of H_2 to moles of N_2 = 50:10 = 5:1

Molar ratio required by the equation = H_2:N_2 = 3:1
Therefore, there is an excess of hydrogen, and nitrogen is the limiting reactant.
(b) Moles of N_2 admitted = 10 mol (found above)
(c) Moles of H_2 required = 3 × 10 = 30 mol
Moles of excess H_2 = 50 − 30 = 20 mol
(d) Since each mole of N_2 produces 2 mol NH_3, 10 × 2 = 20 mol NH_3 can be produced.
(e) Molecular mass of NH_3 = 14 g/mol + 3 g/mol = 17 g/mol
Weight of NH_3 that can be produced = 20 mol × 17 g/mol = 340 g.

Problem Set No. 6

1. °F = 9/5 C + 32; $F = \frac{9 \times 30.°C}{5} + 32$; F = 86°F

2. °C = 5/9 (°F−32);
°C = 5/9 (68°F−32; C) = 20°C

3. °C = 5/9 (F−32);
C = 5/9 (95−32); C = 35°C;
K = 35°C + 273 = 308K

4. $V_2 = 500.\text{ cm}^3 \times \frac{770.\text{ Torr}}{1540.\text{ Torr}} = 250.\text{ cm}^3$

5. $V_2 = 350.\text{ cm}^3 \times \frac{461\text{ K}}{320.\text{ K}} = 504\text{ cm}^3$

6. $P_1/T_1 = P_2/T_2$ (by canceling constant V from combined gas law)

$$P_2 = 750.\text{ Torr} \times \frac{525\text{ K}}{293\text{ K}} = 1340\text{ cc}$$

7. 750. Torr × 0.108 = 81 Torr of CO_2; 750. Torr × 0.022 = 16.5 Torr of CO; 750. Torr × 0.045 = 33.8 Torr of O_2; 750. Torr × 0.825 = 618.7 Torr of N_2

8. $V_2 = 150.\text{ mL} \times \frac{273\text{ K}}{295\text{ K}} \times \frac{740.\text{ Torr}}{760\text{ Torr}} = 135\text{ mL}.$

9. $V_2 = 330.\text{ mL} \times \frac{303\text{ K}}{273\text{ K}} \times \frac{29.92\text{ in.}}{30.40\text{ in.}} = 360.\text{ mL}$

10. $V_2 = 400.\text{ mL} \times \frac{273\text{ K}}{295\text{ K}} \times \frac{(767.4 - 22.4)\text{ Torr}}{760\text{ Torr}}$
= 362 mL

Problem Set No. 7

1. Volume of gas at standard conditions:

$$V_2 = 555 \text{ mL} \times \frac{273 \text{ K}}{295 \text{ K}} \times \frac{740. \text{ Torr}}{760 \text{ Torr}} = 560 \text{ mL}$$

$$\text{Density} = \frac{0.6465 \text{ g}}{500 \text{ mL}} = 0.00129 \text{ g/mL}$$

2. $$\frac{\text{Helium rate}}{\text{Chlorine rate}} = \frac{\sqrt{70.9}}{\sqrt{4.0}} = \frac{\sqrt{17.73}}{1} = \frac{4.2}{1}$$

   Helium diffuses 4.2 times faster than chlorine.

3. Molecular mass of hydrogen is 2.016 g/mol. Mass of a hydrogen molecule is

$$\frac{2.016 \text{ g}}{\text{mol}} \times \frac{1 \text{ mol}}{6.022 \times 10^{23} \text{ molecules}}$$

$$= 3.34 \times 10^{-24} \text{ g/molecule}$$

   A hydrogen atom thus weighs half this, or 1.67×10^{-24} g.

4. Using the ideal gas equation; $PV = nRT$;

$$n = \frac{PV}{RT} = \frac{(740. \text{ Torr}/760 \text{ Torr/atm})(0.0822 \text{ L})}{(0.0821 \text{ L} \cdot \text{atm/K} \cdot \text{mol})(390. \text{ K})}$$

$$= 0.0025 \text{ mol}$$

   Molecular mass = g/mol = 0.24 g/0.0025 mol = 96 g/mol

5. Volume of gas at standard conditions:

$$V_2 = 523 \text{ mL} \times \frac{273 \text{ K}}{302 \text{ K}} \times \frac{733 \text{ Torr}}{760 \text{ Torr}} = 456 \text{ mL}$$

   Molecular weight would be:

$$\frac{456 \text{ mL}}{1} \times \frac{1 \text{ mol}}{22{,}400 \text{ mL}} = 0.0204 \text{ mol}$$

   0.855 g/0.0204 mol = 42.0 g/mol

   Simplest formula of compound would be:

$$\text{Carbon: } \frac{85.72 \text{ g}}{12 \text{ g/mol}} = 7.143 \text{ mol}; \quad \frac{7.143 \text{ mol}}{7.143 \text{ mol}} = 1$$

   Hydrogen:

$$\frac{14.28 \text{ g}}{1 \text{ g/mol}} = 14.28 \text{ mol}; \quad \frac{14.28 \text{ mol}}{7.143 \text{ mol}} = 2$$

   Simplest formula is therefore CH_2. Formula mass of simplest formula is 12 g/mol+2 g/mol =14 g/mol. Multiplier of subscripts must be 42 g/mol/14 g/mol = 3. True formula must then be C_3H_6.

6. The equation for the reaction is

$$Zn + 2\,HCl \rightarrow ZnCl_2 + H_2$$

   Thus each mole of zinc produces 1 mol H_2. Moles of zinc = 10. g/65.4 g/mol = 0.153 mol. Moles of H_2 = 0.153. Volume of H_2 at standard conditions:

$$0.153 \text{ mol} \times 22.4 \text{ L/mol} = 3.43 \text{ L}$$

7. The equation of the reaction is

$$CaCO_3 \rightarrow CO_2 + CaO$$

   Thus 1 mol $CaCO_3$ produces 1 mol CO_2. Moles $CaCO_3$ = moles CO_2 produced = 5 L/22.4 L/mol = 0.223 mol. Molecular mass of $CaCO_3$ = 40 g/mol + 12 g/mol + 48 g/mol = 100 g/mol. Mass of $CaCO_3$ required = 0.223 mol × 100 g/mol = 22.3 g

8. The equation for the reaction is

$$2\,C_2H_6 + 7\,O_2 \rightarrow 4\,CO_2 + 6\,H_2O$$

   Since volumes are in same ratio as coefficients,

$$\frac{15 \text{ L}}{1} \times \frac{1 \text{ mol } C_2H_6}{22.4 \text{ L}} \times \frac{7 \text{ mol } O_2}{2 \text{ mol } C_2H_6} \times \frac{22.4 \text{ L}}{1 \text{ mol } O_2}$$

$$= 52.5 \text{ L}$$

   x = 52.5 L of oxygen.

9. In PH_3, 31 parts by mass of P are combined with 3 parts by mass of H. Therefore equivalent mass of phosphorus is:

$$\frac{31}{3} = \frac{x}{1}$$

$$x = 10.3$$

Problem Set No. 8

1. All substances listed in Table 7.1 are gases at room temperature except ethyl alcohol and water.

2. (a) $$\frac{-78.5 + 273}{31.1 + 273} = 0.64$$

(b) $\frac{-34.0 + 273}{144.0 + 273} = 0.57$

(c) $\frac{78.5 + 273}{243.1 + 273} = 0.68$

(d) $\frac{-252.8 + 273}{-239.9 + 273} = -0.61$

(e) $\frac{-183.0 + 273}{-118.8 + 273} = 0.58$

(f) $\frac{100.0 + 273}{374.0 + 273} = 0.58$

3. Despite the extensive variation in the properties of the substances considered in Problem 2, the ratio is in remarkable agreement for each substance. The absolute normal boiling point seems to be approximately two-thirds of the absolute critical temperature.

4. The temperature indicated on the thermostat is 180°F. The boiling point of water at the top of Pike's Peak is about 62°C, or on the Fahrenheit scale:

$F = 9/5C + 32$

$= (9/5 \times 62°C) + 32 = 144°F$

Thus, the water would boil below the operating temperature of the thermostat. Therefore, the thermostat should be removed to permit circulation of the water at this high elevation.

5. To melt ice:

$25 \text{ g} \times 79.7 \text{ cal/g} = 1992.5 \text{ cal}$

To heat water to 100°C:

$100°C \times 25 \text{ g} \times 1 \text{ cal/g°C} = 2500.0 \text{ cal}$

To boil the water:

$25 \text{ g} \times 540 \text{ cal/g} = 13500.0 \text{ cal}$

Total 17992.5 cal

Problem Set No. 9

1. (a) $\frac{0.02 \text{ mol}}{0.080 \text{ L}} = 0.25 \text{ M}$

(b) $\frac{0.234 \text{ g}}{58.5 \text{ g/mol}} \times \frac{1}{0.050 \text{ L}} = 0.08 \text{ M}$

(c) $\frac{222 \text{ g}}{111 \text{ g/mol}} \times \frac{1}{4 \text{ L}} = 0.5 \text{ M}$

2. (a) $\frac{0.2 \text{ mol}}{0.200 \text{ L}} \times \frac{2 \text{ equiv}}{1 \text{ mol}} = 2.0 \text{ N}$

(b) $\frac{0.2 \text{ equiv}}{0.200 \text{ L}} = 1.0 \text{ N}$

(c) Equivalent weight of K_2CO_3 is half the molecular mass, or 69 g/eq.

$$\text{Normality} = \frac{2.76 \text{ g}}{0.400 \text{ L}} \times \frac{1 \text{ equiv}}{69 \text{ g}} = 0.1 \text{ N}$$

(d) Equivalent mass of $Al_2(SO_4)_3$ is one-sixth of molecular mass, or 57.

$$\text{Normality} = \frac{6.84 \text{ g}}{0.250 \text{ L}} \times \frac{1 \text{ equiv}}{57 \text{ g}} = 0.48 \text{ N}$$

3. (a) $\frac{4.0 \text{ g}}{40. \text{ g/mol} \times 0.400 \text{ kg}} = 0.25 \text{ m}$

(b) $\frac{333 \text{ g}}{111 \text{ g/mol} \times 6 \text{ kg}} = 0.5 \text{ m}$

4. (a) $\frac{6.0 \text{ g}}{256 \text{ g}} \times 100 = 2.3\%$

(Note that in dilute solutions of solids in liquids, the solid dissolves without appreciably changing in the volume of the solution.)

(b) $\frac{6.0 \text{ g}}{120. \text{ g/mol} \times 0.250 \text{ L}} = 0.2 \text{ M}$

(c) $0.2 \text{ mol/L} \times 2 \text{ equiv/mol} = 0.4 \text{ N}$

5. $\frac{10 \text{ mL}}{50 \text{ mL}} \times 100 = 20\%$

6. $\frac{34.0 \text{ g}}{170. \text{ g/mol} \times 0.750 \text{ L}} = 267 \text{ M}$

7. The equivalent mass of $AlCl_3$ is one-third the molecular mass, or 44.5 g/equiv

$$\text{Normality} = \frac{2.67 \text{ g}}{44.5 \text{ g/equiv} \times 0.400 \text{ L}} = 0.15 \text{ N}$$

8. $C_1V_1 = C_2V_2$

$0.8 \text{ M} \times 35 \text{ mL} = 0.5 \text{ M} \times x$

$V_2 = x = \frac{0.8 \text{ M} \times 35 \text{ mL}}{0.5 \text{ M}} = 56 \text{ mL}$

56 mL − 35 mL = 21 mL water to be added

9. $N_1V_1 = N_2V_2$

$0.1 \text{ N} \times 24.0 \text{ mL} = (25.0 \text{ mL})\, x$

$x = \frac{24.0 \text{ mL} \times 0.1 \text{ N}}{25.0 \text{ mL}} = 0.096 \text{ N}$

10. $N_s \times V_s$ (in liters) = # of equiv

$$x \times 0.020\text{ L} = \frac{0.285\text{ g}}{143.5\text{ g/equiv}}$$

$$x = \frac{0.285\text{ g}}{143.5\text{ g/equiv} \times 0.020\text{ L}} = 0.099\text{ N}$$

11. Reaction: $H_2SO_4(aq) + 2\,NaOH\,(aq) \rightarrow Na_2SO_4(aq) + 2\,H_2O(l)$

$$\frac{0.1000\text{ mol } H_2SO_4}{\text{L}} \times \frac{0.0250\text{ L}}{1} \times$$

$$\frac{2\text{ mol NaOH}}{1\text{ mol } H_2SO_4} \times \frac{1}{0.0385\text{ L}} = 0.1299\text{ M}$$

Problem Set No. 10

1. From Equation 15:

$$m = \frac{(w) \times (M) \times (P)}{(W) \times (P-p)}$$

$$m = \frac{23\text{ g} \times 18\text{ g/mol} \times 17.5\text{ Torr}}{500\text{ g} \times (17.5-17.34)\text{ Torr}}$$

$$= 90.56\text{ g/mol}$$

2. From Equation 17:

$$B - b = \frac{0.52\,w}{m}$$

$$w = 12.4\text{ g} \times \frac{1000\text{ g}}{200\text{ g}} = 62\text{ g}$$

For $C_2H_6O_2$ m = 12 g/mol + 6 g/mol + 32 g/mol = 62 g/mol

$$\text{So, } B - b = \frac{0.52°\text{C/mol} \times 62\text{ g}}{62\text{ g/mol}} = 0.52°\text{C}$$

Therefore, the solution boils at 100.52°C.

3. From Equation 18:

$$m = \frac{0.52\,w}{B-b}$$

$$w = 15.5\text{ g} \times \frac{1000\text{ g}}{100\text{ g}} = 155\text{ g}$$

So, $B - b = 101.3 - 100.0 = 1.3°\text{C}$

$$m = \frac{0.52°\text{C/mol} \times 115\text{ g}}{1.3°\text{C}} = 62\text{ g/mol}$$

4. From Equation 20:

$$m = \frac{1.86\,w}{F-f}$$

$$w = 11.5\text{ g} \times \frac{1000\text{ g}}{100\text{ g}} = 115\text{ g}$$

So, $F - f = 0-(-2.325) = 2.325°\text{C}$

$$m = \frac{1.86°\text{C/mol} \times 155\text{ g}}{2.325°\text{C}} = 92\text{ g/mol}$$

5. The ratio of the weight of glycol to the weight of water is

$$\frac{1.26}{1} \times \frac{6}{12} = \frac{1.26}{2}$$

Therefore, $w = 1.26 \times \frac{1000}{2} = 630\text{ g}$

From Equation 20:

$$F - f = \frac{1.86\,w}{m}$$

$$F - f = \frac{1.86°\text{C/mol} \times 630\text{ g}}{62\text{ g/mol}} = 18.9°\text{C}$$

The solution freezes at −18.9°C. We then convert this to °F:

$F = 9/5\text{ C} + 32$
$F = (9/5(-18.9)) + 32 = -2°\text{F}$

Problem Set No. 11

1. Each mole of $CaCl_2$ produces 3 mol of ions; therefore,
$F - f = 3\text{ mol} \times 1.86°\text{C/molal} \times \text{m};$

$$m = \frac{44.4\text{ g}}{1} \times \frac{1\text{ mol}}{111\text{ g}} \times \frac{1}{0.200\text{ kg}} = 2.0\,m$$

$= 3\text{ mol} \times 1.86°\text{C / molal} \times 2.0\text{ m} = 11.16°\text{C}$
Therefore the solution freezes at −11.16°C.

2. (a) $MgCl_2 \rightarrow Mg^{2+} + 2\,Cl^-$
$[Mg^{+2+}] = 0.05\text{ M}$
$[Cl^-] = 0.10\text{ M}$
(b) $H_2SO_4 \rightarrow 2\,H^+ + SO_4^{2-}$
$[H^+] = 0.50\text{ M}$
$[SO_4^{2-}] = 0.25\text{ M}$
(c) $Fe_2(SO_4)_3 \rightarrow 2\,Fe^{3+} + 3\,SO_4^{2-}$
$[Fe^{3+}] = 0.2\text{ M}$
$[SO_4^{2-}] = 0.3\text{ M}$

3. (a) $K_2CO_3 + CaCl_2 \rightarrow CaCO_3 + 2\,KCl$
(Forms precipitate)

(b) $2\ HNO_3 + Na_2SO_3 \rightarrow H_2SO_3 + 2\ NaNO_3$
(Forms weak electrolyte)

(c) $H_2SO_4 + 2\ AgNO_3 \rightarrow Ag_2SO_4 + 2\ HNO_3$
(Forms precipitate)

(d) $HNO_3 + Na_2SO_4 \rightarrow$ No reaction

4. $HNO_2 \rightarrow H^+ + NO_2^-$

$$\frac{[H^+] \times [NO_2^-]}{[HNO_2]} = K$$

$$\frac{[0.0063]^2}{[0.1]} = 4.0 \times 10^{-4}$$

5. $$\frac{[x]^2}{[0.1]} = 1.8 \times 10^{-5}$$

$x^2 = 1.8 \times 10^{-6}$

$x = 1.3 \times 10^{-3}$ M

6. $$\frac{1.3 \times 10^{-2}}{10^{-1}} \times 100 = 1.3\%$$

7. $$\frac{[H^+] \times [CN^-]}{[HCN]} = 4 \times 10^{-10}$$

$$\frac{[x]^2}{[0.1]} = 4 \times 10^{-10}$$

$x^2 = 4 \times 10^{-11}$

$[H^+] = x = 6.3 \times 10^{-6}$

$pH = -\log[6.3 \times 10^{-6}] = 5.2$

8. $$\frac{[NH_4^+] \times [OH^-]}{[NH_4OH]} = 1.8 \times 10^{-5}$$

$$\frac{[x]^2}{[0.1]} = 1.8 \times 10^{-5}$$

$x^2 = 1.8 \times 10^{-6}$

$[OH^-] = x = 1.3 \times 10^{-3}$

$pOH = -\log[1.3 \times 10^{-3}] = 2.89$

$pH = 14 - 2.89 = 11.11$

9. $$\frac{[H^+] \times [C_2H_3O_2^-]}{[HC_2H_3O_2]} = 1.8 \times 10^{-5}$$

$$\frac{[x] \times [0.01]}{[0.05]} = 1.8 \times 10^{-5}$$

$[H^+] = x = 9.0 \times 10^{-5}$

$pH = -\log[9.0 \times 10^{-5}] = 4.05$

10. (a) $KNO_3 \rightarrow K^+ + NO_3^-$

$H_2O = OH^- + H^+$

All strong electrolytes; no hydrolysis. Solution remains neutral.

(b) $KNO_2 \rightarrow K^+ + NO_3^-$

$H_2O = OH^- + H^+$

Forms HNO_2; therefore solution becomes basic.

(c) $NH_4NO_3 \rightarrow NH_4^+ + NO_3^-$

$H_2O \rightarrow OH^- + H^+$

Forms NH_4OH; solution becomes acidic.

(d) $Na_2SO_3 \rightarrow 2Na^+ + SO_3^{--}$

$H_2O \rightarrow OH^- + H^+$

Forms H_2SO_3; solution becomes basic.

(e) $FeCl_3 \rightarrow Fe_{3+} + 3\ Cl^-$

$3\ H_2O \rightarrow 3\ OH^- + 3\ H^+$

Precipitates $Fe(OH_3)$; solution becomes acidic.

11. Solubility = 0.0009 g/100 mL = 0.009 g/liter =

$$\frac{0.009\ g}{58\ g/mol} \times \frac{1}{L} = 1.6 \times 10^{-4}\ M$$

$Mg(OH)_2 \rightarrow Mg^{2+} + 2\ OH^-$

$[Mg^{2+}] \times [OH^-]^2 = K_{sp}$

$[1.6 \times 10^{-4}] \times [3.2 \times 10^{-4}]^2$
$= K_{sp} = 1.6 \times 10^{-11}$

12. $AgC_2H_3O_2 \rightarrow Ag^+ + C_2H_3O_2^-$

$[Ag^+] \times [C_2H_3O_2^-] = 2 \times 10^{-3}$

$[x]^2 = 2 \times 10^{-3}$

$[Ag^+] = [C_2H_3O_2^-] = x = 4.4 \times 10^{-2}$ M

13. $MgCO_3 \rightarrow Mg^{2+} + CO_3^{2-}$

$[Mg^{2+}]\ [CO_3^{2-}] = 1 \times 10^{-5}$

$[x]\ [0.05] = 1 \times 10^{-5}$

$[Mg^{2+}] = x = 2 \times 10^{-4}$ M

14. (a) $MgCO_3 \rightarrow Mg^{2+} + CO_3^{2-} \cdots + 2\ H^+ =$
H_2CO_3. So salt dissolves.

(b) $AgCl \rightarrow Ag^+ + Cl^- \cdots + H^+ =$ nothing.
Salt does not dissolve.

(c) $Cu(OH)_2 \rightarrow Cu^{2+} + 2\ OH^- \cdots + 2\ H^+$
$= 2\ H_2O$. So salt dissolves.

(d) $BaSO_4 \rightarrow Ba^{2+} + SO_4^{2-} \cdots + 2H^+ =$
nothing. Salt does not dissolve.

15. $Ca(OH)_2 \rightarrow Ca^{2+} + 2\ OH^-$

$[Ca^{2+}]\ [OH^-]^{2-}\ 8 \times 10^{-6}$; let $x = [OH^-]$

$$\frac{[x]}{2} \times [x]^2 = 8 \times 10^{-6}$$

$$\frac{[x]^3}{2} = 8 \times 10^{-6}$$

$$[x]^3 = 16 \times 10^{-6}$$

$$[OH^-] = x = 2.5 \times 10^{-2}\ M$$

$$pOH = -\log[2.5 \times 10^{-2}] = 1.60$$

$$pH = 14 - 1.60 = 12.40$$

Problem Set No. 12

1. (a) Oxidation–reduction.
   (b) No valence change.
   (c) Oxidation–reduction.
2. (a) $H_2S + I_2 \rightarrow S + 2\ I^- + 2\ H^+$.
   (b) Oxidizing agent is I_2.
   (c) I_2 is reduced.
3. (a) $2\ Cu^{2+} + 2\ I^- \rightarrow I_2 + 2\ Cu^+$.
   (b) Reducing agent is I^-.
   (c) I^- is oxidized.
   (d) This would not balance net charge.
4. (a) $2\ MnO_4^- + 5\ Sn^{2+} + 16\ H^+ \rightarrow 2\ Mn^{2+} + 5\ Sn^{4+} + 8\ H_2O$.
   (b) The oxidizing agent is MnO_4^-.
   (c) Sn^{2+} is oxidized.
5. (a) $Cr_2O_7^{2-} + 6\ Fe^{2+} + 14\ H^+ \rightarrow 2\ Cr^{3+} + 6\ Fe^{3+} + 7\ H_2O$.
   (b) The reducing agent is Fe^{2+}.
   (c) $Cr_2O_7^{2-}$ is reduced.

Problem Set No. 13

1. (a) The more positive half-reaction proceeds in each case.

   At anode: $2\ H_2O \rightarrow O_2 + 4\ H^+ + 4\ e^-$
   At cathode: $2\ H_2O + 2\ e^- \rightarrow H_2 + 2\ OH^-$

   (b) Net reaction is sum of half-reactions with electrons balanced. Doubling the coefficients in the second reaction and adding both reactions in part (a), we get:

   $$2\ H_2O \rightarrow 2\ H_2 + O_2$$

2. Noting that the anode reaction has already been turned around, the minimum voltage will be the sum of the two emfs.

   (a) $(-1.358) + (-2.711) = -4.07$ V. It would then require a voltage of 4.07 V or greater to cause the electrolysis to take place.
   (b) Since a dry cell produces only 1.5 V, it will not do.
   (c) Since an automobile battery produces 12 V, it will cause the electrolysis to take place.

3. $$\text{Mass} = \frac{30\ \text{a}}{1} \times \frac{\text{C}}{\text{a–s}} \times \frac{3600\ \text{s}}{\text{hr}} \times \frac{1\ \text{mole}^-}{96{,}500\ \text{C}} \times \frac{1\ \text{mol Ca}}{2\ \text{mol e}^-} \times \frac{40.08\ \text{g}}{1\ \text{mol Ca}} = 22.43\ \text{g Ca}$$

4. $$\text{Time} = \frac{1\ \text{a–s}}{\text{C}} \times \frac{1}{20\ \text{a}} \times \frac{96500\ \text{C}}{1\ \text{mol e}^-} \times \frac{1\ \text{mol e}^-}{1\ \text{mol Na}} \times \frac{1\ \text{mol Na}}{22.99\ \text{g}} \times \frac{40\ \text{g}}{1} = 8{,}400$$

5. $$\text{Volume} = \frac{15\ \text{a}}{1} \times \frac{1\ \text{hr}}{1} \times \frac{3600\ \text{s}}{1\ \text{hr}} \times \frac{\text{C}}{\text{a–s}} \times \frac{1\ \text{mol e}^-}{96500\ \text{C}} \times \frac{1\ \text{mol Cl}_2}{2\ \text{mol e}^-} \times \frac{22.4\ \text{L}}{1\ \text{mol}} = 6.3\ \text{L}$$

6. (a) $Mg + NiCl_2 \rightarrow MgCl_2 + Ni$
   (b) $3\ H_2 + 2\ AuCl_3 \rightarrow 6\ HCl + 2\ Au$
   (c) $Cu + ZnCl_2 \rightarrow$ No reaction
   (d) $Ag + HCl \rightarrow$ No reaction
   (e) $2\ Al + 3\ CuSO_4 \rightarrow Al_2(SO_4)_3 + 3\ Cu$
   (f) $Cu + 2\ AgNO_3 \rightarrow Cu(NO_3)_2 + 2\ Ag$
7. (a) Half-reaction: $Al^{3+} + 3\ e^- \rightarrow Al\ (-1.66\ V)$
   $Pb^{2+} + 2\ e^- \rightarrow Pb\ (-0.126\ V)$
   (b) Total emf : $-0.126 + 1.66 = 1.534$ V
   (c) Oxidizing half-reaction loses electrons. Therefore:
   $$Al \rightarrow Al^{3+} + 3\ e^-$$
8. (a) Half–reactions:
   $$Pb^{2+} + 2\ e^- \rightarrow Pb\ (-0.126\ V)$$
   $$Ag^+ + e^- \rightarrow Ag\ (+0.7996\ V)$$
   (b) Total emf: $+0.7996 + .126 = 0.9259$ V
   (c) Reducing half-reaction accepts electrons. Therefore:
   $$Pb^{2+} + 2\ e^- \rightarrow Pb$$
   (Contrast this with the results in Problem 7.)
9. (a) At the anode, electrons are given off (oxidation). Therefore, iron is the anode in the Edison cell.
   (b) Total emf: $0.877 + (+0.49) = 1.367$ V
10. (a) Half reactions:
    $Fe^{3+} + e^- \rightarrow Fe^{2+}\ (+0.770\ V)$
    $Cr_2O_7^{2-} + 14\ H^+ + 6\ e^- \rightarrow 2\ Cr^{3+} + 7\ H_2O\ (+1.33\ V)$
    (b) Total emf: $1.33 + (-0.770) = 0.560$ V

Problem Set No. 14

1. In each case divide the density by 1.293. N_2: 0.9675. O_2: 1.1052. CO_2: 1.5290. H_2O: 0.6218. H_2: 0.0696. O_3: 1.6582. Air: 1.0000
2. From Table 6.2, vp at 20° = 17.5 Torr; vp at 10° = 9.2 Torr

   $$\frac{9.2 \text{ Torr}}{17.5 \text{ Torr}} \times 100 = 52.6\%$$

3. From Table 6.2: vp at 29°C = 30.0 Torr; 30.0 Torr × 89% = 26.7 Torr. From Table 6.2, 26.7 Torr gives a dew point of 27°C or 80.6°F.
4. On p. 66.
5. +7
6. Nitrogen, oxygen, hydrogen, helium, argon, neon, krypton, xenon, and possibly a bit of radon.
7. He, because molecular mass equals atomic mass.
8. A person, and the parts of his body, are cooled by moving through air because of increased rate of evaporation. Fast-moving bodies are actually heated, by friction with the air. Meteorites, on striking the atmosphere, are heated until they burn brilliantly with oxygen.
9. The dissolved carbon dioxide is liberated more readily by the shaking.
10. CO_2 comes off as a result of the action of citric acid on sodium bicarbonate.

Problem Set No. 15

1. Relatively inert—less active than iodine.
2. $2KHF_2 \rightarrow 2\,K + H_2 + 2\,F_2$
3. In solution, HF is a weak electrolyte and therefore supplies fewer hydrogen ions per unit concentration than the strong electrolyte HCl. The specific property of HF to attack glass has nothing to do with its acid characteristics.
4. At anode: Br_2. At cathode: H_2. Resulting solution: NaOH.
5. Probably the HCl, which forms in light or under the heat of the iron, attacks the fibers to produce the stains.

Problem Set No. 16

1. (a) ZnTe (b) H_2Se (c) H_2SeO_4 (d) $Na_2S_2O_3$
2. $Zn + Te \rightarrow ZnTe$

   $ZnTe + 2\,HCl \rightarrow ZnCl_2 + H_2Te$
3. Two or more forms of the same element having different properties as a result of different molecular or crystalline structures
4. (a), (c), and (e) are weak. (b), (d), and (f) are strong.
5. (a) S is oxidized and O_2 is reduced.
   (b) S in SO_3^{2-} is oxidized and O_2 is reduced.
   (c) S in SO_3^{2-} is oxidized and free S is reduced. (*Note*: In this case the free S behaves just like O_2 in the previous equation and has an oxidation number of −2 in the thiosulfate ion. The S in the sulfite ion is oxidized from a +4 to a +6 oxidation number in the thiosulfate ion.)

Problem Set No. 17

1. Fish is a source of phosphate, which is required in brain and nerve tissue.
2. $2\,Sb_2S_3 + 9\,O_2 \rightarrow 2\,Sb_2O_3 + 6\,SO_2$

   $Sb_2O_3 + 3\,C \rightarrow 3\,CO + 2\,Sb$
3. $4\,As + 5\,O_2 \rightarrow 2\,As_2O_5$

   $As_2O_5 + 3\,H_2O \rightarrow 2\,H_3AsO_4$
4. (a) H_3AsO_3 (b) H_3SbO_3 (c) H_3SbO_4
5. Bicarbonate ions from the sodium bicarbonate react with hydrogen ions from the monosodium phosphate to produce carbon dioxide, which forms throughout the batter and causes it to rise. The equation is

   $$H^+ + HCO_3^- \rightarrow H_2O + CO_2$$

Problem Set No. 18

1. (a) Carbon monoxide (b) Carbon dioxide
2. Both producer gas and water gas are made from steam and coke, but air is also used in making producer gas. This introduces large amounts of incombustible nitrogen, which lowers the heating value of the producer gas per unit volume.

3. Elements present in window glass are oxygen, silicon, sodium, magnesium, calcium, aluminum, and iron.
4. Peach kernels contain deadly hydrogen cyanide. If a sufficient number of them are eaten, violent illness or even death can result.

Problem Set No. 19

1. $2\ KOH \rightarrow H_2 + O_2 + 2\ K$
2. The alkali metals react chemically by losing the outermost electron. In the heavier members of this family, this negative electron is farther removed from the positive nucleus and thus is given up more readily.
3. The equation is

   $2\ Na + 2\ H_2O \rightarrow 2\ NaOH + H_2$

   $$\frac{4.6\text{ g Na}}{1} \times \frac{1\text{ mol Na}}{22.99\text{ g}} \times \frac{1\text{ mol }H_2}{2\text{ mol Na}} \times \frac{2.02\text{ g}}{1\text{ mol }H_2}$$

   $= 0.20$ g
4. Ca^{2+} from *limestone* goes into the formation of $CaCl_2$. Cl^- from *salt* goes into the formation of $CaCl_2$. *Water* is not recovered in the process.
5. Ammonium hydroxide is a solution of the gas ammonia in water. The solubility of all gases decreases as the temperature is increased. Therefore heating ammonium hydroxide will drive off ammonia gas from the solution.

Problem Set No. 20

1. (a) Heating carnallite to its melting point will drive off the water of crystallization as steam, so no hydrogen will be present in the fused bath to be a by-product.
   (b) KCl: $2.924 + (+1.358) = 4.282$ V
   $MgCl_2$: $2.37 + (+1.358) = 3.728$ V
   (c) If the voltage during electrolysis is held below 4.28 V, potassium cannot deposit.
2. $Al_2(SO_4)_3 \rightarrow 2\ Al^{3+} + 3\ SO_4^{2-}$

   $6\ H_2O \rightarrow 6\ OH^- + 6\ H^+$

   ‖

   $2\ Al(OH)_3(s)$
3. (a) $NaAl(SO_4)_2 \cdot 12\ H_2O$
   (b) Alkaline earth metals are divalent. Alums contain only monovalent and trivalent ions. Thus, no alkaline earth metals are found in alums.
4. Beryl is $Be_3Al_2Si_6O_{18}$. Molecular mass is 537 g/mol
   Be: 3 mol × 9 g/mol = 27 g
   Al: 2 mol × 27 g/mol = 54 g
   Si: 6 mol × 28 g/mol = 168 g
   O: 18 mol × 16g/mol = 288 g

   $$\%\ Be = \frac{27\text{ g}}{537\text{ g}} \times 100 = 5\%$$
5. The chemical activity of metals depends on their ability to lose electrons. As is the case with the alkali metals, the heavier alkaline earth metals lose electrons more easily because the electrons are farther from the nucleus of the atoms. Therefore, radium, barium, and strontium will be *above* calcium in the activity series in this order: radium, barium, strontium, calcium, magnesium.

Problem Set No. 21

1. Hematite – 70%; magnetite – 72%; pyrite – 47%; siderite – 48%
2. Refer to corrosion section in Chapter 11
3. (a) 3; (b) 1; (c) 2
4. $Cr_2O_3 + 2\ Al \rightarrow Al_2O_3 + 2\ Cr$
   $3\ V_2O_5 + 10\ Al \rightarrow 5\ Al_2O_3 + 6\ V$

Problem Set No. 22

1. Malachite, $Cu_2(OH)_2CO_3$. The copper is dissolved from piping by the water containing dissolved carbon dioxide.
2. By using the impure copper as the anode, copper and all metals above it in the activity series present in the anode go into solution as ions. All metals below copper in the activity series simply fall to the bottom of the bath as the anode decomposes. At the cathode, only copper, which is the least active metal in solution, can deposit. Thus copper is effectively separated from all other metals.
3. $2\ ZnS + 3\ O_2 \rightarrow 2\ ZnO + 2\ SO_2$
   $ZnO + C \rightarrow CO + Zn$

4. (a) Copper—wiring and electrical equipment
   (b) Tin—babbitt and bearing metals
   (c) Mercury—mercury fulminate
5. Because it is a source of poisonous lead salts formed when water containing dissolved oxygen dissolves lead.

Problem Set No. 23

1. A sacrificial anode must be chemically more active than the metal it protects. Since the noble metals are very inert, they would not serve as sacrificial anodes.
2. 1 part concentrated nitric acid + 3 parts concentrated sulfuric acid
3. Fineness – parts by weight of silver per 1000 parts alloy; carats – parts by weight of gold in 24 parts alloy
4. Silver is removed from photographic film and printing paper as sodium silver thiosulfate in the fixing process. Since silver is much less active than sodium, silver would easily be obtainable as a cathode deposit by subjecting this solution to electrolysis.
5. Lanthanides: lanthanum, cerium, praseodymium, neodymium, promethium, samarium, europium, gadolinium, terbium, dysprosium, holmium, erbium, thulium, ytterbium, lutetium.

   Actinides: actinium, thorium, protactinium, uranium, neptunium, plutonium, americium, curium, berkelium, californium, einsteinium, fermium, mendelevium, nobelium, lawrencium.

Problem Set No. 24

1. (benzene ring)—OH

2. $C_2H_5OH + CH_3\overset{O}{\overset{\|}{C}}OH \rightleftharpoons CH_3\overset{O}{\overset{\|}{C}}OC_2H_5 + H_2O$

3. $\dfrac{1\text{ L}}{22.4\text{ L}} = \dfrac{1.88\text{ g}}{x}$

   $x = 42.1$ g in 1 mol

   $0.856 \times 42.1 = 36.0$ g carbon
   $0.144 \times 42.1 = 6.1$ g hydrogen
   $36.0\text{ g}/12\text{ g/mol} = 3$ mol atoms carbon
   $6.1\text{ g}/1.01\text{ g/mol} = 6$ mol atoms hydrogen
   Therefore, the molecular formula is C_3H_6.

4. 1. d 2. e 3. c 4. a 5. b

5.
```
    H   H   H
    |   |   |
H—C—C—C—H.
    |   |   |
    H  Cl   H
```

Problem Set No. 25

1. Both the mass numbers and the atomic numbers on both sides of the equation must balance. Therefore x must be written ${}^1_1x$. According to the periodic table, the element with atomic number 1 is hydrogen. Therefore x = hydrogen, H.
2. In fusion, nuclei collide and fuse. In fission, nuclei split apart or disintegrate. In an ordinary chemical reaction the nuclei stay intact and the outer electrons surrounding the nucleus (valence electrons) are involved in chemical reaction.
3. (a) from the nucleus.
4. Neutron-induced reactions may result in both heavier and ligher elements.
5. The energy produced by the Sun involves fusion or thermonuclear reactions.

Problem Set No. 26

1. (a) Hydroelectric and (c) wood
2. True. They contain carbon and hydrogen, which oxidize or burn to CO_2 and H_2O.
3. Solar energy strikes silicon photovoltaic cells, which provide the electricity to electrolyze water.
4. Refer to Table 24.1.

   $\dfrac{(21.6 + 7.2)\text{ quads}}{94.3\text{ quads total}} \times 100 = 30.5\%$ in 1980

   $\dfrac{71 + 18}{143} \times 100 = 62.2\%$ in 2020

   Supplies of petroleum are dwindling.

5. $\dfrac{34.2\text{ kcal}}{1\text{ g}} \times \dfrac{2.02\text{ g}}{1\text{ mol}} \times \dfrac{1\text{ BTU}}{0.252\text{ kcal}}$

   $= 274$ BTU/mole

Problem Set No. 27

1. 0.05 ppm $= 0.05 \times 10^{-6} = 5 \times 10^{-8}$ g lead per gram of water.
2. (b) automobiles
3. The rain is acidic; it contains sulfuric acid from stack gases, much of which is from coal-fired electric plants.
4. It is stable. It is situated away from rivers, lakes, and (it is to be hoped) underground water streams, and it is immobile.
5. True. CaO is basic, like MgO, and neutralizes acidic oxides such as SO_2 and SO_3.

Problem Set No. 28

1. (a) Carbohydrate
   (b) Fat
   (c) Protein
   (d) Fat
   (e) Fat
2. (a) Polymer
   (b) Monomer
   (c) Monomer
   (d) Polymer
   (e) Monomer
3. $2\ C_{17}H_{35}COONa + Ca^{2+}$
   $\rightarrow (C_{17}H_{35}COO)_2Ca + 2\ Na^+$

   $(C_{17}H_{35}COO)_2Ca$ is insoluble and forms a ring around the bathtub.
4. (c) Animals and plants.
5. $-NH_2$. Remember that NH_4OH is a base. It can be written as $NH_2^-\ H_3O^+$. Therefore, NH_2^- can be considered to be the basic portion of NH_4OH as well.

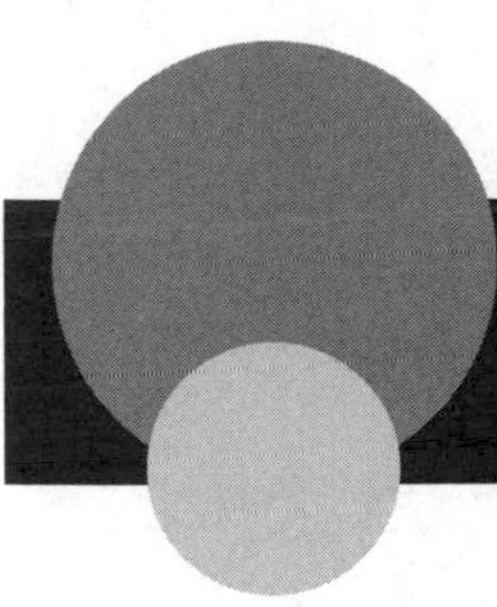

CONCISE GLOSSARY OF CHEMICAL TERMS

acid
a substance that ionizes in solution to produce hydrogen ions (H^+)

acid rain
rain of a low pH formed by the dissolving of air pollutants, such as the oxides of sulfur, nitrogen, and carbon

activity
the ease with which an element loses or gains electrons

activity series of metals
a list of metals in their order of reactivity with water and some acids, with the most reactive metal at the top

activity series of nonmetals
a list of nonmetals in their order of reactivity with water and some acids, with the most reactive nonmetal at the top

aliphatic
a compound of carbon that does not possess a benzene ring or the series of alternating double and single bonds

allotropic form
the same substance in a different molecular form and possessing different properties

alloy
a mixture of two or more metals

alkali metals
the elements of group IA (1) in the periodic table.

alkaline earth metals
the elements in group IIA (2) of the periodic table

alpha rays
nuclei of helium atoms, consisting of particles containing 2 neutrons and 2 protons.

amorphous solid
a solid having a completely random arrangement of particles

amphoteric
hydroxides that behave as acids toward bases and as bases toward acids

anode
in an electrolytic cell, the positively charged electrode that attracts negative anions

aromatic
a carbon compound that contains a benzene ring or a series of alternating single and double bonds

atmosphere
the "sea" of gases that surrounds the Earth

atom
the smallest particle of an element that has all the properties of the element

atomic mass
the sum of the number of protons and the number of neutrons in the nucleus of an atom

atomic number
the number of protons in the nucleus of the atom

aufbau principle
the rules electrons follow when they fill the various energy levels

Avogadro's law
the principle that equal volumes of gases measured at the same temperature and pressure contain equal numbers of molecules

Avogadro's number
6.022×10^{23}, which is the number of atoms in one mole of an element or compound

base
a substance that ionizes in solution to produce hydroxide ions (OH^-)

beta ray
a stream of negatively charged particles, similar to electrons, that move almost at the speed of light

biological oxygen demand (BOD)
a measure of the amount of organic material in waste water

biotechnology
molecular engineering to modify or create parts or products of living organisms

boiling point
the temperature at which the vapor pressure of a liquid equals the pressure of the atmosphere about it

bonding
the tendency of elements to form compounds through a shift of electronic structure

Boyle's law
at a given temperature, the volume occupied by a gas is inversely proportional to its pressure

BTU (British thermal unit)
a unit equal to the amount of heat required to raise one pound of water one degree Fahrenheit at one atmosphere pressure

carbohydrate
a compound of carbon, hydrogen, and oxygen that usually has a hydrogen to oxygen atomic ratio of 2 to 1

cast iron
a form of iron obtained from cooling pig iron, which contains more than 2% carbon

cathode
in an electrolytic cell, the negatively charged electrode that attracts positive cations

chain reaction
a reaction in which one of the agents necessary to the reaction is itself produced by the reaction, thus causing like reactions to continue indefinitely

Charles's law
at a given pressure, the volume occupied by a gas is directly proportional to the Kelvin temperature of the gas

chemical property
the ability of a substance to change into a new and completely different substance

colligative property
a characteristic of a solution that depends on the number of solute particles, but not their type

colloid
an intermediate between a true dissolved particle and a suspended solid that will settle out of solution

combustion
the rapid combination of a substance with oxygen, releasing energy in the forms of heat and light

common ion effect
the principle that the addition of an ion in common with an ion of solute represses the dissociation of the solute, thereby reducing the solubility of the salt

compound
a pure substance made of elements that are chemically combined in a specific whole-number ratio

corrosion
the redox process that oxidizes metals, forming oxides and sulfides

covalent bonding
the process of forming a compound through the sharing of pairs of electrons

crystalline solid
a solid in which the particles are arranged in definite geometric patterns

Dalton's law of partial pressures
the total pressure of a mixture of two or more gases that do not chemically combine is the sum of the partial pressure of each

density
the mass-per-unit-volume of a substance

deoxyribonucleic acid (DNA)
the nucleic acid that carries the genetic code

destructive distillation
heating in the absence of air

dew point
the temperature to which air must be lowered to saturate it with water vapor

electrode
a solid piece of conducting metal used in a battery or electrolytic cell

electrolysis
the passage of an electric current through a solution of an electrolyte

electrolyte
a substance that conducts an electrical current in the molten state or when dissolved in water

electromotive force (emf)
a measure of the electrical driving force needed to complete an electrochemical reaction

electron
a subatomic particle possessing a negative (−) electrical charge

electron configuration
a notation that shows the type of orbital at each energy level that contains electrons and the number of electrons in each orbital

electronegativity
the attraction that an atom has for a bonding pair of electrons

element
the simplest form of matter that cannot be formed from simpler substances or decomposed into simpler substances

empirical formula
the simplest formula of a compound, indicating only what atoms are present in the lowest whole number ratio

energy level
a volume of space at a definite average distance from the nucleus of an atom and containing a definite number of electron orbitals

environment
everything that surrounds and affects living things on Earth

equation
a concise statement of a chemical reaction using the symbols and formulas of the reactants and products

equilibrium
the state in which two opposing processes take place simultaneously and at the same rate

equivalent mass
that amount of the element that has combined with or displaced an atomic mass's worth (1.0 gram) of hydrogen or 8 parts (by weight) of oxygen

family
a vertical column in the periodic table

fat
an ester of glycerin and an organic acid

fission
the decay of a nucleus by breaking apart into two or more lighter isotopes

flux
a material that aids in the melting on another material

formula
the composition of a substance indicated by symbols of each element present and subscript numbers showing the number of each type of atom involved

formula mass
the sum of all the atomic masses of the elements present in the formula of the compound

fossil fuels
fuels, including coal and petroleum, that formed millions of years ago from the remains of plants and animals

fractional distillation
the separation of substances by their boiling points

freezing point
the temperature at which a substance solidifies from the liquid to the solid state

functional group
the part of an organic molecule that gives that molecule its specific reactivity

fusion
the combining of two light isotopes into a heavier one

gamma rays
represented as γ, photons (small packets) of energy

glass
a rigid, supercooled liquid made from molten silica

Graham's law of diffusion
the rates of diffusion of gases are inversely proportional to the square roots of their molecular weights, when the gases are at the same temperature and pressure

greenhouse effect
the warming that results when solar radiation is trapped by the atmosphere

group
a vertical column in the period table

halogen
an element of Group VIIA (17) of the periodic table

hard water
water that contains calcium and magnesium ions in solution

heat of fusion
the number of calories required to melt 1 gram of a solid to liquid at the transformation temperature (melting point)

heat of vaporization
the number of calories required to change 1 gram of the liquid to a gas at the transformation temperature (boiling point)

hydrocarbon
a compound of hydrogen and carbon

hydrogen halide
a compound of hydrogen and a halogen

hydrolysis
the reaction of a substance with water

hypothesis
a possible explanation for a phenomenon; a "best guess"

inner transition elements
the 28 metals of the lanthanide and actinide series of the periodic table

ion
an atom that has acquired an electrical charge due to the loss or gain of electrons

ion exchange
a reversible process in which ions are released from an insoluble permanent material in exchange for other ions in a surrounding solution

ionic bonding
the process of forming a compound through the transfer of electrons

ionization
the process by which electrolytes dissociate into positively and negatively charged ions in solution

isomers
different compounds possessing the same empirical formulas

isotopes
atoms of an element with different atomic masses

kinetic-molecular theory
the principle that explains the behavior of gases in terms of particles in motion

law
a theory proven to be a nonvarying performance in nature

law of conservation of matter
the principle that matter is neither created nor destroyed during chemical change

law of definite proportions
the principle that a given compound always contains the same elements combined in the same proportions by mass

law of mass action
the law that the velocity of a reaction is proportional to the product of the molar concentrations of the reacting substances

Le Châtelier's principle
a principle stating that a system in equilibrium, when subjected to a stress resulting from a change in temperature, pressure, or concentration, and causing the equilibrium to be upset, will adjust its position of equilibrium to relieve the stress and reestablish equilibrium

Lewis structural formula
a formula that indicates the bonding pattern by using a line to represent a shared pair of bonding electrons in the molecule and showing electrons not involved in bonding as dots

mass
a measure of the quantity of matter

matter
anything that takes up space

metal
an element that tends to lose electrons in reactions, has a luster, and is malleable and ductile; also a good conductor of heat and electricity

metalloid
an element with some of the properties of metals and some of the properties of nonmetals

mixture
a combination of pure substances held together by physical rather than chemical means

molality
the number of moles of solute per kilogram (1000 grams) of solvent

molarity
the number of moles of solute per liter of solution

mole
an amount of a compound expressed in grams equal to its formula mass

molecule
a group of covalently bonded atoms that possesses all the properties of the compound

neutron
a subatomic particle with no electrical charge found in the atomic nucleus

noble gas
any element of group VIIIA (Group 18) of the periodic table

noble metal
any metal of low chemical activity that is found naturally in the free state

nonelectrolyte
a substance that does not conduct an electrical current in the molten state or when dissolved in water

nonrenewable
an energy source that will be exhausted in supply at some time

normality
the number of equivalents of solute per liter of solution

nucleic acid
any one of a group of polymers found in all living cells made of subunits containing a nitrogenous base, a phosphate molecule, and a sugar molecule

nucleus
the center of an atom composed of one or more neutrons and one or more protons

octane number
in gasoline, the percentage of isooctane that must be added to n-heptane, C_7H_{16}, to cause a standard engine to operate with the same characteristics as the gasoline being tested

orbital
a volume of space in which the probability of finding an electron is high

organic chemistry
the chemistry of carbon compounds

osmosis
the process of a solvent moving through a semipermeable membrane

oxidation
a process involving the loss of electrons

oxidation number
the charge that would be present on an atom if the compound in which the atom is found were ionic

oxidizing agent
in a redox reaction, the substance that gains electrons

period
a horizontal row of elements in the periodic table

periodic table
a chart showing the atomic number, symbol, and atomic mass of each element

pH
the negative logarithm of the molar concentration of the hydrogen ion in a solution

phenomenon
a particular natural event

phosphate
a compound that contains the phosphate ion, PO_4^{3-}

phosphorus family
the elements below nitrogen in group VA (15) of the periodic table

photovoltaic cell
a device that emits electrons (electricity) when exposed to sunlight

physical property
a characteristic of a substance as it is

pig iron
solidified iron obtained from a blast furnace

polarization
the reduction in voltage of a dry cell caused by the accumulation of hydrogen gas

pollution
hazardous substances put into the air, water, or land

polymer
a large molecule composed of many repeating smaller units called monomers

protein
a complex polymer of amino acids that is the building block of all plant and animal tissues and organs

proton
a subatomic particle consisting of a positive (+) electrical charge equal in magnitude (but opposite in type) to the charge on the electron

quantum mechanical model
a theory of atomic structure that describes electrons as being found in energy levels and being contained in orbitals

radioactivity
the property that some elements or isotopes have of spontaneously emitting energetic particles (radiation) by the disintegration of their atomic nuclei

reaction
a chemical change or change from one substance to another

redox reaction
an oxidation-reduction reaction in which one substance gains electrons and another loses electrons

reducing agent
in a redox reaction, the substance that loses electrons

reduction
a process involving the gain of electrons

reduction potential
the potential for a reduction reaction in which electrons are gained

relative humidity
the ratio of the partial pressure of water vapor in air to the vapor pressure of water at the temperature of the air expressed as a percent

renewable
an energy source that can be replenished

salt
a substance that ionizes in solution, but produces neither hydrogen nor hydroxide ions

saturated
a hydrocarbon that contains only single bonds

silicate
an ion containing silicon and oxygen (in any proportion); present in many rocks and minerals

smelting
the process of heating a metal oxide with some reducing agent such as coke (carbon) to produce an elemental metal

solubility
the maximum amount of a substance that can be dissolved in a given amount of solvent at a specified temperature and pressure

solubility product principle
at saturation, the product of the molar concentrations of the ions of the electrolyte, raised to proper powers, is a constant

solute
the substance that is dissolved

solution
a homogeneous mixture consisting of two components, a solvent that is the dissolving medium, and a solute that is the substance being dissolved

solvent
the dissolving medium

specific gravity
the ratio of the mass of a given volume of a substance to the mass of the same volume of water at the same temperature

specific heat
the ratio of the heat capacity of a given substance to the heat capacity of water

steel
an alloy of iron that is less than 2% carbon

structural formula
a formula that shows the actual number of each type of element in a compound and also indicates the bonding pattern

sublimation
the passage of a solid directly to the gaseous state as it warms

sulfur family
the elements below oxygen in Group VIA (16) of the periodic table

Système International d'Unités
the international system of units (abbreviated SI)

theory
a hypothesis supported by the weight of the evidence

transition element
an element in which the d orbital is not full of electrons

unsaturated
a hydrocarbon that contains either double or triple bonds

valence electrons
the electrons in the outermost energy level (particularly the s and p electrons) that are involved in chemical reaction and bonding

vapor pressure
the pressure exerted by a vapor when the liquid and its vapor are in dynamic equilibrium

weight
a measure of the gravitational force on an object or mass

wrought iron
a soft, tough, and somewhat corrosion-resistant form of pig iron that contains some slag

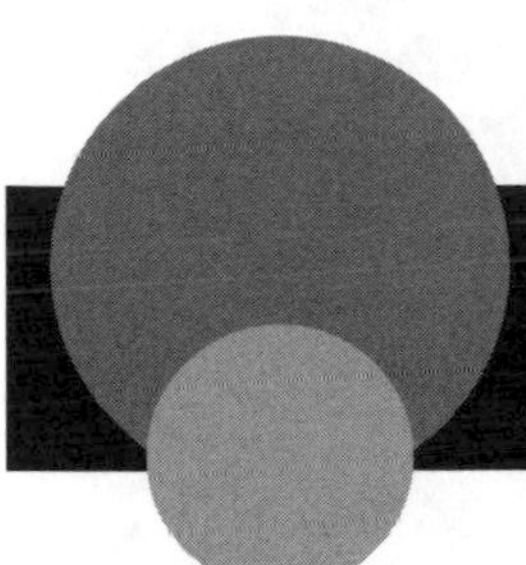

INDEX

N

NOTES

NOTES